The Internet President:
None of the Above

P.G. Sundling

Adaptonium   Inglewood California

Mobi ISBN: 978-1-948786-01-0
Paperback ISBN: 978-1-948786-00-3
Library of Congress Control Number: 2018904496

Table of Contents

Dedicated to those who see not just what is but imagine what could be.

01. War Paint

If Maria Cortez hears the word "gorgeous" again, someone's getting it in the nuts. Never been kissed and never will. She hides behind grandma glasses and unflattering makeup, but not even clothes that cover every inch of skin, besides her hands and face, can save her from the curse of beauty.

Maria sits in an executive suite with the feel of a teenager's bedroom. A half-built Lego battle cruiser and robotic components clutter the elegant decor. Science fiction posters wallpaper the room. Near the door, a couple of dozen ants tunnel through space-age gel, in an ant habitat lit up like an alien world.

At the rear of the room is an enormous oak desk. Alvin Renquist presides in the seat of power. He is all teeth, a shark eater. If greed is good, he's a saint. Six chairs surround the desk, for those who might seek an audience with him.

Maria wonders why James picked Renquist to buy the company. He acts like he owns the place, but he doesn't yet. "The purchase closes today. Why haven't your funds hit escrow?"

Renquist slips into the adjoining seat. "I want you as interim CEO, under my guidance."

Maria eyes him suspiciously. "If I'm CEO, what about James?"

"I'll make you the face of the company." Renquist slides his hand onto her thigh.

She retreats a seat closer to the door. "Answer my questions."

Renquist pursues and sits next to her.

Maria's leg quivers.

He reaches towards her thigh again but pauses.

"Of course, a makeover!" Renquist undresses her with his eyes, imagining every hidden contour. "Take off those glasses. Add some makeup. Put on a big smile. I can see it now. Billboards. TV. You'll be *everywhere*." The words slither from his mouth as he tries to caress her cheek.

Renquist grazes her face before she swats his hand away.

1

Maria tries to stay professional and composed, but her whole body shudders. Is it fear, or rage, that stirs within her? She fends off his advances and stands. "Stop! Let's pretend we have a safe word, and I used it."

He closes in.

Maria won't be a victim. She takes a karate stance.

Renquist withdraws, with his hands raised. "Fine. Grab James, and we'll sort this all out."

She flings the door open and strides outside. Tiny helicopters, inflatable sharks, and small drones buzz around an unremarkable cubicle farm. Nerf Darts whiz by Maria's head. A six-inch helicopter crashes to the carpet. She smirks, but her apprehension remains.

James Wong advances down the aisle with a cardboard tower shield and homemade lightsaber. Software developers shoot him with Nerf guns. Their projectiles bounce off his tower shield.

More creative supernova than businessman, it's moments like these where Maria isn't the only one to see his inner child.

When he sees Maria, James drops his tower shield and sprints towards her. "Do you remember our first hit app?" He clicks a button on the hilt of his sword and thrusts. Fart sounds change pitch as he swings his weapon.

"May the fart be with you, always," James says. Maria and James chuckle together. "From fart app to billion-dollar company. We did it. Today is the best day of my life." James fist pumps.

Her smile crumples. "The buyer wants to see us."

The intercom turns on with a hiss. "James Wong to my office now." Over the speakers, Renquist's voice sounds like an angry school principal.

James puts down his lightsaber. "Why does the Eye of Sauron have to summon me every time I'm having fun?" He grins. "At least it's good news."

The employees scatter. Cubicle airspace clears. The programmers put away their toys, slink into their Aeron chairs, and return to work.

Distress fills Maria with every step. It's like a roller coaster rising farther and farther into the sky.

James seems too preoccupied with his perfect day to notice. He struts into the executive suite.

Maria follows closely.

Two burly bodyguards flank Renquist.

James checks his bank on his phone. "What is your bidding my…" He points to his empty bank account on the phone. "Oh wait, you're not my master."

Renquist rips an *Aliens* movie poster off the wall. James lunges at Renquist, but the security guards intervene.

"Every time you make a movie reference, I destroy a poster." Renquist rips the poster into strips, and then into oversized confetti.

"This isn't your office." James takes a deep breath. "Not until the money clears."

Renquist squints at the ant habitat. "Everything will be clear in a minute, but first, what kind of freak has ants in his office?"

James says, "Ants are my reminder that with determination and teamwork, anything is possible. As many as a million ants can work together in a colony. Imagine if humans could accomplish—"

"I didn't ask for the whole Wikipedia entry." Renquist shakes his head.

Maria retreats to the back of the room. She can feel her roller coaster about to plunge.

"Well James, I've got good news. You're fired." Renquist throws up poster confetti.

James does a double take. "I think I misheard you."

"I didn't say the good news was for you," Renquist says. "I'm taking Adaptive Unlimited in a new direction, without you."

"You said you wanted me, along with my company!"

"I lied. I tend to do that."

James puffs out his chest. "You'll never find a replacement who knows this company as well as I do."

"I already have." Renquist smiles.

"Who could…"

James looks at Maria. How could she? Shock. Fear. Betrayal. "Maria? Maria."

There's an eon of silence.

Maria bolts from the room without a word. The door slams behind her.

She zips into the adjoining office and dashes straight to her desk. Monitor on. Keys clack. Live surveillance of the executive suite displays on the monitor. Maria listens, as she pulls a dusty box from the bottom of her cabinet. She opens the box and smiles at the provocative clothes and boots inside.

James pleads, in the other room. "This company is my life. I sleep more in the office than I do at home. Don't do this!"

"You couldn't even lead a Boy Scout troop. You've only muddled this far because Maria covers for you."

Maria monitors events in the next room as she undresses. She flings her conservative outfit on the floor one item at a time.

"I built this company one crazy idea after another, a whole trainload of them," James says. "Maria follows me. She's the caboose on the crazy train."

"Is that what you think?" Maria grumbles at her screen and scrunches her nose in anger.

"If you, Maria, or anybody, doesn't want to ride the crazy train, this is your stop." James points to the door.

Maria slinks into knee-high boots with stiletto heels. It's the first step of a sensual self-makeover. She puts on her power suit. Form-fitted leatherette pants and a scandalous front-zip black bustier hug every curve. Her personality shifts, molts, leaving her old self on the ground. Confident. Powerful. Invincible.

James says, "I'm not selling!"

"Just hand over your $150 million cancellation fee, and you can have your company back," Renquist teases.

"Fine. When does our $1.2 billion arrive?"

Renquist says, "The money's never coming. You didn't catch the resale clause. I can resell your company to another buyer before the purchase completes, extending the contract. Profits and control in the meantime go to the seller."

"Right, and I'm the seller."

"Wrong. During a resale, I'm redefined as the seller. So, your company will go through one resale after another, for years, decades, if it's worth the trouble. I get the profits and control. You get nothing. Where did you get your lawyer? A park bench?"

"You reassigned variables on me," James murmurs. He slumps to his knees like a deflated tire.

Maria springs up and pounds the table. She gets self-conscious, worried they might hear her in the next room. She scrutinizes the screen for a reaction to the noise.

He spreads out the fingers in his hand and stares at them, as he imagines all that cash slipping away. "Nuked from orbit. Game over."

"That sounds like movie talk to me. Were you born in a theater or something? Talk normal." Renquist rips an *Avatar* movie poster off the wall. He tears through it like an impatient child unwrapping a present.

Renquist circles James, a predator finishing off his prey.

Maria roots for him, riveted to the monitor. "Fight him, James. Don't let him do this to you. To us."

Renquist leans towards James. "Crazy smart works for startups, but not in Corporate America. You should know your place on the org chart. If I say wear a pink dress, or eat out of a dog bowl, that's what you do. You don't respect the chart. You don't belong on it. That's why security is gathering your toys as we speak."

The bodyguards snicker and high five. Their body language suggests they no longer consider James a threat.

Renquist grabs his briefcase from under the desk. He opens it and removes a can of Raid Ant and Roach Spray.

Maria pulls a mirror from her desk. She fixes her makeup and applies ruby lipstick with deliberation as war paint.

James eyes the spray can. He pushes off the floor to lunge at Renquist. The guards tackle James before he reaches his target. Renquist steps over James, approaches the ant habitat, and flips open the lid. James watches helplessly on the ground, pinned underneath 500 pounds of hired muscle.

"You're an ant, and I'm top of the food chain." Renquist sprays into the ant habitat. The deadly liquid flows down the gel tunnels. Chemical odors waft through the air. The ants scamper no more.

Renquist crouches down towards James. "I'd say you were a worthy adversary, but even I have a limit on how many lies I can tell in a day." He gets back up and motions to his guards. "I want people talking about it for weeks. Remove him."

Each guard grabs an arm and drags James backward, beyond the edge of her monitor. Maria can no longer see him. She rushes to her doorway, just in time to see him dragged into the hallway.

James glares at Maria. He fixates on her lipstick and snickers. "You're going to pay for your betrayal." He tries to kick free, but only kicks metal filing cabinets along the hallway.

Bystanders peek their heads out of their cubes like gophers to catch the spectacle.

The guards yank James around the corner, out of sight.

Maria creeps back into her office. She pulls three external solid-state drives from under the desk. She covers them with bubble wrap and slips them into pre-addressed packages. Maria peers out to check if Renquist has his door closed. Confirmed.

Thelma, the company's only executive assistant, sits at the reception desk across from the executive suite.

Maria slips Thelma the packages and nods.

Thelma nods back. She pushes the intercom button. "Maria Cortez is here to see you."

"Thanks, Thelma, buzz her in," Renquist says.

The door opens. Maria slinks in, fully transformed from drab female to sex goddess. She's on the prowl for big game.

Renquist forgets who she is for a moment. The whole day is forgotten. He focuses on her.

"How's this for a makeover?" she purrs. "I hope you brought a pill. I might need a full four hours."

All the blood rushes from his brain. If he were any more brain dead, he'd require life support.

She saunters to his desk, one sensual step after another.

He fumbles for the intercom key. "Cancel all my meetings. I don't want to be disturbed."

"Is this seat taken?" Maria asks innocently, as she slides onto his lap. Her hand explores the contours of his chest. He gently inhales the scent of her perfume. His smile barely fits on his face.

She yanks at his tie and whispers. "Do you like it rough?"

He nods vigorously.

She chokes him with the necktie.

He turns red. A gasp. Panic. Renquist slams the table with his flailing right arm.

Maria loosens the tie. With a soft caress, she lulls him back to the fantasy.

Thelma buzzes the intercom. "Are you alright, Mr. Renquist?"

"Disregard any disturbances. It's going to get noisy."

"Understood." Thelma sighs, with disgust. "Since I have you, the 11 o'clock mailman is here if you—"

"I don't want to be disturbed. That includes you, the mailman, and everyone else. I don't care if your hair is on fire. Don't bother me." Renquist disconnects the intercom.

"Nibble on my ear. Lick me like a dog." He sticks out his tongue to the max and pants.

She strokes his ear and pats him on the head. "Good doggy."

Maria rises off his lap and pulls him up. He attempts to embrace her, but she backpedals. The desk blocks her retreat.

Renquist shoves everything off the desk. His water pitcher shatters, creating a puddle. "I hope the carpet isn't the only thing that's wet."

It's too much. It's all Maria can do to stifle a snort of contempt. She turns away and rolls her eyes. She turns back to Renquist and smiles seductively.

He heaves Maria into the air and plops her on the table. He leans in for a kiss. Maria shakes her finger no and points to the floor.

Renquist throws off his suit jacket, kneels down, and imitates a dog begging for food.

She tips him over onto his back. He loosens the top buttons on his shirt.

Hostility seeps through her facade. "I'm sorry about earlier. I must have forgotten to take off my 'please fuck me' sign."

Renquist looks confused. He laughs nervously and unbuttons down to his navel, slower with each button.

Maria jabs Renquist playfully in the thigh with the toe of her boots. She circles him counterclockwise and kicks as she goes along as if tenderizing meat.

His face reflects a battle between a libido clinging to a fantasy and his brain attempting a reboot.

Sexy ends, anger starts. In an instant, fire in Maria's eyes. "You want to fuck me? Am I meat, for you to devour? Did you think I was yours? I'm not for you. I have feelings. I. Am. A. Person."

Her words hit him like an instant cold shower. He's back. "You're too hot to be a person."

The assault escalates. Repeated blows crush his ribs. Crunch.

Renquist groans in pain. He scoots across the floor to flee. "I've never hit a woman, but there's always a first time!"

"Security!" They aren't coming. He eyeballs the intercom. Disconnected. He fishes his cell phone from his left pants pocket.

Maria punts the phone from his hand. It flies across the room and hits the wall. Shards break off, but it's intact. She feigns an attack.

He cowers and holds hands up defensively.

"For months, I made vids of every dirty thing you've done in this office." Maria towers over him with a cocky smile. "I'll ask you for a favor someday. You pay up. I destroy everything. Until then, you owe me one."

Renquist rises to his feet. "The only thing I owe you is an unmarked grave!"

Maria does a roundhouse kick to his head. Her right stiletto heel slices his cheek. The heel breaks on impact but remains attached to the shoe. She wobbles, then regains balance. "I'd better not get hurt. If this evidence gets out, they'll bury you next to me."

Renquist breathes heavily through gritted teeth. He clenches his fist and strains every muscle to hold back. Blood trickles down his cheek.

Maria stumbles out the door, with one high heel dangling. She leaves a small trail of bloody heel prints. After a few yards, Maria groans and intentionally breaks off the other heel. Maria lumbers away on her tiptoes. The heels flop with each step.

02. Justice Drive By

In an affluent Santa Monica neighborhood, well-manicured lawns and a contrived floral ecosystem maintain the illusion of an oasis from the crowded streets.

Maria strides through afternoon shadows towards a postmodern house with sharp right angles and an overabundance of transparent glass exteriors. In the driveway sits a yellow Lamborghini with rear spoilers. James notices Maria and rushes to get inside the car.

"Talk to me," Maria implores, as she moves into position a few feet behind the Lamborghini.

James revs the engine. The car lurches back a foot.

Maria jumps back, startled. She continues to block the car exit. Would he run her over? She's less certain. Maria grits her teeth and doubles down. She eases up to the rear of the car. "Come on bro. We've been best friends since before you liked girls. That's got to mean something."

The engine shuts down. James exits the Lamborghini without a word. He glares at Maria.

She stares back with pleading eyes. "The last thing you said to me was 'You're going to pay for your betrayal.' I didn't betray you. It was a ruse to get Renquist alone. After what he did, revenge was my only thought."

James stares with angry eyes.

Maria says, "Don't shut me out. It was my company too. Are you just going to pout?"

James turns towards the house.

"OK. OK. I'm sorry," Maria says with outstretched arms.

He swings back towards her. "How could you let me think you betrayed me, even for a second? Do you know how much I went through, just thinking that?"

"I'm sorry. Are we cool?" Maria inches her hand out.

He stares at her hand, motionless. It's a battle of wills. Maria feels more awkward every second he doesn't shake.

"Bros?" Maria overextends her arm in an awkward pose.

"Bros?" She extends her arm to the very limit. If Maria moves her hand any closer to James, she will fall over.

Maria can't hold the position much longer. Her patience and self-respect are almost depleted. Her muscles twitch.

James nods. "Bros." He steps towards her to take her hand.

They shake and transition to a one-armed hug.

Maria finally relaxes, spent.

"I still need to drive to clear my head." James gets back in the car. The engine hums.

Maria is dejected. Abandoned. All that to be left behind?

James pokes his head from the vehicle. "Are you coming?"

Maria is ecstatic. She dashes to the Lamborghini and gets inside.

James backs the sports car onto the street and peels out. Small talk, banter, then finally the two are old friends again. He accelerates around a Prius and a BMW as if it were a race. James banks around a corner towards Wilshire Boulevard. Ahead lies a solid wall of cars. He hard brakes. The car fishtails just short of the thoroughfare.

"Rush hour. Welcome to L.A., the entertainment capital of the world and traffic nightmare. That's as much clearing as my head is going to get." James pounds the steering wheel and repositions to merge into traffic.

Maria says, "Then let's work, James. What's your plan to take back the company?"

"A better lawyer might find a contract loophole." James drives aggressively. He forces his way onto Wilshire.

"Yeah, a lawyer off a park bench would be an improvement," she teases.

James eyes her suspiciously.

Maria tenses up. Did he notice her park bench comment? He doesn't need to know about her cameras, not yet. How can Maria distract him? An apology. "Sorry for insulting your brother like that."

He shrugs. "He did blow it. I'm waiting to hear back from half a dozen law firms. The lawyers aren't calling me back. I feel helpless like I can't move forward."

"The phone works both ways. Call them," Maria says.

James dials a number on his phone. A dial tone drones on the car stereo system.

A receptionist picks up. "McClintock and Associates, how may I help you?"

"This is James Wong. I need to follow up with Jim Forrester."

"Just one moment please." After a long pause the receptionist returns. "Our law firm is not interested in your case. Please do not call again." She hangs up.

Maria gasps. "You'd think we were telemarketers, from her reaction."

James does not take the news well. Gridlock seems to add to his tension. He watches a pedestrian outpace his extravagant sports car. He loses it. James pounds on the steering wheel.

"I'm going to try side streets," James growls. He fights his way off the boulevard, returning to reasonable traffic.

James dials again. It rings once, then goes straight to voicemail.

"Let me try with my phone," Maria says.

"Stevenson Horowitz Partners, how may I direct your call?"

"I've got a ten-figure contract dispute, and I'd like to discuss the possibility of your firm representing me," Maria says.

"Let me check if any of the partners are available," the receptionist says.

The phone picks up quickly. "Horowitz speaking. Just to confirm, that's a billion-dollar contract?"

"Yes. Do you take cases on a contingency basis?" Maria asks.

"We rarely do contingency on business litigation, but the potential payout is high enough, so we'll consider the case. Our fee is 40% of any settlement."

"Almost half?" James chokes on the words.

"Who is opposing counsel?"

"Briggs, Hultz, and Cantaker," Maria says.

The lawyer says, "Never mind. No lawyer would take a contingency case against them. I'd get better odds with the lottery. If you pony up two, or three million, and wait a few years, I'll see what I can do. Don't expect much."

"Two...or three...million?" James looks at Maria, almost hyperventilating. The Lamborghini drifts in the direction of his eyes. The sports car smashes into a parked Lexus. The airbag deploys. The right front wheel crunches to the sounds of shearing metal. The impact sends the Lexus hurling like a lumbering domino into a Civic, which hits a Camry, which taps the bumper of a Mercedes.

The lawyer ignores the noise. "Will you be putting us on retainer?"

"That was the sound of my life crashing. All my money is tied up in a company, which I have no access to. I can't afford justice! I'm not going to give you millions of dollars when I'm not sure how I'll even make rent."

"You could sue them for five grand in small claims." The lawyer laughs.

James hangs up the phone. "I always find a solution. I can't accept this is out of my hands. My brother is my only shot now. After the verbal assault I unleashed on him this morning, I'm not sure he'll talk to me, much less take the case. Maybe we can't get the company back."

Maria cringes. "At least the speakers still work."

"Can you just let me stay angry for ten minutes before you try cheering me up?" James asks.

"Call roadside assistance, and I'll give you some space. I'll be outside when you're ready to talk." Maria pushes the airbag out of the way. The door won't budge. She crawls out the window. Maria grabs a pen and paper from the glove compartment.

Vintage heavy metal blares from the car. James bangs his head into the airbag repeatedly in time with the music.

Maria steps over a bumper and car shrapnel. She leaves notes on each damaged car.

Five minutes later, a tow truck drives up behind the Lamborghini. A Persian tow truck driver yanks chains from the back.

James turns off the music and greets the driver. "That was quick. I'm glad I upgraded to the executive package. Here's my card."

The driver lowers the winch. "I'm not here for that."

Maria looks at the street sign and approaches. "Why are you here then? Street sweeping is Thursday on this side. Today is Friday."

"Don't worry," the driver says. "You're not getting a parking ticket. I'm just taking your car."

The driver puts chains around the right rear tire. He notices the damaged right side. The driver grumbles. "You got in a crash? They'd better not charge me for this." He takes photos of the accident scene on his phone.

James says, "You can't take my car! Banged up or not, it's mine. What? Are you a repo man, or something?"

"Your employer informed us you're no longer working for them."

"I don't have an employer."

The driver puts away his GPS locator. "So, it's confirmed then."

"No, I mean it's my company."

The driver shakes his head. "I hear this every time with these fancy cars. It's my company. We're not bankrupt. It was in the divorce settlement."

The driver positions wheel lifts under the tires. He gets more agitated as he talks. "I've got to take the car. I've got two ex-wives. Look, I'm a bit behind myself. Can you imagine my shame if I have to repo my own car? Not that I'd turn down the job, work is work. What would I do on my weekend? My tow truck doesn't fit a car seat."

"Please, there has to be something you can do to help us," James says.

"Look, here's what I can do. Can you cover your next payment today?"

James gives Maria puppy dog eyes.

"Don't look at me," she says.

"Well…no." James shrugs.

The driver says, "I don't care if it was mommy, daddy, or the board of directors that cut you off. It's not your car anymore."

The wheels lift off the pavement. James watches the Lamborghini disappear. The sky looks artificial. Red hues announce the arrival of night.

James sulks. "You can cheer me up now."

11

"We'll get something to eat, change clothes, and head over to Beer Beakers, my treat." Maria gets him to smile.

The two comrades stroll off into the sunset.

03. **Steamed Punks**

Beer Beakers is the fantasy bar of a mad scientist. Laboratory equipment functions as glassware. Tubes and exaggerated plumbing adorn the walls. An emergency chemical shower in one corner emphasizes the scientific theme. Only unattended musical instruments and gear on a small stage give away the bar's real purpose.

They call him The Blacksmith. The lanky white bartender with a leather apron looks like a blacksmith but acts like a chemist. He serves beverages in Erlenmeyer flasks, each ingredient precisely measured.

The majority of clientele don various steampunk fashions with goggles and top hats being the most common accessories. Dozens of anachronistic pirates, cowboys, pilots, and engineers enjoy themselves, among those too lazy to dress up.

Maria sports cowboy boots with spurs, a Victorian corset with an intricate system of buckles and chains, topped with a fedora augmented with clockwork pieces. Goggles hang around her neck, more a fashion statement than anything practical. Her makeup is low key except for ruby lipstick.

James wears a utility belt, vest, and slacks, all in muted earth tones. Wrenches, dual replica pistols in holsters, and a welding helmet dangle from his belt.

The Blacksmith places an Irish coffee on a stand at their table. A portable Bunsen Burner heats the drink.

James and Maria sit with body language more appropriate for a funeral than a bar. James talks to his beer glass more than anyone in particular. "I'm a billionaire…without any money."

A downpour of bad news broadcasts on the bar TV. Unemployment. Protests at the European Central Bank. Protests at the Federal Reserve. Riots at the G20 summit. Death. Fires. It's a money Apocalypse. How could the whole world go broke? That doesn't make any sense.

James stares at his beer, downbeat. "I filed for unemployment. It's all I have left. I just…" James trails off without enough energy to finish his sentence.

Bad news drones on, the perfect companion for their day so far. Head lice are now five times more popular than Congress. The Take Back America Movement is trying to take down every incumbent, Democrat or Republican. With everyone running scared, a schism in the Democratic Party created the Occupy Party, and The Tea Party split from the Republicans.

The political landscape is so toxic and splintered that "None/None of the Above" is one of the more popular election choices, and that poll was taken before a sting operation indicted five senators on corruption charges.

"We've got enough of our own problems." Maria stomps over to the bar. Her spurs jingle. She climbs on top of the countertop and manually changes the TV channel. She struts back to the table and sits down.

"How did your supermodel take the news?" Maria exaggerates the sssss in supermodel with a hissing sound and matching scowl.

James struggles to get out each word. "She dumped me."

Maria smirks at the news.

"She knew. Someone told her," James says.

Maria's eyes go wide, as her smirk vanishes. She stirs her drink nervously.

James eyes the Irish coffee, as steam wisps off the inviting beverage. The drink is there to be taken. Instead, he nurses his empty glass down to the last bits of beer foam. Nothing left. His head droops. "She didn't even bother to say goodbye. She sent another guy to pick up her stuff."

"I told you she was a gold digger." Maria's words hang in the air.

James lifts his head back up, incensed. "I told you so? Today? That's what you're going with?"

Maria sits straight, defiant. "You go after girls for the wrong reasons, and this is what you get. At least make some new mistakes."

"This is not how to cheer me up."

"No, but it is how I snap you out of mopey-mopey."

James rolls his eyes. "You could have tried good news. Tell me everything you did to Renquist."

"I tracked his blood all the way to my car," Maria says.

"Go, Maria! James Bond wouldn't get very far in high heels. Imagine how much ass you could kick in sensible shoes. Details."

Maria says, "Right after you threatened me in the hallway—"

James makes a stop sign with his hand. "I did not threaten you. When I said 'You're going to pay now,' that was for Renquist. I knew you were going to attack him."

"How?"

"Every time you wear lipstick, someone gets it in the nuts."

Maria purses her lips self-consciously. "You knew I didn't betray you, and you still put me through all that grief?"

James smiles and pretends to beg. In a shrill voice, he pleads with her. "Bro? Bro? Please, bro?"

Maria laughs. "I did not sound like that."

James continues in character, "Please. Please. I'll do anything, just be my friend." He mimes wiping tears from his eyes.

"Don't make me punt you." She pretends to kick him.

His smile disappears quickly.

Maria watches the mood change with concern.

James reflects, his expression somber. "I trust in us, but not much else. Before I saw the lipstick, that moment of doubt shook me. "

She already explained her reasons for the ruse. What else can Maria say?

The television blares louder, catching their attention. On TV, a young man in shabby clothes watches in frustration as steam comes up through the hood of an old Oldsmobile. "Tired of filling out job applications? Is your boss a jerk?"

Maria raises her voice and talks over the TV. "Why do they have to make commercials louder? It irritates me."

An overly energetic narrator continues. "Come find the perfect job at SlamDunkJobs.com."

Maria shuts up abruptly. She and James stare at each other. Their jaws drop at the same time. They turn back and watch the ad intently.

The young man sits at a computer, while the narrator explains. "We assess your skills to find the perfect job. Bid on the ones you want. If you win, the job is yours. Slam dunk! Then let us know if you want to keep the job."

The job seeker is wearing a suit. "I got a big raise! Thanks, Slam Dunk Jobs." He fist pumps and drives off in a yellow Lamborghini.

James almost falls out of his chair, as they both get to their feet. They high five and jump around. The excitement lasts for a minute before they calm down.

He slumps back into his seat. "Our new brand launches, and the first I hear about it is a TV commercial featuring my car? Aiya! I should be there."

Maria says, "Slam Dunk Jobs launched. That's good news, whether we're running the company, or not."

James shakes his head. "They didn't mention anything about training. I bet Renquist cut that part, just to save money. Better jobs were the whole point of Slam Dunk Jobs."

Maria looks deep into his eyes. "Renquist might be untrustworthy, ruthless, and evil, but he's a good businessman. When we get the company back, it will probably be worth even more."

James waves her off. "His money was supposed to power *my* vision, but he used it against me. The only way to fight power is more power. I need a power-up."

"James, this is not a video game. Life doesn't have power-ups. In case you forgot, we came here to have a good time. At least try."

James stares at the bar TV.

Maria gives James a dirty look. "Sulk if you want. You're too much work tonight." Maria reads emails on her phone.

They ignore each other, lost in separate worlds.

An old movie washes over James, punctuated by the occasional set of commercials. He finally comes back to life during a political ad. "I found my power-up," James says.

Maria looks up from her phone, puzzled. "I've seen that look. Here comes the crazy."

James is ecstatic. "I've decided I'm going to be president."

Maria slaps the table and cracks up. "You don't decide to become president. That's not how it works."

James ignores Maria's derision.

She sips her drink, but it goes down the wrong pipe. Maria grasps her throat as she chokes. Cough. Laugh. Cough and Laugh.

James looks concerned. "Are you alright?"

Maria takes a deep breath and tries not to chuckle. A few snorts escape. "You're serious."

"People want something different, why not the truth?"

Maria raises her hands in frustration. "You're too honest. You can't even bluff. You want to stroll into the most dishonest place on Earth? How's that supposed to work? Even the backstabbers have backstabbers."

"I can do a better job."

"You're a stress case, and you can't even stand up to your mother. In what world can you handle being president?"

James nudges closer to Maria. "The one where you're beside me as VP."

Maria puts her hands on her hips. "Maybe I should be president, and you should be VP."

"I call dibs." He grins.

"Did I forget to mention immature?"

James says, "The last time I had this look, I made us a billion dollars."

Maria snorts. "I'd be more impressed if we still had the money."

"I concede all of your points, but I can do this. I'm what this country needs."

She points her finger at him. "The whole world is going to hell, and there's nothing you can do about it."

James says, "If you want to be my VP, help me with reasons I can, instead of reasons I can't."

"I'm not your VP." She waves him off. "I'm not getting on the crazy train here. Stop wasting time on this ridiculous idea and figure out how to get the company back."

The waiter sees her waving and puts an ornate clockwork box on the table. James tosses a credit card in the box. The waiter takes it away.

James leans in towards Maria, trying puppy dog eyes. "Come on. I need you. What would it take to convince you?"

"OK. OK. Prove to me you can bluff. Impress me in a big way. Then I'll think about it."

"Impress you how?"

"I'll know it when I see it."

The waiter returns with the box. "How was everything?"

"Happy hour wasn't that happy." James signs the receipt and takes his card.

Maria holds out a Visa. "James, this was supposed to be my treat."

"I'm sorry, it's too late. The card has already gone through." The waiter takes the box away.

Maria glares at James, but he stares at his credit card. He looks left, then right, to see if anyone is nearby.

James whispers to Maria. "They didn't cancel the corporate card. Time for revenge spending."

He strides across the bar to the band setup. Maria follows at a leisurely pace, her spurs jingle with each step. She draws lecherous glances as she strolls across the bar.

He turns on the amp and speeds to the microphone. He taps the mic to make sure it's on. "I'm James Wong. Tonight, in this bar, I decided to run for president."

The crowd reacts with laughter, claps, and jeers. The Blacksmith fiddles with his leather apron, shrugs, and returns to work. Twenty partiers converge in front of the barely raised stage.

James yanks his biggest wrench from his utility belt and thrusts it up. "Do you like steampunk?"

The audience cheers.

"Who wants a steampunk president?"

Excitement grows, as people filter from their seats to join the commotion.

"To make democracy work, we need participants, not spectators."

A portly aviator waves at James. "Buy me a beer, and I'll vote for you."

James points his wrench at the crowd in a widening arc. "I'm going to start a conga line, so that everyone can participate in the celebration of my candidacy. Dance with me, and I'll buy everyone a beer."

Free beer earns a standing ovation, which would be more impressive if more of the audience were still sitting.

Electronic music plays on the speaker system. James chants with the music, "James Wong, steampunk president. That's right." He raises his arms repeatedly in encouragement. The chant spreads across the room.

"James Wong, steampunk president. That's right," the audience shouts.

At critical mass, James gets the courage to leave his microphone. He steps off the stage. A flash of doubt. Will anyone join him?

James waves his wrench overhead with the beats of the music. He kicks out on the fourth beat. Two kicks later, the portly aviator is the first to join. The conga line extends every second as others join the fun.

He looks back and sees Maria standing with her arms crossed. He leads the dance in a circle around the bar, back to Maria. "Don't be a spectator." He snakes the line around her so he can talk longer.

"I told you. I'm not getting on the crazy train," Maria says.

Over half the bar patrons join the dance. The chant dies down, replaced by a murmur of happy party goers. The train of people is too long for everyone to see the wrench, so kicks are out of sync.

Another revolution brings James back to Maria again. He tilts his head and gives her puppy dog eyes.

"OK. OK. I'm in front. I don't want anyone getting handsy." She cuts ahead of him to be first in line.

James puts one hand on her back. "Are you impressed now?"

Maria reaches back and grabs his wrench. She leads, waving his wrench, and kicking in time.

James stumbles, misses a beat, and shakes his head. He grabs the second biggest wrench from his utility belt. He attempts to take charge on a different beat. The kicks are haphazard in the line as dancers pick from the dueling metronomes.

Maria looks back with a smirk.

James fumes. He puts his wrench back and roots around in his utility belt pockets. He pulls out his rave whistle. It's an echo of a vintage club scene when suspenders and mushroom hats were fashionable.

"Rock, paper, scissors. Whistle beats wrench." He double whistles for each kick. The whistle makes a much better metronome, so the entire crowd syncs with it.

Maria laughs and glances back. She syncs her wrench to his beats too. The dance is better than ever.

After another lap, the front door opens. Eight Latino men with matching "Los Muchos" tattoos enter. They wear jeans and white wife beaters, except for their shirtless leader, Jaguar. Tattoos of Mayan glyphs cover him, with his fists inked to look like jaguar mouths. Gangbangers. The conga line near the door momentarily collapses like an accordion, before resuming its flow.

The skinniest gangbanger, Flaco, approaches the conga line and pushes to insert himself between James and Maria.

James blocks him, but Flaco pries his hand off Maria. Flaco puts both hands on Maria's back.

Maria kicks left.

Flaco's eyes glance down, admiring the view.

Her hips sway side to side with the music. Maria kicks right.

Flaco's hands drift down to her ass.

Maria kicks back. Her spurs catch Flaco hard in the groin.

He slumps to the ground with a shriek, clutching his privates.

She doesn't miss a beat with her wrench.

James leaps over Flaco to rejoin Maria. "See what I mean? Lipstick."

Maria snickers, but doesn't turn around.

Behind them, people step over Flaco, one by one, like a trail of army ants.

When the song finishes, the conga line disperses. The thirsty herd converges at the bar. The Blacksmith marches towards James.

James holds up the corporate card.

The crowd cheers as The Blacksmith takes the card.

"I'm going to check on my roadkill," Maria says.

James backpedals towards his table. "I'll grab my drink and meet you over there."

An albino waiter approaches Jaguar. "Sorry sir, no shirt, no service. I'm going to have to ask you to leave."

Jaguar lunges towards the waiter. "Vanilla, I suggest you bounce."

The waiter retreats behind the bar.

Flaco points at Maria. His other hand still clutches his privates. He struggles to speak. "That's the bitch."

Maria strolls towards the gangbangers, enjoying each clank of her spurs. She stops ten feet short of them and eyes Flaco. "My ass is a hands-free zone."

Jaguar struts towards Maria and motions his posse to follow. "If you'd hit him any harder we'd be calling him Flaca." Jaguar laughs at his own joke.

The crowd gives a wide berth to the area, content to watch. The circle of spectators expands as if this were an impromptu sporting arena.

James notices trouble brewing and hurries to Maria.

Jaguar shifts his attention to James and makes an anxious laugh. "You know martial arts?" Jaguar sizes him up. "You don't look like a Kung Fu."

James forces a smile and asks Jaguar. "So, what brings you here? A love of chemistry?"

"Why the fuck would I care about chemistry?" Jaguar lunges the last few feet until he is face to face with James. His men stay two strides back.

James backs off, hands raised. "Sorry. I just meant, look around. It's a themed bar. There's even a safety shower in the corner for chemical burns."

Jaguar swings his head around, mocking surprise. "I just noticed." He glares at James. "You don't think I have two eyes in my head?"

James grabs Maria's arm to pull her into retreat with him.

Maria yanks her arm back. She's not backing down. "Why didn't you ask if *I* knew martial arts?"

Jaguar shifts focus to Maria. "What's your name, chica?"

She stares at Jaguar and clanks her spurs together. "My name's not Benjamin."

Jaguar turns back toward his friends. He shakes his head with a confused expression. "What?"

James takes advantage of the distraction and books it to the edge of the encircling crowd. He flags down the portly aviator and a few other bar patrons. "I need your help."

"I'm not fighting those guys," the portly aviator says.

James pulls the welding helmet from his utility belt and puts it on with the faceplate open. "No fighting. Just follow my lead."

The portly aviator does a salute. "Goggles at the ready, Mr. President."

A pirate with a clockwork parrot glued to his shoulder adjusts his eyepiece. "Goggles at the ready."

James swipes a Bunsen Burner and a beaker of clear liquid from a nearby table. He runs towards Maria.

Both the aviator and pirate grab the same supplies and follow.

"That bitch is crazy," Jaguar says to his amigos. He swivels and returns to Maria. "You must be loca to mess with Los Muchos."

"Suck my dick, assholes." Maria readies for a fight, hands open.

Hearing those words from a woman confuses the gangbangers. They hesitate but surround Maria anyway.

James attempts to flank them with his two new friends. James thrusts out his beaker. "Stand back. I've got a beaker of dihydrogen monoxide."

The pirate holds out his beaker.

Jaguar and his posse back away, so they can keep an eye on James and Maria from the same direction.

James says, "With second-degree burns, your tattoos would be cooked right off your burnt flesh. This chemical is used to distribute pesticides. It's also fatal when inhaled."

Jaguar notices his friends are getting skittish. He assesses the threat James poses. Undecided.

"If you get third-degree burns, it won't hurt…because burns scorch off your nerves. Are you ready for a taste of dihydrogen monoxide, punk?" James makes a menacing step forward. "It's found inside nuclear power plants and cancer patients. Dihydrogen monoxide was used in torture in Iraq."

Like an afterthought, the aviator also holds out his liquid.

Jaguar smiles. "Bullshit. If it were as dangerous as you say, you'd protect yourself."

"You want to mess with science?" James puts down the faceplate on his welding helmet. The aviator and pirate put on their goggles in sync a second later.

Jaguar looks around at the bar. "Oh shit! That's what all the goggles are for."

James motions with the portable Bunsen Burner at the clear liquid. "What kind of burns do you want? Are you ready for the third degree?"

The gangbangers panic and jet for the door. The sea of spectators parts to allow the gang to evacuate.

Jaguar throws Flaco over his shoulder as they hurry out the exit.

The crowd cheers. The aviator and pirate high five.

Maria slow claps for James. "Dihydrogen monoxide. Two hydrogens, one oxygen. H_2O. You were bluffing with a glass of water?"

James says, "If they'd remembered high school chemistry they would've kicked my ass. See? I can bluff without lying."

"You bluffed to scare off some punks, using steam, in a steampunk bar. Wow! OK. OK. Vice President sounds good. I'll finally get a vacation. Just don't call me your sidekick." Maria tilts her left leg up. "Or I'll introduce you to *my* side kick."

James takes a sip of his clear liquid and spews some out. A small fireball erupts over the Bunsen Burner for a second, then dissipates. "This is vodka. I almost served myself a Molotov cocktail."

Maria snickers and turns the flame off on the Bunsen Burner.

"What?" James says. "How was I supposed to know it wasn't water?"

Maria grabs an ice cube from an abandoned drink and plops it in his beaker. The ice cube drops to the bottom. "If it were water, the ice would float. You're not the only one who got A's in school."

James moves quickly to change the conversation. "Are you excited? I'm going to be president."

Maria rolls her eyes. "Yep. Making great life decisions in a bar, how could we lose?"

"If we can get the Take Back America Movement to back us, I'm sure we'll win. We have work to do." James downs the vodka.

04. Unleash the Internet

An overgrown lawn surrounds a two-bedroom yellow stucco house with a clay tile roof. Three political posters jut out of the grass. Maria uproots them one by one, tapping the grass back in place with her boots.

James claps as Maria removes each political opponent from her lawn.

Maria throws the posters on the porch, into a stack with another half dozen. She stomps her boots and opens her front door.

A rack with dozens of shoes, slippers, and a backpack sits near the entrance. They change into slippers at the doorway.

The living room has the feel of a theater inside a library. Cabinets filled with Blu-rays, DVDs, CDs, and books line the living room walls, meticulously organized by genre and alphabet. A 90-inch LCD TV is the centerpiece of the room. Two couches and adjoining tables are arranged for optimal viewing angles on a white tile floor.

Maria heads toward her bedroom. "I'm going to change into my power suit." She closes the door behind her.

James sets up a camera on a tripod in the front of the TV, facing behind him. He turns on the TV and camera. "We should be able to get some good practice footage."

One too many drinks have given James rosy cheeks. He moves behind the camera and waves at the camera. James watches his actions play out on the TV. He strokes his cheeks and makes funny faces.

Maria strides confidently into the living room in her curve-hugging power suit, with fresh lipstick and new knee-high boots. Her form enlarges on the TV as she approaches.

James pretends to protect his balls with a playful expression. "Uh-oh. Lipstick."

Maria socks him lightly on the arm from behind. She stands by his side towards the camera.

"You know what I learned today?" James shakes his head. "The best criminals wear suits. You invest your whole life in something. Then, a lawyer comes by, and they want a 40% cut. That's the price of justice?"

Maria shifts the camera to point at her. "Why do men think you want to sleep with them if you're the slightest bit friendly?"

James nudges Maria aside. "Justice is something for the rich. The rule of law can be bought and paid for. If you're poor, you can't get justice on layaway."

Maria flirts with her hips and strokes her leg seductively.

James shares the spotlight with her. His mouth gapes open.

Maria compromises between a pout and a smirk. "Why are we prisoners to how people react to us? I love the way this outfit makes me *feel*, but I hate the reactions I get. Why is it not OK for me to feel powerful *and* sexy?"

The distraction of her outfit deflates his rant. James nudges his way towards the camera. "Lift that rug and see what we swept under there. Can we solve problems, instead of fighting?"

Maria crosses her arms with a serious look. "My ideas should not be ignored or sidelined, regardless of what I wear. I want to be heard." She looks at James. "But you don't always listen."

James stares at the camera, oblivious to Maria's glance. "Exactly. Politicians are talking past each other, not listening to each other."

She points at James. "You're not listening to me."

"You tell them, Maria." James fist pumps, still oblivious.

Maria facepalms and shakes her head.

"The Take Back America Movement shouldn't just vote against incumbents, they should vote none of the above," James says.

Maria nods twice and rushes to the door.

"These people don't represent me. Do they represent you? Why not an Asian president?" he wonders aloud.

Maria returns with the posters from the porch. She plops them on a table. "Or a Latina president?" She winks at the camera.

"These candidates are all more of the same. Demand better choices," he says.

Maria yanks open a table drawer. She retrieves a black marker and grabs a "Garfield for President" sign.

James scoffs. "Senator Garfield is so in bed with the banks. I'm surprised they don't already have a love child."

Maria draws a large circle with a line through it across the sign. "None of the above," she shouts. Maria flings the sign across the room.

She picks up a "Danvers for President" poster.

James says, "Danvers is CEO for a defense contractor. Give him the power to start a few wars. Good idea."

Maria defaces his poster with the universal no symbol. "None of the above," she shouts, even louder. Maria splinters the poster over her knee. She lets the broken pieces flop to the ground. "To all the other candidates, get off my lawn."

James cheers. "When polls show none of the above is popular, people want real change. It's not a wasted vote. Vote none of the above," he says.

Maria chimes in. "And vote more jobs."

"It's a message they need to hear. I'm James Wong. I'm none of the above."

"I'm Maria Cortez. I'm none of the above."

James growls. "You didn't even want to run for president." He pushes her, but Maria doesn't budge. She turns it into a tug of war.

He tries running in place, but his slippers are no match for the traction of Maria's boots. It's like watching a car stuck in mud dig itself into a deeper hole. With alcohol-impaired balance, he slips and grabs Maria to stay upright.

She pushes back. "If you could be president, I could be president. I'm just as capable."

He slides backward until he's no longer visible on the TV.

"Do you accept me as an equal?"

"I wouldn't do it without you," he says.

Maria drags him back in view and lends James a hand.

James stands. "Working together is the only way. America needs problem solvers, not more lawyers and politicians. We'll fix campaign finance. I know how to make it more transparent."

Maria shakes her finger no. "Focus on one issue first, more jobs."

"We'll fix the education system. I have a plan to create rock star teachers," he says.

"Rock stars teaching our kids, that's a scary thought. Smashed guitar, sold separately. We need more jobs." Maria says.

"Fine, we need more jobs. We'll fix that too." James says.

Maria grabs a bottle of vodka and pours two shot glasses.

James and Maria make a toast. "To America. Let's fix it." The glasses clink.

James tilts his head back. A shot of vodka vanishes. He wobbles as he speaks. "We can fix America. We *will* fix America. Internet, how would you fix America?"

"Unleash the Internet!" James jumps in the air with a double fist pump. He belly flops onto the table. The particle board splinters under his weight.

He lifts up his head and looks at the broken table sections. "And we'll fix...the table." His eyes flicker. James slumps over, passed out.

Maria turns off the camera and surveys the carnage. "I'm so done with your campaign." She checks on James. Maria shakes her head and uploads the unedited video to the Internet.

24

She stomps off towards her bedroom, pauses at her doorway, and looks over her shoulder at James. He lies sprawled out on broken drawers and particle board shards. Maria sighs and trudges back to him. She yanks his arm, lifts James out of the table wreckage, and throws him on the couch. Maria brushes the dust and debris off his clothes with harsh strokes. Her strokes grow slower and gentler, as she calms down.

Maria takes off his utility vest, gun holsters, and anything uncomfortable. The wrenches clank as they slip onto the tile floor. She cringes at the noise. James remains a silent action figure, in the last pose Maria left him in. She heaves him off the couch, drags him to her bed, and tucks him in. Maria puts on pajamas and retires for the night.

05. **Presidential Hangover**

Morning light pours through the windows of Maria's bedroom. A katana and two smaller swords with matching water serpent scabbards hang on the wall next to posters of karate stances and striking points. Martial arts trophies and rainbow knick-knacks that could just as easily belong to a 12-year-old schoolgirl fill a display case next to the bed.

The open closet door reveals wardrobes from the exotic to the mundane. Steampunk, goth, and kimono ensembles clash with outfits Aunt Opal might wear lounging around the old folks' home. The conservative versus the decadent. Even Maria's clothes are ready to fight.

Four pillows prop up James in bed. Maria sleeps next to him with a single pillow. She stirs first.

Maria turns to James in bed. "Thanks for not puking on my sheets, Mr. President."

James clutches his head with both hands. "Last night is a little fuzzy." He rubs his eyes. His eyes go from open to bulging. Maria lies next to him. In bed. "Last night must have been wonderful!" He leans in a few inches towards Maria.

Maria looks up, pondering the night. "Wonderful is a strong word. I'd rather not go into all the ins and outs."

James lights up at the mention of the words in and out. He breathes deep and stares at Maria with a lifetime of desire. "I've liked you for a long time." He scoots a couple of inches closer.

Maria smiles at him. "There's no one I like spending time with more than you." She socks him playfully on the arm.

His voice quivers with emotion. "I've wanted something to happen for so long. I'd lost hope." He leans in further towards Maria.

She notices him invading her personal space. Her eyes go wide. Red alert! He's going for a kiss.

She evades his lips, does a rolling tumble onto the floor, and rotates to face the bed in a crouched fighting stance. "I'm your bro, not your girlfriend." Maria stares down James. "Are we clear?"

James shakes his head, confused. "Yeah, but…"

Maria repeats the words slower. "I'm your bro, not your girlfriend. Are we clear?"

James stands up on Maria's mattress. "I never thought we'd be in bed together, but here we are. Are you saying we didn't sleep together?"

She rolls her eyes. "OK. OK. Literally, we slept together, but there was no sex and never will be."

James raises his arms and looks around. "I'm in your bed." James bounces a few times on the mattress and points down. "I wouldn't be here if you didn't want me here."

Maria says, "You were in bad shape. I wanted to make sure you didn't puke up a lung. Chivalry isn't just for boys."

He points down again. "This is not the couch. This is your bed."

"James, if something happened last night, then why are you fully clothed?"

James looks down at his steampunk clothes from last night and blushes. He steps down from the bed. "We can pretend today never happened."

Maria springs up and strolls over to James. "Oh, no. I'm going to tease you for weeks over this."

James laughs halfheartedly. He retrieves a wad of paper from his pocket. He holds one end and flicks. The receipt is long enough to hit the floor. "This is worse than a CVS receipt. Is this a bar tab, or a novel?"

James follows Maria into the living room. He surveys it with fresh eyes. "You need some more self-control, Maria. Who pissed you off this time? That table is totally smashed."

Maria retreats to her room in a huff and slams the door.

"The tech community should love having one of their own as president. Easy sell," James war dials every tech CEO in his contacts. He leaves dozens of messages. Not even long-term friend, Rick Slater of Slater Dynamics, returns his calls. Déjà vu. It's just like the lawyers.

Maria changes from her pajamas and returns to the living room.

"Maria, can I borrow your phone?" James asks.

She hands him the phone.

He dials. "I'm the assistant to Maria Cortez of Adaptive Unlimited. I need to connect her to Rick Slater."

"One moment," the receptionist mumbles.

Maria chuckles. "Hey *assistant*, can you get me some coffee?"

James gives her a dirty look.

Rick comes on. He speaks with urgency. "Maria, I hope you're staying away from James. He's a black hole. Anything he touches disappears."

"Thanks, Rick," James says. "Good to know I can count on you."

There is only a deep breath on the other end, then an exhale.

James says, "How long have we been friends?"

"If you're asking that, you obviously know how much trouble you're in," Rick says.

"Maybe I don't."

Rick says, "There isn't a powerful VC that Renquist isn't invested with. Nobody by the name of James Wong is raising money in tech, or anywhere else, until you get Renquist off the warpath."

"I'll handle it," James says.

"You know, you're not the only James Wong," Rick says. "Two companies lost late round funding because they had a James Wong on the payroll. Stay away from me. I prefer not to be collateral damage. You're the Titanic. Put Maria in a lifeboat. Don't sink her into the depths with you." Rick hangs up.

James gives Maria a puzzled look. "Why aren't they coming after you? You're the one that kicked his ass."

"I doubt Renquist wants to trumpet the circumstance of our last meeting," Maria says.

He says, "Without donors, that campaign video is more important than ever. Let's work on it."

"It's already done."

Concern creeps on his face. "I was so wasted. There is no way it doesn't need major editing."

"I uploaded the raw video last night," she says.

His jaw drops. "You did what? Aiya! That was only a rehearsal."

"You were out of control. I retaliated. Sorry. We haven't promoted it yet. I'm sure nobody even noticed."

James grabs a backpack from the rack by the front entrance. "You're probably right. I'll look at it when I get home." He shoves a laptop and everything Maria took off him last night into his backpack. James pulls out his car keys. "Right…my car."

"You need a ride, just ask," Maria says.

James opens up the Uber app on his phone.

She puts her hand on his phone. "It's my way of apologizing."

He reluctantly accepts her ride home. The conversation in the car is like a frigid wasteland, filled with icy stares and desolation.

06. Shambles

James and Maria approach his apartment. Vertical blinds conceal the contents of the home. James opens the mailbox outside his apartment, while Maria waits silently.

He ruffles through his mail until he sees a California EDD logo. "I didn't expect my unemployment check yet. Finally, some good news."

James opens the letter and reads. "My unemployment claims are contested for work-related misconduct. That's so petty."

Maria notices the front door is slightly ajar. She steps past James to investigate.

"I'm rich on paper, but two steps from the curb. I'm going to be homeless, and now I can't even live out of my car." James realizes her attention is elsewhere. "Maria?"

She pushes the door. It creaks as it swings open to an apartment in shambles. Maria barges in. She checks each room, but the vandals are long gone.

James treads lightly, his steps shaky. He surveys the devastation with lost eyes. Nothing inside the apartment is undisturbed. Holes in the walls. Ripped up floorboards. Upended cabinet drawers, with contents sprawled across the floor. Slashed couch cushions. Even the refrigerator door is left wide open.

"Why is my apartment ransacked? What did you do?" James asks.

Maria eyes the ground and mutters inaudibly.

James takes on a mocking tone under his breath. "Don't take down Maria. Pfft. I'm the collateral damage." He raises his voice. "What's going on, Maria?"

She speaks much louder. "I might have put Renquist under surveillance and maybe…threatened him."

"Maria, where did you put the camera?" James cringes, afraid to hear the answer.

"Your office. It's the only place big enough for his ego."

His gaze sears into her.

29

She wears her guilt like a scolded child. "I recorded every office visit for the last two months. I knew Renquist was up to something. Today proved it."

James blushes. "Months? Sometimes I do private stuff in my office."

Maria rolls her eyes. "I know. I fast forward through those parts."

"Please tell me you didn't make recordings." James looks at the ground.

She says, "I never trusted Renquist. I had to know what he was up to. Someone has to protect you, while your avatar floats off in imagination land. That someone is me."

James rushes to the refrigerator and slams the door. "I don't feel protected." He twists a kitchen cabinet drawer right side up and places kitchen utensils inside. His eyes tear up.

Maria watches quietly from a safe distance.

James jolts up. "What does it matter? As a last resort, I was going to sell my furniture to make rent. Who's going to buy a slashed-up couch?" He flings the drawer across the room. The kitchen utensils scatter from the point of impact.

"The hard part is those moments when I don't feel you believe in me. They aren't just taking my company. They're taking my life." Tears stain his cheek.

"I believe in you," she says.

"You say the words, but your actions don't always agree. You even fought me for who leads a dance. A dance."

"I challenge you to bring out the best," Maria says.

"I get that enough from my mother. Can't you just support me?"

Maria gives James a remorseful look. "What would make you feel supported?"

His tears stop. James considers the price to quell Maria's guilty conscience. What would prove her support? "I need a hug."

They both remain silent for a few moments. Did James ask too much? It's emotional, but also awkward.

Maria opens her arms. "I know it's been rough, but I didn't realize…"

James embraces her in a bear hug.

She winces. Maria looks like she's enduring a root canal, but she softens. It's not that bad.

He rests his head on her shoulder.

She's somewhat puzzled by it. The expressions on her face reflect her rating the experience in real time.

James takes relaxing deep breaths.

Maria's breaths follow his.

He rocks back and forth.

She syncs with the rhythm. Comfort. It's a safe place, a meditation. Maria nods in approval. She nestles her neck on his shoulder like a cat getting comfortable. They smile in deep contentment.

Evidence of intrusion at their feet can't penetrate the happy bubble. Everything is OK. Calm. Sleepy. Safe. Spiritual. Safe. Comforting, like a mother's womb. Safe.

Or is it safe? James jerks back, awoken by an unwelcome realization. "They repossessed the car but didn't bother closing the credit card. This isn't about money. What if they're after your surveillance recordings? They wanted to search my car. They're probably searching your apartment right now. Let's go." James grabs his backpack.

They sprint to the car and speed off to Maria's house.

Maria runs to her front door, leaving her keys in the ignition.

James grabs the keys and follows.

Her door is ajar. She shoves it open, mid-run.

A man and a woman wearing dark clothes, ski masks, and gloves take DVDs and Blu-rays from her wall. They toss the disks in a box and throw the containers on the floor.

James peers inside and ducks back out. He calls 911.

Maria rushes into her bedroom.

Caught. The intruders scramble towards the bathroom.

She yanks her water serpent katana from the wall display and charges after them.

The female intruder pulls a gun. Her accomplice deflects the weapon down, as it fires. The bullet blows through the floor. "He said alive," the man says. The intruders enter the bathroom and lock the door after themselves.

Maria thrusts the sword through the door and quickly retracts. Blood. A second stab encounters only air. Maria kicks a gaping hole near the lock but gets stuck. She drops the sword, frees her leg from the door, and reaches into the hole she made, to unlock the door. Maria flings the door open.

A red trail leads to the man near the bathroom window, as he wraps his left arm with a bloody towel.

She picks up her katana and charges again.

The wounded intruder steps on the toilet and launches himself through the open window, letting himself fall. He lands on the grass with a thump and a scream of pain.

His accomplice, already outside, helps him to his feet.

He stumbles away, favoring his wounded arm.

The intruders can get to the alley from Maria's backyard. She can take a short-cut through her neighbor's yard to catch them. She darts out the front door, bloody katana in hand.

James stands behind two empty police cars parked on the street. He gives a worried look and sprints towards her.

Maria veers right across her porch. Her boots stomp with loud thuds.

The bushes next to the porch rustle, both behind and to the left. Footsteps and creaking wood follow her on the porch.

Maria doesn't look back. She hops over the porch railing and runs across the neighbor's lawn.

When she crosses the property boundary, two police bark orders. "Police! Don't move. Drop the weapon."

Maria sputters to a halt. She peers over her shoulder.

Both police officers point guns at her. Officer Samuels aims from behind the porch railing. He seems reluctant to leave the cover and height advantage of his position.

Officer Monroe approaches Maria across the grass with his gun drawn. James races after him.

Maria remains stationary in an awkward position, with only her neck and head facing the police. "My katana has DNA from my assailant," she says.

At 15 feet, Officer Monroe creeps closer. "Drop the weapon."

"That might contaminate the evidence. Take it from me," Maria says.

Officer Monroe cocks his gun to intimidate Maria.

"Don't hurt her!" James screams, a few feet from Officer Monroe.

Officer Monroe holds his hand out at James. "Stand back. Remain calm." Officer Monroe is anything but calm. He motions for James to stop and shifts between pointing his gun at James and Maria.

"OK. OK." Maria drops the weapon onto the neighbor's sidewalk with a clank. "You're letting them escape."

"Hands behind your back. On the ground now."

Maria complies.

Officer Monroe jerks her arm back and places handcuffs on her.

James says, "What are you doing? I asked for your help."

"As per California Penal Code Section 417, I am placing you under arrest for brandishing a weapon."

"It was self-defense, they shot at me," Maria says.

Officer Monroe scoffs. "If it was self-defense, you're lucky. Bringing an oversized knife to a gunfight usually doesn't end well."

The policeman pats down Maria, walks her to the car, and shoves her in the back seat. Officer Monroe places the bloody katana in plastic inside the trunk and drives off.

Maria yells at Officer Monroe. "My front door was open. You didn't cordon off the area. What kind of cops are you?"

"I'm so glad you watched enough TV that you know how we should do our jobs," Officer Monroe says.

07. **Trench Coat**

James feels claustrophobic in a small precinct interrogation room. He occupies an uncomfortable metal chair seated behind the solitary table. Barren walls surround James, except for the one-way mirror that occupies half of one wall, and the door out.

Samuels paces on the other side of the table. "Was there really a home invasion?" He pauses and leans over the table. He lowers his voice like he's sharing a secret. "Don't be embarrassed. It happens. Just tell the truth."

"We caught them in the act," James says. "Why else would I call you?"

"I know it's hard to talk about." Samuels nods at James sympathetically. "Has Maria ever hit you?"

James pushes off the table to stand. He notices Samuels putting his hand closer to his weapon and sits back down. "What? Why are you...well...not really."

"That doesn't sound very convincing. Yes or no? Be straight with me." Samuels channels more cop, less friend.

"Just friendly taps on the arm," James says.

The cop leans in further and stares eye to eye with James. "Hard enough to cause bruises?"

James squirms in his seat. "Sometimes she plays rough. She forgets her own..." This cop doesn't know Maria. He won't understand. Sweat glistens on his forehead. "No. She doesn't hurt me."

"James, you know how that sounds? Can you lift up your shirt?"

"Whatever, I have nothing to hide." James unbuttons his vest and shirt. Bruises and numerous scabs scatter across his chest.

Samuels cringes. "Yikes, she really did a number on you."

James relies on instinct and a half-remembered impression. "Oh, right. I think I fell."

The cop interprets his insecurity as lying. Samuels pounds the table. "Don't make excuses for her. Look, your old lady came looking for you with a bloody sword. I'd be freaked out too. You were right to call us."

James shifts in his chair. "She's not my old lady. She's my bro."

"She didn't look like a bro to me." Samuels plunks down pictures from a folder showing both sides of the bed slept in. "Do you usually sleep with your *bro*? Tell me about last night."

"Last night is a little fuzzy."

"So, you admit, Maria might have hurt you."

"I don't remember." James wonders about last night. He tries to dredge up any memories he can.

Samuels says, "Did you cut yourself shaving with the samurai sword? That's the only excuse I haven't heard yet."

"It's not my blood."

"We'll find that out soon enough. Can we get a DNA sample to prove it?"

James folds his arms. "Go ahead. Am I being charged?"

"No," Samuels says. "We'll release you after we finish your statement."

"Where's Maria?"

"By law, we have to arrest a suspected batterer, when we see visible marks." Samuels points at James' chest. "We'll hold her 72 hours, more if you press charges. I suggest filing a restraining order."

James grows impatient. "What about my apartment?"

"Wait here, while I check on your break-in investigation." The door buzzes and Samuels leaves. The lock clicks behind him.

If he's just taking a statement, why is the door locked? James tilts his head down and closes his eyes. He attempts to center himself, to reconstruct the last 24 hours.

Before long, a door buzz interrupts his thoughts. James opens his eyes. It's not Samuels.

A blond male with gloves and a trench coat enters. Wraparound sunglasses and a hat conceal sections of his face. His clothes seem out of place on such a hot day. "I'll give you a choice. Very rich, or very dead?" He speaks with a thick Russian accent.

"Hmm, hard choice." James tests with a laugh, but Trench Coat Man doesn't even smile. "What's the fine print?"

"Police think Maria hurt you. What if new clues point to…murder instead? Like, say, a body to match Maria's bloody weapon."

James breathes unevenly. Chills travel down his spine. He thinks back to all the episodes of violence he's seen or heard about. "Maria didn't kill anyone…did she?"

Trench Coat Man grins at his doubts. "Kill today, or later, that rose has razor thorns. Get too close, and she cuts you. Renquist has scars to prove it."

"I don't hear choices," James says.

"Recover our secrets and testify against Maria for murder. You get your life back, and the Adaptive Unlimited sale goes through. You've got a billion reasons to help us."

What secrets? What was on Maria's surveillance recordings? James tries to hide his worry with a smirk. "What's behind door number two?"

"We give you reasons to take your own life. Once we tire of waiting, we help you to the great beyond."

"I pick none of the above." James stands, defiant. "I pick Maria. I pick taking my company back."

Trench Coat Man cracks his knuckles. "Next time I see you, the only choices will be what bones you want broken next." With a swirl of the coat, a buzz, and a click, he vanishes like an apparition.

Who opened the door? At least one other cop must be in the booth conspiring with Trench Coat Man. Why does Renquist need to own cops?

James stares at the one-way mirror. Maybe if he bores them, they'll leave. His new staring strategy only makes him sleepy. The only off switch his brain has is alcohol, which is not normally found in police interrogation rooms. He zones out, thinking, thinking, thinking.

He awakens from his trance with a memory. James strides to the door and pounds on it. "Officer Samuels?" James continues pounding and calling out for him until he hears voices outside.

Samuels opens the door with a surprised look. "Sorry Mr. Wong, they told me you already left." He holds the door open for James and motions for him to follow. "We have everything we need. You are free to go."

James emerges into the wide-open spaces of the police precinct. There are no cubicle walls, just an open floor plan arrangement of desks and people. James follows Samuels back to his desk.

"Officer Monroe is looking into your break-in." Samuels hands James contact info.

"I can prove Maria didn't hurt me." James doesn't remember what's in the video, but Maria wouldn't hurt him, at least intentionally. "She posted a video online that night."

Navigating the web flusters Samuels.

James reaches across Samuels' chest to type. He finds Maria's post. The thumbnail for the video features Maria doing a sensual pose in her curve-hugging power suit.

Samuels points to his monitor. "Two million views. Wow. You're a virus." He tilts his head with a thoughtful expression. "Yeah. I get those on my computer all the time."

James chooses not explain the difference between virus and viral. He watches the video intently.

35

Other cops loiter behind Samuels to watch. One of the youngest cops blurts out, "Hey, it's None of the Above Guy. It's so long. This must be the extended version. How does it feel to be an Internet celebrity?"

James shakes hands. "I just found out. It feels surreal." He cringes at the embarrassing parts of the video. When he splats straight into the table, James points excitedly, "See I fell. That's where the bruises came from."

Samuels narrows his eyes. "What a jerk. You get drunk and let this woman take the fall for hurting you? You should be ashamed. I'm getting her released right now. If she smacks you, don't come crying to me. You deserve it."

James sighs and rolls his eyes.

08. **Stolen Memories**

Maria browses the contents of her bookcases along the walls of her living room. Movies and music blend on her shelves.

James sits on a couch and sets up his laptop at the unbroken table. He eyes the damage on the other table with guilt and embarrassment. The legs on the broken table hold up two triangular sections on either side of the crushed center. They point up like tents for two opposing camps. "It's weird they put everything back. Did they stage the scene for the cops?"

He checks out their YouTube video. "Maria! 2.5 million views! O.M.G." He fist pumps.

Maria inventories her media collection with vacant eyes lost in her own world.

James checks for trending topics. Not there. He scowls at the laptop screen. James searches and reads recent tweets for his name. "Only 50 tweets about James Wong and most aren't about me. I don't understand." His positive attitude wears down.

Frantic. James reads through the thousands of text comments on his video. The most popular topic is Maria's curve-hugging power suit. Posters applaud the clothes as a statement not to judge women based on their appearance. Opposing comments discuss her hotness, including ratings in the 8 to 11 range. Vulgar suggestions and a feminist backlash round out the debate.

James talks at his screen. "Maria, your outfit got us a lot of attention. I hope that translates to votes."

Maria ignores James and mutters to herself. "Wrong genre, wrong shelf." She separates music and movies into different bookcases with mechanical movements.

Many posts have the phrase "Unleash the Internet," not always spelled correctly, with links to YouTube videos. James clicks.

Two Italian men in New York Yankee caps talk in a thick accent. "It's not enough to bring down the two-party system. We need to give them the finger." They flip off the camera in unison. "I'm voting none of the above," says the guy on the left. "And more jobs. Fuhgeddaboudit." says his friend. The video ends.

A man from Nebraska rambles for five minutes about how teaching prayer in schools would fix America. He ends his video by saying, "Vote none of the above and more jobs. I'm Alan Perry. I'm none of the above."

Hundreds, maybe thousands, are claiming to be none of the above. Right, left, or middle, the videos represent every view. Tax the rich. Tax the poor. Abolish the IRS. Kill the EPA. Protect the environment. It's a firestorm, every possible argument on the Internet distilled into one location. Everyone agrees to fix America. *How* is another story.

His agitation bubbles higher with each video. "Just what the Internet needs, more rants."

Maria stares at a DVD case for *Moonraker*. Perhaps irritation is contagious. "Duplicates? I never." She checks the disk for scratches. Panic and shock roll across her face. She inspects one after another in desperation.

James inhabits a separate hell, losing energy with each post that ignores him. The pranksters support the idea of a protest vote as the ultimate lulz. You can almost hear echoes of them laughing in the shadows. The Internet trolls pile on, just to rile up anyone they can. None of the above has become the Internet meme to end all memes.

He feels like he set off an atomic bomb by mistake. "I can't even go viral right." He lowers his head, dejected. The firestorm he ignited goes on without him. He slumps, a lonely ant in an anthill. He wallows in his disappointment until a scream startles him. He rushes to Maria, the source of the shriek.

Maria screams again. "Who the hell puts *Hellraiser* in a case for *Forrest Gump*?!" She positions her hands as claws with wild eyes, a feral animal ready to pounce.

James says, "Life is like an evil puzzle box. You never know what you'll get." His joke is no more successful than a deer staring down a predator.

Her lip twitches, like a dog ready to bark. "Out of order. Duplicates. Disks in the wrong cases. All my UFC fights are missing. Being shot at is one thing but violating me like this…" Maria's voice trails off.

James holds his hands out to calm her. "You can put them back. I'd offer to help, but my ears were ringing for a week the last time I mishandled one of your disks."

Maria takes a step towards him, angry, but in control. "They came into my home and had their way with my entire collection. I spent decades curating that collection." Wetness trickles down her cheeks. Tremors rumble through her hands. "What if they do it again?"

James says, "Go digital. You don't need the Dewey Decimal System. The computer sorts any way you want." Perhaps he's right, but she's not ready to hear it.

"It's not the same. Computers don't solve everything. These aren't even my disks."

James takes in her meaning. Perhaps the police interrogation was a ruse for more time to switch cases. "Did you record your surveillance on disks? I assume that's what they were after."

Maria's ears pick up. "What's this sudden interest in the recordings?" She eyes James suspiciously. Her breath quickens.

"They threatened me in a police station!"

"They offered you something, didn't they? What was their price?"

James pauses, unsure how Maria will react. "The price was you."

Her body convulses. Tears stream down her cheeks. "Did you pay it?"

"Of course not," James says.

Maria waves James off. "Don't worry about the packages. It's better you don't know." Maria calms and wipes tears from her eyes, angling away from James.

James expected her to be pissed, but crying? Maria is built of stone and steel. "Is this really about some disks?"

She walks to a bookshelf with her back to James. "Caring about people and things is a vulnerability. I'm not fond of being vulnerable." She grabs a box set and returns with *Robotech: The Complete Set*. "Decades ago, this is what we bonded on. I liked the soap opera romances and—"

James finishes her sentence. "I liked the space battles. *ThunderCats* and *He-Man*, whatever. *Robotech* is what I ran home for."

"What *we* ran home for." She corrects him. They smile at each other and giggle like kids, a release from a difficult day.

She pulls out a *Top Gun* DVD from a shelf. "Do you remember when I introduced you to ice skating?"

"You tricked me," James says with a laugh. "I thought we were going to watch ice skating. I didn't figure it out until I noticed people in line with ice skates. My ankle has never been the same."

Maria flips the case over. She shakes her head and points on the back. "This isn't the same case. You left a coffee stain, a ring, right there."

James gasps. "You went so volcanic that I thought you'd never speak to me again."

"Each one of these is a memory," Maria says. She puts *Top Gun* down and grabs a music CD, Enya's *Watermark*. "This was playing the first time you invited me to join your Dungeons and Dragons group. I was the first girl." The room is silent for a long pause as they reflect on happy times.

Maria opens the case. Empty. A stolen memory. "They rummaged through my past. This isn't my collection anymore. This isn't *my* past. They tore it away and left me with a forgery."

James offers Maria a hug. She hesitates, but then accepts.

"I didn't know you were sentimental. You're such a girl." James grins in spite of himself.

Maria pulls away. She playfully punches his shoulder.

James pretends to be scared. "Don't hurt me. I might have to call the cops."

She glares at him. "Too soon."

James says, "Sorry. But, you're right. These memories are good reminders that life used to be more than a job or a company. Thank you. I'm sorry this happened to you."

Maria looks James straight in the eye. "I hate politics. It's quicksand for your soul." She growls in frustration. "I want to hurt Renquist, hurt him bad. That requires money and power." Maria groans softly, then sighs. Acceptance. "Your president idea. I'm all in, for real this time."

09. **Remixed**

Maria wheels a dusty whiteboard into her living room. It partially obscures a few bookcases along the wall. She wipes dust off the board.

James looks at the board with a gleam in his eye. "You still have it? Like old times, you, me, and a lot of hard work." He tests the markers. They're dried out.

"We'll need lots of coordination when we get serious about the campaign." James takes a deep breath, to gather strength for what he knows is a big ask. "We should co-locate." He exhales, still tense.

"I have some big cardboard boxes in the garage you can live in." Maria holds a serious expression until a smile breaks through. "You could do odd jobs for room and board."

James assumes a mock fight pose with a grin. "I'll be your bodyguard."

"The thought that I would need a—" She can't finish the sentence without laughing.

"It wasn't that funny," James says. "You're not invincible."

Maria averts her eyes and motions for James to stop. "OK. OK. The guest bedroom is all yours, *bodyguard.*"

She helps James move anything he can salvage to her place. A lifetime of possessions condenses into a carload. A few trash bags of clothes, broken computer equipment, and random odds and ends isn't much to remember his old life by.

After the move, Maria and James work at adjoining laptops on the unbroken table.

James sifts through search results. "We went viral because of another video with 40 million views linked to us." He clicks. The video plays on the TV.

The mega-viral video remixes their footage. It begins with the brief moment in the original where Maria's outfit distracted James. A single

41

lecherous gaze draws out with camera tricks for eternity, or at least that's how five seconds feels like when your dignity is roasted on a campfire spit. A poorly imitated Asian accent is dubbed in. "Oh. Very clever outfit, sexy girl." A loud gong reverberates.

James looks horrified, then sneers.

Maria giggles.

"Aiya! You'd think I'd never seen a girl before. I'm a player, not a loser," James' face turns red from embarrassment and anger.

The main section of the original upload is heavily edited into sound bites. Every time the words "none of the above," or "more jobs" are used, a cartoon version of Maria flies through the air to dropkick an election poster.

One section remains unedited. James wobbles in the video as he speaks. "We can fix America. We will fix America. Internet, how would *you* fix America?"

"Unleash the Internet!" Cats and animated babies stampede through the living room in the video, as he belly flops onto the table in slow motion. Splat. The sound is sped up, high pitched, like drunk chipmunks, as he says "fix the table." The video ends as he passes out to another gong sound.

James says, "A gong? I could do without the *Sixteen Candles* homage. That's a part of the eighties I'd rather forget." He stares at the screen.

"I liked it. I'm a superhero." Maria kicks, imitating her cartoon self. "It was funny and short. They didn't mess with the core message, voter engagement. People want to feel heard. It's the whole point of billions on social media."

James paces in frustration. "Thousands have taken the mantle of none of the above. I can't outshout all of them. This guy has his own gong. I've lost control of the message. It's a runaway train."

Maria says, "No, James, you did it. The Take Back America Movement just changed their endorsements for president and vice president to 'none of the above' and 'more jobs.' They took us literally."

James points his finger at Maria. "If there were a little more James Wong, and a little less none of the above we wouldn't be in this situation."

Maria purses her lips. "You're right. Otherwise, the video would never go viral. Twenty people would click on it and ask 'who's James Wong?'"

Reality is a slap in the face. "I'll tell you who James Wong is. I know what to do."

10. None of the Above

Maria pulls into the courthouse parking lot with James. She smirks at him. "OK, Crazy Train, what are we doing here? What's the surprise?"

"You're right. Nobody knows who James Wong is. More people know me as None of the Above Guy. So, I've decided to change my name...to None of the Above."

An involuntary laugh takes hold of Maria.

"You're laughing at how brilliant the idea is." James undoes his seat belt, eager to go.

"You're serious?" Maria rolls her eyes. "Of course, you are. I should have seen that surprise coming."

James says, "Oh, that's not the surprise."

Maria quiets and stares intently at James. She's almost afraid to ask. "What is?"

James bounces in his seat. "You're going to change *your* name to More Jobs." He cackles like a cartoon villain and rubs his hands together.

Her eyes turn murderous. Maria releases her seat belt.

James knows that look. He shoves open the passenger door and stumbles onto the asphalt.

Maria erupts from her car.

James says, "You wanted more jobs. There you go."

These are not the words to calm Maria down. She chases James around the car.

James tries to reason with her as he circles the car. "Our names have to match the Take Back America endorsement. It was our idea. It's the only way we can take back *our* message."

Maria gives up the hunt and glares at James across the car hood. "I ought to kick your caboose. This is my stop off the crazy train." Maria stomps on the asphalt. "I'll never change my name."

11. **Ring Not Included**

James and Maria face Judge Pittman, a white lady in her early forties.

Maria struggles not to pout. She grumbles under her breath. "I always figured if I changed my name for a man, I'd at least get a ring out of it."

James says, "Your Honor, I can't wait four weeks for newspapers. I've told millions of people my new name. If we have to wait six weeks for a court date, our campaign is over."

Judge Pittman says, "I understand the urgency of an expedited hearing. James Wong, 'none of the above' has a special meaning in politics, a no-confidence vote, and does not endorse any actual candidates. Wouldn't your name change cause confusion in the election and a possible source of fraud?"

"I'm not James Wong anymore. I'm None of the Above. I've publicly shared my new identity over 50 million times. With this name change, I accept who I am, your Honor. I'm different than all the other candidates. I'm no politician. I'm a problem solver. I will prove I'm different with my actions."

"If you're just trying to be different, you could just go to Burning Man or something." The judge smirks. "We have some exceptions to announcing a name change in a newspaper, but none of them seem to apply."

James says, "Your Honor, the point of four weeks of newspaper ads is to make a public announcement so that anyone could object to a name change. Who reads the newspaper anymore? I announced my new identity online. I reached more people online than I'd ever reach with a newspaper."

His argument seems to sway the judge. "I'm going to allow it. I archived a copy of your video into evidence. Your name change is approved. Your name is now legally None of the Above."

With a stroke of a judge's pen, he is James no longer. It's one thing to declare your uniqueness. But to change your very name to trumpet that quality is quite another. Is this just an election stunt? No. He's always been None of the Above. He just never accepted it in those terms. Think differently, act differently, embrace your inner geek. James is the little boy who was the bullies' favorite in elementary school until Maria came along. None of the Above is the man. He is a man apart, a leader, a fearless revolutionary. This is a name he *chose,* not the one chosen for him. He can leave James behind with all the things that make him feel small or scared. None smiles and takes a cleansing breath, in gratitude, in peace.

The judge's voice sounds distant to None as Pittman continues, "Maria Cortez, I have no issues with your new name. I can't grant the same waiver based on the viral video because while you say 'more jobs' multiple times, you never directly state your name as 'more jobs.' In fact, you tried to claim the name 'none of the above' from your companion."

Maria grits her teeth and bows her head. "I'm sorry, your Honor. There was some confusion. My name is, indeed, More Jobs." She gives None a sideways glance. His mind is still elsewhere.

"You need to post in the newspaper for four weeks," the judge says.

None blurts out, "How about another video with a million views?"

"Two million and I'll allow it. I'll be watching," the judge says.

None and Maria rush into the hallway.

Maria's mouth contorts to an unhappy shape. "Let's get on with it."

None records video on his phone.

Maria composes herself. "I'm Maria Cortez. Do you believe in more jobs? I do. If two million people view this video, I will legally change my name to More Jobs to highlight its importance in the upcoming election. I'll do it. I will. This is not a scam like baby Megatron. You promised the Internet to name your baby Megatron for a million likes. You got your million likes, and you name the baby Dylan? Where's our baby Megatron?"

12. Escape Clause

None and Maria enter the waiting room for the Law Offices of Shen Wong. A wall mounted TV rotates through pictures of buildings and construction sites. Display cases hold architectural models of shopping malls with miniature trees and cars.

A paralegal, Cathy Morgan, chews on nicotine gum at her desk. "Thanks for three days of hell. Unhappy boss, unhappy office."

None steps towards a closed office door.

"Are you sure you need to be here?" Cathy blocks his path.

"I'm here to apologize," None says.

Cathy crosses her arms and stares None down. "He's not ready for an apology. Come back next week."

Maria shoves Cathy aside, walks past None, and swings the door open. Maria makes a small bow and waves None in as if inviting None into her office.

None strides into his brother's inner sanctum.

Shen Wong scrunches his eyebrows in confusion. "I thought we weren't brothers anymore."

Maria blocks the door frame. Cathy makes multiple fruitless attempts to squeeze past her into the room.

None says, "I came—"

Cathy raises her voice. "He came to apologize."

Shen has a slight accent, a reminder that he's not American born like his brother. "That's all Cathy. Close the door."

Cathy slams the door. Maria surges forward to avoid getting hit.

"I'm sorry for calling you a worse disaster than the BP Oil Spill," None says.

Maria mimics a slap to the left.

Shen says, "I was there. I don't need to hear it again. Why are you really here?"

"I'm sorry I said you'd be outclassed as a lawyer by an ape with a typewriter," None says.

Maria mimics a slap to the right.

"You have a future as a court reporter, but you're terrible at apologies. Please stop."

None trembles as he says the words. "I shouldn't have said Ma only needed one son."

Maria gasps and covers her mouth.

When None lost his company, he lashed out at Shen with every ounce of bitterness. He imagined the worst thing he could say and unleashed it on his brother. He immediately regretted every word. Where's a time machine when you need one? None bows his head in shame. "If I could cram those words back down my throat, I would."

Shen glares at his brother. He shakes his head and bites his lips in anger. "Jamie, stop this character assassination, masquerading as an apology, and I'll revisit my analysis. I assume that's why you came."

"I'm done." None's lips quiver with guilt.

The brothers remain quiet.

None slumps into a chair, staring at his shoes.

Shen forcefully kneads his hair with both hands.

Maria squirms, the third wheel to an uncomfortable exchange and a less comfortable silence. She gives them a few moments of peace before she interrupts. "Shen, how can we get the company back?" After the long pause, her voice sounds louder, startling both brothers from their thoughts.

Shen says, "Jamie, I told you not to ask me. I don't do mergers and acquisitions. I'm a real estate lawyer. If you want to talk cap rates, I'm your expert. I gave you that disclosure up front. I don't see liability on my part."

"We don't have time for cover your ass bullshit," Maria says.

Shen says, "What do you think I do for a living, Maria? Limiting liability is why people hire me."

"Ma meant well when she pushed us to work together," None says. "I'm done with the blame game. I'm not going to sue my own brother."

Shen says, "One of my clients had a bad contract like yours. The investor rents washing machines and dryers from a laundry service. The contract gives the renter more money each year. The renter can extend the ten-year contract another 20 years. There's just one problem. Guess who the renter is. The laundry service defines itself as the renter because it's leasing the land that the machines sit on. Some misleading words almost stuck my client with a 30-year contract, no easy way to cancel and less money every year. Jamie, I caught that one, but it wasn't a 300-page behemoth like yours. Days like that, I hate lawyers too."

"Don't call me Jamie. I legally changed my name to None of the Above."

"I want front row seats when you tell Ma," Shen says.

None shudders at the thought of his mother angry. Shen enjoys None's discomfort.

Maria clears her throat to get the conversation back on track.

Shen says, "I've seen dirty tricks before. Your contract is a whole new level of diabolical. If you try to exploit a loophole, they'll drown you in motions and interrogatories, until you're bankrupt. You'll have to use one of the few direct escape clauses, like the $150 million cancellation fee."

"I thought that clause was to protect us," None says.

"It was," Shen says. "Who knew we'd want to turn down a billion-dollar offer."

None toys with his brother. "Can you lend us the money?"

Shen pulls out his wallet and pretends to count his money. "$150 million. Must be in my other pants."

Maria says, "That can't be our only option."

"If Jamie dies, or is incapacitated, that would also void the contract," Shen says.

None says, "Not helpful. I get rich if I kill myself?"

"No, they thought of that," Shen says. "There's a suicide clause."

"Aiya! Even less helpful," None says.

"Shen, I'm really not impressed with your suggestions," Maria says.

"I saved the best for last. A higher counteroffer scuttles the deal. Make rich friends and develop patience. It could work."

None says, "So, I can die, become super rich somehow, or attempt a hostile takeover of my own company. What a waste of time. All those apologies, so that I could ask you to be my campaign manager."

Shen claps his hands together. "You only visit when you need me. That's why you're here. I rest my case."

Maria nods in agreement. "James, er, None, doesn't do it on purpose. He just waits until he can achieve two goals at the same time." The brothers stare at her, the interloper. She retreats to the rear of the office and sits quietly.

Shen says, "You definitely need a manager. From what I've seen, your core platform for becoming president is making a fool of yourself."

Maria jumps from her seat and closes the distance to Shen's desk.

Shen backs away. "Put a leash on your pit bull. I'm on your side."

None waves Maria away. She stands her ground, ready to pounce.

Shen says. "Maria, what are you going to do when you get to Washington? Beat up Congress? Learn impulse control. They will do whatever it takes to rile you up. I wouldn't even put you on a witness stand. You'd turn into The Hulk under cross-examination."

None says, "He's got you pegged. If you did beat up Congress, you might get a standing ovation, before they threw you in jail."

"You're taking his side?" Maria asks.

"I'm on the side of truth," None says.

Shen shifts his gaze to his brother. "Honesty's your problem. For me to invest my time, I need to know you're ruthless enough to pull this off. What's the line you won't cross?"

"I won't lie, but I can tell the truth in a misleading way." None throws his hands up. "I already went through this with Maria. How many Excalibur tests am I going to have to go through?"

"He's got *you* pegged," Maria says.

"Jamie, does the Take Back America Movement know that you're literally campaigning as None of the Above and More Jobs?"

"Probably not, but we're just taking back the message we started," None says.

Shen says, "You're doing more than that. You're stealing the nomination from them. When the Take Back America Movement says vote for 'none of the above' and 'more jobs,' they don't realize they're voting for James Wong and Maria Cortez, or even your alter egos. They're voting for the concept, the brand."

"Whether they realize it, or not, I'm exactly what they're asking for," None says.

Shen says, "Be careful not to draw too much attention. We have to beat Take Back America to the punch and get you on the ballot in the easy states. We'll rely on write-in votes for the rest. As long as you're just some fringe guy on the Internet, you're no threat. If they find out you're a real campaign, they'll fight."

Maria's phone buzzes. She reads her email. "We passed two million views. The name change went through." Her shoulders slump, her angry eyes stare at None.

None smiles with pride. "We kept our first campaign promise. More Jobs is on the job."

Shen takes a long look around his office and nods. "I could use some adventure in my life."

He sighs. "Even adventures require paperwork." Shen types notes. "Are you running third party, or independent?"

"Put us down as the Independents Party," None says.

Shen chuckles. "It's not hard enough to run for president? You want to start a new party?!"

"No," None says. "That party exists already. It just doesn't know it yet."

Shen says, "Perfect. I'll scrawl that in the margins when I file with the Federal Election Commission. That should help with the crackpot image. The crazier, the better."

Lan Wong, Shen's wife, bursts into the office. Her stance is aggressive, but her feelings hide behind smiles like an emotional Botox. Her upbeat facade would make a game show host proud.

Lan strides to Maria. "Fantastic to see you again. We should find you a property worthy of your success." Lan hands Maria her business card.

Maria ignores Lan and drops the business card on the floor.

"Clumsy me," Lan says. She grins even wider and hands Maria another card.

Maria flings the card to the floor.

Lan picks up her business card and dusts it off. She hands it to None. "I heard you had a recent home invasion." Instead of somber, it comes off as gleeful. She pulls an envelope from her purse and hands it to None.

None opens the envelope. Inside is a "Congratulations on Your Home Invasion" greeting card. A cartoon burglar sneaks through the window on the cover. Inside it says "Don't let it happen again. Stay safe. Ask Lan about security systems and neighborhood crime stats." Sales are her religion. Even during an Apocalypse, Lan could list ten reasons why NOW is the perfect opportunity to buy property, while the city burns in the background.

"Let me show you around to some safer areas later. I wouldn't want anything to happen to my favorite little brother. I'm sure I can find a new home you'll love." Lan doesn't wait for a response before pointing at her husband. "I need to talk to my partner in private."

Shen waves goodbye with an anxious laugh. None and Maria return to the waiting area. Lan closes the door.

Maria puts her ear to the door. Raised voices are too muffled to hear anything. There's an echo to the sound. None swivels his head back and forth to find the source. He approaches Cathy's desk. None overhears Shen's voice on Cathy's headphones. None unplugs her headphone jack. Shen's office can be heard over the speakers. The noise attracts Maria to the desk.

"Partner means you don't make big decisions without consulting me," Lan says over the intercom.

"Is eavesdropping on your boss legal?" None asks Cathy.

Cathy shrugs.

Maria avoids eye contact, trying to look innocent.

Cathy sneers at None. "The NSA spying program is in trouble if you get elected. Lan authorized it, and she's the boss."

None raises an eyebrow, disappointed and surprised. "Why is Lan running Shen's company?"

Lan says, "If you're going to be gallivanting around the countryside with your brother, you still need to keep up with your clients. You can't take a vacation from our company."

"Anything else?" Shen asks.

"You won't be needing Jenny," Lan says. Keys clank together.

"Your brother got beat down. She even took his Corvette," Maria says.

None doesn't want to hear any more. He plugs the headphone back in, to quiet the speakers. "How did she know to come?" He looks at the intercom, then Cathy. None answers his own question and glares at Cathy.

"Stop staring. You're creeping me out." Cathy tosses off her headphones and glares back. "Sure, I called her. If he's going to be in a foul mood, I'd rather it be at home."

Lan bursts into the room. She waves goodbye to None and Maria. "Again, amazing to see you both. I have to rush off to a client." Lan speeds out the door.

None looks out the window. He sees Lan zoom off in a Chevy Corvette with the top down. The tires screech, with a tiny bit of smoke.

Cathy sits up straight in her chair, while None and Maria try to act natural.

Shen shambles in with a Saddleback Leather Briefcase over his shoulder. He sulks, his movements are lethargic. "Unhappy wife, unhappy life," he mumbles, barely audible.

None forces a laugh. "Unhappiness seems to be contagious around here. Don't upset me, or you might set off an epidemic."

Everyone is quiet, the room an empty void. Maria laughs politely to fill the enormous silence. Cathy joins in.

Shen chuckles. His laugh gets louder and larger until it sounds maniacal. The energy returns to his face, as he shakes off the doldrums. "Great news. I was able to negotiate a tour bus for our campaign." He flings the door open and wanders outside. Everyone follows him to a red minivan.

"This is going to be the best road trip ever." Shen turns towards None. "Maybe two Wongs can get it right. Wait a second. You have no last name. Are you a Wong anymore?"

13. **Buying Democracy**

Shen drives, while Maria shoots campaign videos from the front seat.

"Let's do another take." Maria points her camera at None.

None says, "With a $25,000 budget, how much democracy can we buy? Even with a free attorney, is that enough to be heard? With no political party or volunteers, we're going on a month-long road trip to find out."

Maria rotates the camera around to scan the passengers. She poses for a quick selfie.

None says, "What would it take to get an independent candidate on the ballot in all 50 states? Over a million signatures and millions in expenses and fees. Even then, major parties use lawsuits to keep independents off the ballot, like Ralph Nader in 2004.

"You might think that all you have to do to elect None of the Above and More Jobs is write those names on your ballot, called a write-in vote. You should be able to vote for whomever you want, but you can't. Seven states don't allow write-in votes. In most states, write-in votes won't count, unless your candidate registers with the proper paperwork. A few states charge thousands of dollars to be a write-in candidate, someone not even on the ballot.

"So, we'll get on the ballot in states that require the least signatures. Out of the gate, we skip eight states worth 36 electoral votes. The good news is we paid to be on the ballot in three states and registered as a write-in candidate for D.C. plus 31 states. Adding Texas, we've spent about $14,000, and we haven't even left California. Enough math. First stop, Seattle."

Maria puts down her camera and turns forward.

None leafs through a yellow legal pad, back and forth. "Shen, I don't see California on the list."

Shen says, "When I filed your paperwork, I talked to the elections division. They said None of the Above was already on the California ballot. So, Take Back America must have already filed."

"How can we win without California?" None asks.

No one has an answer.

The minivan arrives in Seattle after 20 hours on the road. Thunder rumbles in the distance.

They turn in for the night. The next morning, Shen turns a corner of the hotel room into an office with a printer, scanner, laptop, and office supplies.

None answers a knock at the door. He lets Maria in. She hands out Egg McMuffins and orange juice.

"What's our plan for today?" None says.

Shen says, "I'm following up with election divisions of the states that haven't confirmed your candidacy. I'll also generate billable hours for my company. I'm not on vacation."

"Will that take all day?" None asks.

"I'll be stuck in hotel rooms the whole trip." Shen throws None keys to the minivan. "Pick me up when you need to file petitions."

None drags Maria to a supermarket to collect signatures after the skies clear on a soggy afternoon.

An older man in cowboy boots approaches None. The old man seems to be itching for a fight. "Are you a RepuliCON, or a Democrap?"

"I'm None of the Above. Sign this if you want an independent for president."

"This country's going bankrupt. The major parties will stick us with the bill. Give me that damn paper. I'll show them." The old man jots down his details.

Maria ignores several people as they walk into the store.

None approaches a young man in a hemp shirt. "Will you sign a petition to get More Jobs on the ballot?"

The young man says, "Hell yeah. I'm out of work. Of course, I'll vote for that."

None wonders why Maria isn't helping. Is she feeling shy? Maybe she needs him to break the ice. He hands Maria his clipboard. "This one's yours."

Maria turns to watch the young man write. Another signature.

A middle-aged man in an orange parka approaches. "Aren't you None of the Above Guy?"

"I am. If you want to help fix America, sign the petition," None says.

His eyes bug out. "You're running for President, for real?"

"Definitely."

"Awesome! All politicians are clowns. At least you two entertain me." The man reads the petition. "More Jobs? She did change her name. Very cool." He gives Maria a thumbs up and grabs a pen.

Maria runs off into the parking lot.

None flashes an uncomfortable smile. "She must have left to run errands." Once the signature is done, None leaves to find her.

He finds Maria crying next to the minivan. None sneaks towards her like a timid animal. "Are you crying?"

Tears stream down Maria's cheek. "No. It's the fucking rain."

None double checks the clear skies. His eyes return to Maria.

"We're a sideshow!" Maria wails.

None says, "I'm sure he meant the clown comment in the kindest way. Look how happy your new name made him. He's your biggest fan."

"My name is Maria Cortez."

"Not anymore."

"My name is Maria Cortez. Are we clear?" She purses her lips and advances into his comfort zone.

"You can't go back to Maria." He not only stands his ground but inches forward.

"I never stopped being Maria. I don't care what a piece of paper or some Internet video says. This is ridiculous. I want to take out my anger on someone, but I'm the one that let you talk me into this cluster-fuck."

None says, "Bro, I thought I had an original idea. I looked it up on Wikipedia. There have been at least four candidates in various countries that have changed their name to None of the Above. There is even a None of the Above political party in Serbia. Even Maria Cortez has kind of been done. Scott Fistler changed his name to Cesar Chavez to draw Latino voters in Arizona. More Jobs is the true original. You are the original."

Her sniffles quiet. Maria crosses her arms with a smug expression. "I succeeded where you failed?"

"Yes. People only pick None of the Above when they're pissed, but they'll always want More Jobs. You were right."

Maria smiles at the acknowledgment. "I just want to be heard."

"More jobs is the most important issue. Your name is our political platform."

"Oh." Maria nods.

None says, "Be More Jobs. Embrace it, and we can do this."

"Vice President is like being an appendix. I don't feel important."

"I'll treat you almost like co-president."

"Co-president." Maria nods.

More Jobs doesn't fit her identity like None's name did his. That's not who she is. Maria knows that more jobs, better jobs, are the key to a better America. How strange is it to condense the essence of a political platform into one name and wear it as your own? It's the change she wants to make. "Be the change," Maria whispers to herself.

Her eyes widen in recognition. Maria accepts her new name. "Be the change that you wish to see in the world. I will make more jobs happen. Nice to meet you. My name is More Jobs."

They shake hands.

None says, "We could call you Jobs, like Steve Jobs, or MJ. MJ is pretty close to your original initials."

"Steve Jobs and me in the same sentence? That kind of heresy could send half the Internet into a nerd rage. MJ works."

They smile at each other.

"You can always change it back in eight years, 16 if you get elected," None says.

MJ pokes him on the arm. "*If* I get elected? I finally accept my name, and you're already talking about changing it back?"

14. Lawnmower Madness

After a half day circulating petitions, None and MJ drive off for an early dinner.

MJ says, "We haven't made a video in a couple of days. What makes you think of Seattle? The Space Needle?"

"Lawnmowers." None drives into a residential area with verdant lawns and lush vegetation.

"Lawnmowers are nothing special. We have those in L.A.," MJ says.

None says, "Look how green everything is. Every inch without concrete bursts with life. How often do they mow? Probably enough to justify a riding lawnmower. I've never ridden one. I want to."

MJ nods. "That's our video."

None finds a lawn with five-inch grass, closed off with a black wrought iron fence. Rows of purple crocus, yellow tulips, and white daisies grow along the periphery of the lawn next to the porch.

None opens the iron gate, walks down the cement walkway, and knocks on the door. A gruff man in his forties opens the door. "Sir, if you have a riding lawnmower, I'd like to mow your lawn," None says.

"How much?" asks the homeowner.

"Free, if we can shoot a campaign video," None says.

The homeowner sizes None up and walks out to open his garage. The riding lawnmower resembles a miniature tractor with yellow and green paint. "Don't break anything." He hands the keys to None.

None has the excitement of an unlicensed teenager taking his first drive. "Woohoo!" The lawnmower surges down the driveway.

MJ dangles the camera from her neck. She climbs the hood to the top of the minivan for a good angle.

Camera ready. She waves at None.

None mows in a big circle. He raises his voice to be heard over the lawnmower. "Seattle is alive with all this green. It rains so much here. I'm surrounded by grass. It's fitting that Washington was one of the first states to legalize a different type of grass, marijuana."

None leans over to reach grass clippings from the previous time around.

His right shoe wedges between the brake and gas levers during the awkward reaches.

None lets go of the steering wheel. He leans over precariously and grabs the grass shavings. None slips towards the left but grasps a handhold and regains his balance.

He jerks the wheel left to avoid crashing into the fence. None mows across the middle of the circle.

"Here's to Washington." None salutes with one hand and throws up the clippings like confetti. "Vote None of the Above and more grass! I mean More Jobs. Sign our petition. Bring the protest vote to Washington!"

He double fist pumps and stands on the mower. "Unleash the Internet." His foot stomps on the gas.

The mower jolts forward. None falls off backward, leaving his right shoe behind.

He brushes off his ego and chases the mower. He falls on the slippery grass but scrambles on all fours to return upright.

With one shoe, his cadence is uneven. He run-walks with his feet farther apart for better traction. His lumbering gait is apelike.

The mower has a ten-foot lead. It reaches the end of the circle. From above, MJ can see the universal no symbol mowed into the lawn. The mower continues towards the flower section.

Five-foot lead.

He grabs a handhold on the mower and skips on board with long strides. The mower devours the flowers, leaving behind a mix of purple, yellow, and white petals.

None turns left to avoid crashing into the porch.

The brakes don't work. The wedged shoe blocks the brakes.

None kicks at his wedged shoe to break it free.

The shoe thuds to the ground.

None stops the mower and reaches for the shoe.

The homeowner burst through the front door. "You murdered my garden. I'll turn you into mulch!" He grabs a shovel from the porch and sprints towards None.

None leaves the shoe and stomps on the gas. He aims towards the gate.

The homeowner is on an intercept course, approaching on the left.

None shifts his left leg onto the right side, riding the mower side saddle. "We can fix it," None says.

"I'll fix *you*." The homeowner is almost within attack range. He swings and hits the seat.

MJ points the camera in the general direction of the action. She slides down the side of the minivan.

The homeowner swings again. None extends his body to the right, far enough to dodge the blow, but slips and lands on his butt. The mower slows.

MJ opens the passenger side door and moves to the driver side, still filming.

The homeowner drops the shovel. He grabs a handhold and works his way onto the mower. He brakes and turns off the ignition.

None gets up and makes a beeline for the gate. The homeowner is two strides behind.

None unlatches the gate and vectors towards the minivan, the homeowner on his tail.

"Keys," MJ says.

None fumbles in his pockets as he runs. He pulls out the keys and passes them to MJ. None jumps into the passenger seat.

MJ starts the car and passes the camera to None.

"You're still filming?" None asks.

"No, you are," MJ says.

None pulls to close the door, but the homeowner grabs it.

The homeowner slips as the vehicle pulls away. He slashes the side of the minivan with his keys as it escapes.

The keys scrape off paint with a loud screech. None points the camera out the window for a parting shot of the homeowner flipping him off. He turns off the camera and catches his breath.

MJ laughs. "Maybe you need a stuntman."

15. **Gardening**

Shen watches lawnmower footage on the laptop at the hotel. He yells at None. "You're risking this campaign with your behavior." Shen shakes his head at MJ. "And you uploaded that video? It's evidence."

"Cops don't watch YouTube for evidence," MJ says.

None says, "Shen's right. He might sue us or call the cops. We have to make it right. How can we fix America, if we can't fix a flowerbed?"

MJ says, "OK. OK. We'll eat and make a Home Depot run."

They eat takeout. With only one shoe, sit-down restaurants aren't an option.

After eating in the minivan, MJ uses the video to pick out the right type of flowers. They load crocus, tulips, and daisies into the minivan.

"Is Washington a 'stand your ground' state?" MJ asks.

"We'll all carry flowers," None says. "Nobody shoots the flower delivery guy."

Shen says, "You two poked a bear. Let's return to its lair in daylight."

None nods. "First thing tomorrow morning."

Back at the hotel, Shen inputs receipts into a spreadsheet. MJ sits on the bed, watching None pace. No one speaks.

After an hour, Shen breaks the silence. "Those flowers were the last straw. We're burning through money too fast. We have to stick to fast food. No drinks. Maria, no more separate rooms."

"I'd rather live out of the van than share a room," MJ says.

"Good idea," Shen says.

"I wasn't offering," MJ says.

Shen says, "It might come to that. We also need to decide if we're skipping the ballot for Utah or Iowa. We can't afford both."

"What's the difference?" MJ asks.

"Utah requires 500 fewer signatures, but petition circulators have to be state residents. Iowa, we can circulate petitions ourselves," Shen says.

"Skip Utah," None says.

Maria nods.

"We still have to stop in Utah because they require write-in candidates to apply in person," Shen says.

Maria groans. "Why does it have to be so complicated? Let's get some sleep."

The next morning, they eat Egg McMuffins and drink complimentary hotel coffee.

Dread of the unknown permeates the room. None rushes everyone into the minivan. "No matter what happens, we don't hurt this guy. If there's a fight, we get out of there."

"I don't hurt anyone, except in court," Shen says.

MJ snickers. "Good Shen, because you look so dangerous."

None glares at her.

"OK. OK," MJ says. "I won't hurt anyone, as long as they stay hands-free."

In sunlight, the minivan scratches are obvious. Shen mutters to himself just a bit too loud. "I'll have to let Lan keep the Corvette, or I'll never hear the end of this."

None gulps. He knows how much Shen loves that car.

Shen gets in the minivan with a deep sigh. He drives to the scene of the crime.

On the way over, another lawn has a circle with a line through it mowed into the grass.

The minivan screeches to a halt. Shen stares at the lawn in disbelief.

None and MJ high five. The video went viral.

The minivan pulls up in front of the house. None removes his socks and remaining shoe. MJ and Shen grab flower containers. None opens the gate, and they follow apprehensively.

The front door swings open, and the homeowner sizes each one up as he struts out. "Well, well, well, if it isn't the jackass and his friends."

The trio stop in their tracks. They lower the flower pots to the ground, not making any sudden moves. "I brought my lawyer," None says.

Unfazed, the homeowner charges towards None.

Shen and MJ back away.

None hesitates a few seconds. He barely turns around before the homeowner catches up with him.

The homeowner gives him a bear hug and twirls him around.

Maria sees an attack and pivots back towards the homeowner.

Shen continues to the minivan and starts it.

The homeowner says, "You made me realize how boring my life had become. Yesterday, that changed, thanks to you." He sets None down and laughs. "You should see your face."

Maria approaches in attack mode. She assesses him as a non-threat and relaxes.

None blushes. "I brought you flowers."

"Yes, I will go to prom with you." The homeowner laughs. "I haven't gotten this much attention in years."

"We came to fix the damage from yesterday," MJ says.

Shen turns the engine off and walks over.

"Your crocus and tulips aren't the right color, but I'll let it go. Today I insist on paying you. A shoe for Cinderella here." The homeowner retrieves None's shoe from the porch and hands it to him.

None sets up the camera to take time lapse photos while they plant flowers.

When they finish, Maria twirls to film a donut selfie. "Fixing America, one flowerbed at a time."

None looks at his phone. "11:20 a.m., I posted we'd be at the car wash by 11. We're late. Aiya! We probably had easy signatures from going viral, and now they'll be gone."

At least a hundred people wait in the parking lot. They clap and hoot when the minivan pulls up.

None and MJ work fast to set up the chairs. Shen hands out clipboards. Within minutes, they gather signatures.

One of the girls in front unzips her rain jacket seductively. "I'll sign yours if you sign mine." Underneath she wears an "Unleash the Internet" shirt. It has a cartoon version of None with his fists raised, lightening behind him, and motion lines to denote falling. None signs the shirt and hands her a clipboard.

MJ has fans of her own. "More Jobs, way to support the cause. Changing your name, that's dedication," says a man wearing flannel. He unbuttons his flannel to a shirt drawn by the same artist. A cartoon MJ flies wearing a cape, with the words "More Jobs" on her chest, wielding a sign that has a circle with a line drawn through it. MJ autographs his shirt. He autographs her petition.

None and MJ feel like rock stars. Even Shen senses the energy of the crowd.

After a hectic day, they have more than enough signatures for Washington.

16. Groceries

Utah. Iowa. Ohio. Eleven days of endless campaigning. The Internet yawns.

Another day, another supermarket. MJ wears a Catholic schoolgirl outfit with knee high socks and high heels. She stands listlessly with a clipboard next to the entrance.

None pounces on shoppers within seconds, like a piranha. He asks a young guy wearing Birkenstock Sandals and socks, "Can you support the protest vote?"

"I'm Libertarian. Every vote I make is a protest vote." The man enters the supermarket.

Two sisters in Angel Scout uniforms run to the store entrance, giggling. Their mother sets up a folding table while the sisters hock cookies.

A tall man with a lecherous smile struts to MJ. "Can I get the mint cookies?"

MJ rolls her eyes. "I'm not an Angel Scout."

"You look so young." He winks. "Oh, you're an Angel Scout's mom."

MJ purses her lips. "No." She's no one's mom. "Will you sign—"

"Give me your number and I will."

"You don't even know what it is, and you'll sign?"

"If I can get your number, for sure."

MJ smirks. "My number is 42."

He clenches his fist. "You stuck up bitch. If you were going to give me a fake number, you could at least give me the right number of digits."

"It's Hitchhiker's Guide—"

"I'm gone." He flips her off and struts into the parking lot. He gets to his car and groans. Groceries. The man blushes as he sneaks back past MJ into the store.

A woman wears a shirt with an American flag ripped in half, frayed in the center, covering two parts of a broken table. The caption is "Fix The Table." The woman flashes her eyebrows at None. "My shirt is pretty great, huh?"

"What does it mean, um, what's your interpretation?" None asks.

The woman has a know-it-all smile, as she explains. "It's symbolic of partisan politics tearing this country apart into warring camps."

None nods.

Her enthusiasm drops a notch, unsure of her answer. "The moderates in between have gone missing, or at least unheard from."

None nods.

The woman tries to glean feedback. "We can only fix the country by bringing everyone back to the same table, fixing it by creating a strong center." She's exasperated. "Do you even know what I'm talking about?"

"Spot on. You passed the test. The best answer I've gotten yet." None says. He turns to MJ and mouths the words "no idea."

The star pupil. She signs her name and struts away self-satisfied.

"Leave it to the Internet to take something you did accidentally and give it meaning," MJ says.

"How about a table divided against itself can't stand?" None asks.

MJ says, "My interpretation is 'don't belly flop on my table, or it breaks.'"

None and MJ exchange smiles.

"I am so bored asking everyone the same questions all day, or dealing with attitude and rejection," None says.

"I'm nostalgic about Seattle," MJ says.

None sighs. "Can't I go viral without getting hurt? I get it Internet. You like slapstick."

MJ says, "Maybe we don't need the Internet. Let's play it old school. We can buy yard signs."

"We can't afford any signs," None says. "Maybe there's another way."

17. Signs Point to None

None, MJ, and Shen enter Garfield campaign headquarters, where almost 20 campaign staff mill about. A likeness of Senator Garfield is stamped on every possible form factor. Buttons. Signs. Bumper stickers. Bobbleheads. Impressionable minds.

A whiteboard in the shape of a speech bubble fastens onto a life-size Garfield cardboard cutout. It reads "smaller government."

Shen pulls out the camera. "I'd like to record this so that I can post it online. Is that cool?"

"Sure. Do anything to get the word out. I'm Steve, and this is Ben." The campaign staff shakes everyone's hand.

"We're campaigning. Can we have some signs?" None asks.

"Take what you need," Steve says. None and MJ grab signs.

"Can we keep the signs, as a gift?" None asks.

"Yes," Steve says.

Ben raises an eyebrow.

A jogger enters from the street. He pulls out his phone and records. "Aren't you None of the Above guy?"

"Yes, I'm None of the Above."

The jogger asks, "Are you campaigning for Garfield?"

None says, "No. Garfield has been Senator for decades. How can he pretend to be an anti-establishment candidate?"

While Shen and the jogger record, None and MJ pull out markers and deface their signs. They draw a circle with a line through the middle, turning a Garfield sign into an anti-Garfield sign.

Ben says, "You lied. You said you were campaigning for Garfield."

"I never said who I was campaigning for," None says.

None and MJ write "None of the Above" on top of the line and "More Jobs" below it. His signs are now their signs.

The staffers exchange shocked glances.

Ben says, "Get them!"

The campaign staffers charge the intruders.

None, MJ, and Shen bolt for the door. They make it out the door with an angry mob on their tail.

Shen films the staffers as they chase him.

None weaves through pedestrians on the sidewalk. "Vote None of the Above!"

MJ follows closely. "Vote More Jobs! Why run? Can't I just kick their butts? There can't be more than a dozen."

"Play nice." None turns the corner at the end of the block.

Random onlookers pull out their phones to record the commotion.

Shen lags behind. The mob gains on him. He turns the corner and hides behind a parked car.

The mob passes him.

Shen sneaks back around the corner, the way he came.

None and MJ run around the block, shout slogans, and swerve through pedestrians.

The mob dwindles, as campaign staffers give up.

They finish a lap around the block. None and MJ evade a few tackles, as they pass campaign headquarters again.

Those who gave up give chase again. The mob gets a second wind.

Fatigue takes its toll on None and MJ. The mob gains on them.

Shen opens the back door of the minivan and shoves boxes and chairs further towards the front, to make room. He backs up into the street, with the doors open.

MJ and None throw their signs into the minivan and jump into the rear, to make a narrow escape. The staffers yell and make rude gestures at the escaping vandals.

In the minivan, the pranksters high five one another and laugh.

Shen passes back the camera. None and MJ watch Shen sneak back into Garfield headquarters on the video. He erases "smaller government" and writes "Vote None of the Above and More Jobs" in the speech bubble. On the camera, it appears Garfield endorses them.

None cringes. "Sorry. With all your filming you never got your own sign."

"I know where to find one." Shen's eyes light up. "Next stop, the Robertson campaign."

18. **Homebound**

Video of defaced signs inside Garfield headquarters spreads across the net. Nationwide, voters transform Republican and Democrat presidential signs into None of the Above signs. These personal acts of civil disobedience strengthen the protest vote and make a faction of the Tea Party question if Senator Garfield really is one of them.

None and MJ finish canvassing Ohio within a day. Their anti-Robertson and anti-Garfield signs triple the response rate but increase the hostility they face.

The team stops in Trenton, New Jersey next. New Jersey residents have to collect the 800 required signatures. The firm Shen hires completes the petitions within a single Saturday.

The team loiters by the minivan that evening.

Shen says, "The courthouse opens Monday. We should sleep in the van this weekend if you're serious about your budget. Paying for signatures wiped us out."

MJ loses patience. "I'll cover my own hotel. I won't play this game."

None says, "If I can't handle a small budget, how will people trust me with a large one? I have to keep my word."

"I'm taking the weekend off. See you Monday." MJ grabs her suitcase and leaves the brothers to themselves.

Shen says, "Campaigns cost what they cost. You can use disclaimers."

None shakes his head. "No lawyer tricks."

Inside the minivan, Shen creates makeshift pillows with dirty clothes and plastic bags. "It's like camping."

None says, "You seriously have no money? I thought business was good."

"Profits are up, but so are expenses," Shen says.

"I warned you not to marry Monopoly," None says. "Every hotel on Park Place wouldn't satisfy her thirst. It doesn't matter what you make. What matters is what you keep."

Shen grits his teeth. "You're the last person to lecture me there, *billionaire.*"

None joins Shen in the minivan and closes the door. "I didn't make that mistake alone, *brother.*"

Shen runs his fingers through his hair. "You call MJ, bro, but never me. I'm your actual brother, *bro.* You're not my Jamie anymore. Since we're not married, I'm fine going to bed angry." Shen snuggles under a blanket and sleeps facing the wall.

The tight quarters remind None of his childhood when his family could only afford a studio apartment.

None offers to spend an entire day of family time with Shen to help make up for last night. They check out the Old Barracks Museum, the only remaining colonial barracks in New Jersey.

None ponders history, while he listens to stories about George Washington crossing the Delaware. Taxes to help pay for the French and Indian War helped spark the American Revolution. No taxation without representation. People want to be heard. It's as true today, as it was in the colonies. If average Americans don't feel their leaders listen, could there be another revolution?

Shen says, "I'm thankful I shared in your presidential dream. Can you believe we defaced signs in *their* headquarters? It's the most fun and excitement I've had in a while. I haven't taken real chances like that since I was an undergrad."

The brothers laugh together.

None says, "About last night—"

Shen says, "We each have valid grievances. Leave it at that."

"I might say something different when I'm pissed, but we're OK." None pulls out his phone. "I feel part of history here. I want to share this with MJ."

"Can you spend even one day apart?" Shen covers None's phone with his hand. "What if you and MJ had to be apart for a while?"

None pulls his phone away from Shen. "Nothing could tear us apart." None says the words with absolute faith.

"What if you had no choice?"

"What do you mean?" None gives Shen a worried look.

Shen waves None off the topic. "Never mind."

None brings up MJ on his phone. His thumb hovers over the call button. His eyes linger on his brother.

He puts away the phone and smiles at Shen. "Today, I'm all yours."

The brothers share their day together, the closest to a vacation either of them has enjoyed in a very long time.

The Wongs spend another night in the minivan.

Monday morning, MJ rejoins them. She mails off packages on the way to the courthouse.

They end the campaign with Tennessee and North Carolina.

Next destination, home.

Shen adds up receipts. Only $286.13 of their $25,000 budget remains, not enough for gas money for the 37-hour return drive to California.

"I'll chip in." MJ pulls out her credit card.

None pushes away her card. "I'm not cheating. I guarantee people online will check my math."

Shen says, "If we drive straight through and fast for a day and a half, we'll only be one percent over. That's still damn good."

"How about we be over three percent and enjoy the ride home?" MJ asks.

"That's triple," None says. "I want to prove to voters that I accomplish what I set out to do, no matter the sacrifice. We went through worse when we founded Adaptive Unlimited."

MJ says, "OK. OK. I'll do it, but I reserve the right to bitch."

"You might have to make a bigger sacrifice than that," Shen says.

None's heart races. "Is this what you meant in Jersey?"

"I couldn't find any way around it," Shen says. "One of you moves from California, or we lose the state."

MJ slams her fist on the hotel desk. "That's BS. Who makes up all these silly rules?"

"It's the American Constitution, 12th Amendment. The California electors vote for president and VP, but only one of them can be from the same state," Shen says.

"How serious do we have to take this?" None asks.

"In the 2000 election, George Bush and Dick Cheney both lived in Texas. Cheney sold his house, registered in Wyoming, and moved back there. If he hadn't, they would have lost Texas and the election."

MJ gasps.

Shen says, "MJ, don't you have family in Texas? You could sell your house and move in with them for a while."

MJ glares at Shen. "I'm not uprooting my life."

None says, "I'm practically homeless. I'd do it, but I have nowhere to go."

MJ says, "We'll probably win big, or lose big. I'm sure it won't be close."

"We can't lose an important state, like California, on a technicality," Shen says. "We could end up with a VP from another party."

MJ shakes her head. "So, I'm the one who's supposed to sacrifice. Either I give up my home or let None be president without me."

Shen says, "We'll be going through Texas on the way home. You can think about it."

They talk little as they pack and hit the road. After 19 hours, Shen dozes off for the first time. He stops for gas and buys two large cups of coffee.

None sleeps in the back.

After Shen pulls back onto the freeway, MJ reaches for a coffee.

Shen smacks MJ's hand. "Coffee is reserved for drivers. You want to help me drive?"

MJ shakes her head and yawns.

Shen spits in his cup to mark his territory. "We're in Texas. Do you want me to stop at your parents' place?"

"Keep driving."

Cumulus clouds dot the early afternoon sky. Shen pulls off I-20 close to Dallas, Texas. He opens the door and nudges None awake. "You're up."

None stumbles to the front seat, still groggy. He slaps himself and chugs the remaining coffee. None yawns and pulls back onto I-20, while the others sleep.

Traffic is light on the four-lane highway past Fort Worth, Texas. None nods off. His head drops, then bounces up.

None drifts right, from the fast lane, into the middle left lane. A Ford pickup swerves. Its horn doesn't wake him.

The minivan crosses the line into the right middle lane. A big rig honks, as None veers towards it.

Everyone wakes up.

Shen braces himself.

MJ nudges the wheel to the left.

The minivan swerves. None recovers control a second later.

"This isn't working," Shen says.

MJ says, "Better 10% over budget, than 100% dead. Find a hotel, or I'm walking."

None is wired from the rush of adrenaline. He exits the freeway and pulls up to the nearest hotel. None grabs the door handle.

"Give me a second." MJ sighs and shakes her head. She is silent for a moment.

None gives MJ a confused look.

"We nearly died for this dream. I see how important this is to you, None. My parents are only 30 miles away. I'm driving."

Her sacrifice touches None. Uprooting her life might be one of the biggest things she's ever done for him.

After a short journey, MJ pulls up to an orange stucco house a with clay tile roof, similar to MJ's house, but three times the size with a large yard. A political sign has "Maria Cortez for President" scrawled across it.

"Wait in the car, I'll be right back," MJ says.

None gives Shen a smug look. "See! This is why MJ is my best friend, and you're a distant second place. You'd never make a sacrifice like this for me. You probably wouldn't even stand up to your wife for me."

MJ returns with two packages in her hands. She opens the trunk and puts the packages into her suitcase, using her clothes as padding. She closes the trunk and gets back in the minivan.

She smiles at them both. "Good news. My parents said you could live with them, until after the election."

Stunned, None stares at MJ.

Shen grins broadly.

"I don't belong in Texas. What am I going to do in Texas?"

"Do whatever you were going to do on my couch."

"Aiya!" None considers possible retorts. He looks for holes in her logic. It's bad news, but it's a good plan.

Shen breaks his silence with a smug declaration. "Congratulations! If you stop here, we're under budget. Although, what's left is already in my gas tank."

"All I have to do to balance the budget is move cross country." None grimaces at MJ. "You're enjoying this, aren't you?"

Shen and MJ laugh together.

None says, "Until election day, I'm a Texan. See, MJ. I do take your ideas seriously. I have my laptop, so I'll work on my latest prototype, Adaptonium. We're lucky Renquist showed his true colors before I handed it over."

Shen grabs the camera. "Let's do our final recap so that we can head on home."

None sighs, at the thought of home.

None and MJ hold campaign signs, while Shen films them.

MJ says, "Do you want this election to be remembered? Make your voice heard." None and MJ sway their signs for the camera.

None says, "After 27 days, we ended our journey. How much democracy did we get for $25,000? We're on the ballot in eight states. We're also write-in candidates in D.C. and 34 states. Even to be a long shot write-in, some states make us jump through hoops. We've had to get signatures, show up in person, or even pay, just to be a write-in.

"I ran out of money to reach California, so I'm planting my flag in Texas." None pounds his sign into the lawn. "Democracy doesn't come with a ticket home, at least not for me."

19. **Adaptonium**

It's the longest None has ever been away from his best friend. He video conferences with MJ every night, but still feels alone. MJ's family treats him well, but None locks himself in his room. No distractions. Every minute not used for sleep, or meals, he works on his prototype, Adaptonium. Focus on the code. Change the world. Forget election worries. Forget campaigning. Forget missing MJ. Focus on the code. Adaptonium will change the world.

The months pass. None stays locked in his cocoon, away from the outside world.

The meme machines of the Internet fabricate their own None and MJ antics. The campaign takes on a life of its own, with only the occasional intervention from MJ.

Some states, like Massachusetts, allow stickers to be used for write-in candidates. MJ posts instructions online on how to print your own "None of the Above/More Jobs" stickers. For the rest of the election, they watch on the sidelines.

After every government misstep or economic wrinkle, more lawn signs go up, a manifestation of frustration. Americans vote with their lawns, one by one, a trickle of cynicism and despair. This is the green revolution. Most Americans with "None of the Above" on their lawn have never even seen the campaign videos. Tax evasion charges for a Republican senator, anti-Danvers signs go up. A Democrat House member sex scandal, anti-Robertson signs go up. Unemployment goes up half a percent. All names, from this year and last, are crossed out and replaced with "None of the Above/More Jobs."

Partisan rancor reaches pandemic levels. Even the most basic congressional bills fail to pass for months. Filibusters occur routinely, out of spite. The Republicans and Democrats are a bitter married couple who stopped sleeping together long ago.

Election day arrives. Perhaps the American people will vote for a divorce.

20. **Election Day**

MJ picks up None at LAX. They fight traffic from the airport to MJ's house in Santa Monica. None misses L.A., but not the traffic. Small talk fills the trip back. Texas is never mentioned. Both play it off like they'd never been apart. At MJ's house, they change into steampunk outfits and leave for the bar.

About 8:30 p.m., None and MJ enter Beer Beakers carrying election signs. Shen accompanies them in a well-tailored suit and tie. Dressed so formal, he looks out of place.

The Blacksmith approaches them. "You have a lot of nerve coming here. Because of you, I had to delay payroll a week."

None stammers. "The company didn't pay? I had to move. The bill must have been lost in the mail."

"Your card reversed charges. You're not welcome here until you pay the bill."

"I wanted to hold my election celebration where I started my campaign," None says.

The Blacksmith smirks. "Too bad. I didn't vote for you. I voted None of the Above."

None wiggles his sign. "That *is* me."

"Yeah, right." The Blacksmith puts out his hand. "Show me some ID."

None and MJ pull out their new driver's licenses.

The Blacksmith facepalms. "Texas?" The Blacksmith gives None an odd look.

None ignores the comment.

The Blacksmith chuckles. "If you can pay your tab, I'll be happy I voted for you. Give me a new card, and I'll seat you."

"Shen, pay the man," None says. "I'll reimburse you from my first work check."

Shen hesitates, then hands over his card.

The Blacksmith walks over to the microphone. "Welcome to election night at Beer Beakers. He started his campaign here as James Wong, the steampunk president. He changed his name, but he's still the same genius we know and love. Put your hands together for our very own None of the Above and his VP, More Jobs."

The crowd reacts with loud cheers, several confused looks, and a couple of boos.

None holds his sign with his knees. He pulls his rave whistle from his utility belt and holds his hands over his head. He double whistles, then double claps.

The crowd double claps between whistles.

None picks the sign up and swings in time with the rave whistle.

The Blacksmith takes them to a seat close to a TV. None and MJ dance to their table, waving their signs on the way. Her spurs clank and her boots thump with each step.

Once they reach their table, the claps dissipate, replaced by laughter, then the murmur of the crowd.

The TV is set to LDR News, with anchors Julie Reed and Bruce Cannon.

Julie says, "A record number of write-in votes delayed results tonight. Coming up later, can voting for a write-in candidate get you high?"

None throws up his hands. "Aiya! They'll say anything to scare off the independent vote."

Bruce says, "With 12% of results reporting, we're seeing a rout of incumbents, with a record low 42% re-elected. The Tea Party and Occupy Party are taking huge wins across the country."

A color-coded map of the country shows projected electoral votes for None and each of the four parties.

Americans don't elect the president and vice president directly. They effectively vote for electors, people who promise to vote for their candidate in the electoral college. To become president, 270 of these promises have to be kept.

Julie says, "In the presidential race, a well-organized write-in campaign has the protest vote for None of the Above and More Jobs with a decisive lead in early results."

MJ says, "Well organized? More like chaos."

None, MJ, and Shen share a round of high fives. At least 50 bar patrons clap and cheer.

Bruce says, "We have four parties with double digit voter turnout. This is unprecedented. With None of the Above winning, that will require a new election and a new group of candidates. That will shake things up."

Julie says, "Excuse me. Bruce, why would there be a re-election?"

"None of the Above won," Bruce says. "Voters rejected all the candidates in a protest vote. That means a new election is required."

"None of the Above is a person." Julie almost laughs. "You didn't know?"

Bruce fidgets in his chair. "Oh."

Julie regains her composure. "The campaign began here in Los Angeles."

The TV cuts from the news anchors to a sizzle reel of None's videos.

Footage rolls of None's presidential announcement. None says, "To make democracy work, we need participants, not spectators."

Julie talks over a montage where voters deface lawn signs and put them up. "And participants are what the None of the Above campaign found. Each supporter participates by making their own personalized sign. In North Carolina, it's a Class Three misdemeanor to steal or deface properly placed signs. So, in some places, these individuals braved breaking laws."

Video game clips and Internet remixes show next. Gamers play Capture the Sign mods, where teams can only take a sign to their base after drawing their team logo on it. People backflip off lawnmowers and belly flop into tables. Groups with signs chase Benny Hill style. "This is the first time the Internet has provided such a direct boost to a candidate. None of the Above barely exists in the real world. It's like he only exists online, an Internet President. He is the first presidential hopeful to truly unleash the Internet."

Back in the studio, Bruce is missing. Julie ignores the empty seat. She reports alone. "In Massachusetts, the campaign had its biggest misstep today. The state allows stickers to make write-in votes easier. Officials warned the public of volunteers lacing 'None of the Above' stickers with LSD. If someone asks you to lick a sticker, notify police immediately."

A reporter holds a microphone to an old man in a sunny park. "Can you tell us what happened?"

The old man says, "This nice young man asks if I'm part of the protest vote. I say yes, and he hands me stickers and tells me to lick them to make it stick. I go vote and next thing you know, I'm back in the 60s. Whoa!"

The old man extends his hands toward the reporter. She steps to the left. "I forgot how pretty colors are. I hope None of the Above goes for re-election!" The old man waves a peace symbol and stumbles off.

None talks over a car commercial. "Aiya! Does that hurt us or help us?"

MJ winks. "I think we have the aging hippie vote locked up."

Shen asks, "Who was behind that? Renquist or some Internet trolls?"

"I'm not looking over my shoulder for Renquist after every setback," None says. "The Internet is a capricious place. We can't expect it to always be in our favor, or even to make sense."

"Technology makes everything easier, even being a jerk." MJ hushes up as the news comes back from commercials.

Bruce slouches in his chair, next to Julie. He shakes his head subtly. "We join a live stream where the front-runner celebrates in a local bar."

Video of None's table shows on TV. None waves behind him. Two seconds later, None waves on TV.

Bruce says, "I'm not sure what they're wearing. It looks like a costume party."

Crowds in the bar cheer. Bar patrons point at the televisions. The applause spreads.

"I'm getting word, this is a steampunk bar. Tim Houser will report there live later tonight with the First on the Scene team." Julie says.

The TV screen splits between Bruce on the left and a well-dressed woman on the right. Bruce says, "I have Ann Franklin with the Take Back America Movement. According to exit polls, you waged a very successful campaign against all incumbents in Congress."

Ann says, "Thank you, Bruce. We're very proud of what we've accomplished. The existing parties are too beholden to corporate interests. If we want a party for the people, regardless of politics, we need a do-over. Whether you voted for the Tea Party or the Occupy Party, thank you for taking back America."

Bruce says, "Ann, I understand your group was also first to endorse None of the Above and More Jobs. It's looking like a landslide. You must be very proud."

Ann shakes her head. "No. No, we are not. James Wong and his accomplice, formerly known as Maria Cortez, are common thieves. We're looking at legal remedies and alternatives to block their presidency. This isn't over."

"Thieves?" MJ clenches her fists. She looks ready to fight.

"They tricked you into that endorsement, didn't they?" Bruce frowns.

Ann says, "They actively campaigned. We had no idea. Pure fraud and deceit. They would fit in well with the previous Congress." Ann turns her head and mock spits. "If the presidency gets decided in the House of Representatives, I guarantee None of the Above will not be president."

None asks, "Shen, can they really stop us? What's this about the House?"

"They can try," Shen says. "As long as we win the electoral college, the House doesn't matter."

Julie replaces Bruce on the left screen. "You take credit for electing None of the Above, but your group only canvassed for Congress. Couldn't the Internet videos now seen over two hundred million times be what powered the None of the Above campaign?"

"The Internet?" Ann scoffs. "This is war. You need boots on the ground to win elections."

Julie says, "Those videos spawned their own grassroots movement for a protest vote. Other than the endorsement, did you provide *any* support for the None of the Above campaign?"

Ann fumes silently.

The news anchors return to full screen.

Bruce says, "Let's check on social media. Here are the top five election retweets.

"Five, 'Election hack. Best joke ever.'

"Four, 'Great, we have jackass for president.'

"Three, 'Let's fix America. Unleash the Internet.'

"Two, 'I just got catfished.'

"And one, 'They can't be worse than the other guys.'"

Julie says, "Currently, the top candidate is None of the Above, with 51% of the popular vote. The runner-up is Senator Garfield of the Tea Party with 19%. In spite of only 15% of districts reporting, due to the sizable lead, we are projecting the next president of the United States. Congratulations to None of the Above and More Jobs."

None pulls his whistle out again. He gets the crowd to join another chant. A double whistle, then "We won!" Groups of partiers form small conga lines throughout the bar. The chorus reverberates through the bar for several minutes.

None puts away his whistle. The ambient chatter still drowns out the TV. Others trickle over to congratulate None and MJ.

It takes a while until None can hear the news again.

Julie says, "Let's check in with Corella Fox and the Word on the Street crew."

Corella, an ambitious 20 something with a designer outfit and matching coat, holds her mic out to a man with grimy hands coming her way. "None of the Above just won president. How do you feel about that?"

The guy mutters, "About time. I didn't like none of them." He waves her off and stomps away before she can ask anything further.

A young man skateboards towards Corella. He slows and dismounts off his board next to her.

"Do you think None of the Above is an odd name for a president?" Corella asks.

The skateboarder says, "A musician dude had a symbol for a name, so why not? I didn't want anyone to win president. That's why I voted none of the above. Anarchy, man."

Corella says, "A guy named None of the Above won. It's an actual person."

"Wait, if an actual person won, what's his name?"

"His name is None of the Above."

"You're asking the questions, and you don't even know his name? You only know his name is none of the above? Dude, you must not be too smart."

Corella hisses air and rolls her eyes. She walks past him.

"Dude, what's his name?" He shakes his head and skates off.

Corella interviews a woman in an L.A. Lakers shirt.

"Ma'am, None of the Above was just elected. Your thoughts?"

The Lakers fan says, "I actually voted for More Jobs, the VP. None of the Above seems accident prone. Give it time. One bad slip, and we'll have a woman with strong leadership skills leading this country."

None gasps. "Aiya! She's voting for me to die?"

"You are accident prone," MJ says.

"If I choke, will you do a Heimlich?"

"You'll have to wait and see." MJ snickers. "President sounds better than co-president."

None shakes his finger at MJ. "Hey, that's not funny."

Three-hundred-pounds, with bulging muscles, a man dressed in army fatigues power walks down the sidewalk wearing an oversized military backpack. He shoves the Laker fan out of his way.

Corella runs after him with the mic. "Sir, how do you feel about None of the Above as president?"

He spins around and points at Corella. His arm muscles tense with bulging veins. "He steals my vote. I steal his life." The man pivots back around and sprints to his blue Silverado pickup, parked on the street.

"Make sure you get his license plate," Corella says to the cameraman. The camera captures his license number, right before he zooms off. "I hope the president-elect has enough Secret Service agents to deal with guys like that."

"Back to you Julie." Corella jumps into the news van. She screams at the driver. "Drive!"

The engine revs. The tires screech. The Word on the Street crew chases after the blue Silverado pickup, as the feed cuts out.

Julie says, "Some emotional reactions tonight." She displays updated voting numbers.

None's mind wanders. Why isn't the Secret Service here? Are they stuck in L.A. traffic? If a news van can find us, why can't they? None isn't worried. He already has protection. He has MJ.

MJ interrupts his thoughts. "Take Back America did nothing. What were we so worried about?"

"Then how do you explain None of the Above being on the California ballot?" Shen says.

None says, "It wasn't just president, none of the above was even an option for Insurance Commissioner. Insurance Commissioner! Something else is going on. Shen, how did you vote for president?"

"Don't worry, I voted for you," Shen says.

None asks, "No, how? None of the above was on the ballot. Did you choose that or write-in?"

"Write-in," Shen says. "Not voting for More Jobs could be hazardous to my health."

MJ nods with a grin. "Why worry? They said we won."

None says, "California has 55 electoral votes out of the 270 we need to win. That's big enough to swing elections. What if something goes wrong? Hanging chads anyone?"

"No one else will hit 270," MJ says. "Maybe the Tea Party could hit 120. It's not even close."

Shen says, "It doesn't have to be close. If we don't hit 270, Congress decides. The House of Representatives votes for president, and the Senate votes for VP. Any of the top three candidates would be in play."

"Are you saying even third place Ed Bilsky, from the Occupy Party, could win with 11% of the vote?" MJ asks.

"If he negotiates well enough in the House, yes," Shen says.

"My spider sense is tingling," None says.

The crowd nearest the door cheers. The First on the Scene team with Tim Houser works their way through screaming drinkers.

Bar patrons toast their flasks towards the camera. "To None of the Above."

MJ says, "Here come the media. That must be it. We're really going to be elected."

Julie says, "The First on the Scene team with Tim Houser arrived at Beer Beakers. In just minutes, we'll interview our next president, live. We switch to breaking news from Corella and the Word on the Street crew."

"Always be camera ready," None mutters to himself. He combs his hair using his reflection off the TV.

Corella stands next to the blue Silverado pickup identified earlier. "Death threats were issued against presidential candidate None of the Above. We followed the man in army fatigues to his abandoned vehicle. We're identifying possible destinations."

The camera pans down the street where the pickup is parked.

None freezes up mid comb. He stares at the blue Silverado pickup. Two doors down. Oh no! It's MJ's car.

Freaked. None crouches. He tugs at MJ's ankles.

MJ drops to the floor. "What's wrong?" she whispers.

None whispers back. "He's here, the psycho guy in the military uniform."

MJ plucks the comb out of his hair.

The TV switches back to Julie. "We're going live to meet our new president. First on the Scene, it's Tim Houser."

Tim Houser comes into focus, on TV. "We can't seem to locate the candidate." The bar patrons cheer Tim, as he searches the bar.

None grabs Shen's pant leg, an afterthought.

MJ motions for Shen to be quiet, but he doesn't notice.

Shen looks down. "Why are you on the floor?"

MJ smacks Shen hard on his ankle. He falls, clutching his leg. MJ muffles his shriek with her hand.

Shen grits his teeth. He rocks back and forth nursing his leg.

None glimpses an odd color palette moving through the crowd on TV.

He points at the screen and whispers, "Army fatigues. Psycho guy is inside."

Shen nods and pries MJ's hand from his mouth.

The TV feed jumps abruptly back to Corella without an introduction.

Corella races across busy traffic with her cameraman in hot pursuit. "Oh my god! Someone save the president. Beer Beakers is across the street."

A car slams on the brakes. The impact bounces Corella to the pavement. She struggles to her feet.

Inside. Gunfire.

Erlenmeyer flasks shatter by the bar.

The crowd drops to its feet. Hushed weeping. Urgent whispers.

Tim Houser cowers behind a table. His cameraman films the gunman.

Only the assassin still stands.

The TV feed switches to the First on the Scene team inside. The gunshots replay on TV.

The sound startles the gunman.

He rotates his AK-47 and fires a volley.

A bar TV near the door explodes, with a shower of sparks and glass shards.

The gunman moves towards the TV near None.

None tenses up.

The gunman watches media coverage of his attack. He doesn't flinch at TV gunfire this time. "I want the thief that stole my vote."

TV footage inside the bar continues.

The gunman looks around to triangulate the source of TV footage. "Where's Tim Houser? I know you're here."

The TV footage ends. LDR News shifts abruptly to Corella's camera feed.

Corella peers in a bar window, her back to the camera.

Corella's cameraman says, "You're a beat reporter. This is real news. Shouldn't Tim be covering this?"

She swivels in anger. "I won't be 'Word on the Street' forever." Corella immediately softens when she realizes she's broadcasting live.

Inside, two hostages make a run for the door.

Tim Houser slams a bar stool into a window, making his own exit.

The gunman shoots the two hostages in the back.

Their bodies slump just feet from the door.

"Look what you make me do." The gunman points at his two victims.

Corella helps Tim Houser out the broken window.

"Anyone else want to leave?" The gunman tosses the spent 30 round magazine and inserts a 75-round drum in its place.

Corella mocks Tim on TV. "Last on the scene, it's Tim Houser."

Tim says, "If you're such a hot shot, why don't you go in there?"

The camera pans after Tim as he runs off, badly cut.

Corella grimaces then drapes her designer coat over the glass shards to make the broken window a safer escape path. She shivers from the breeze.

Maybe None and MJ can escape out the window too. MJ looks at her boots and spurs, not exactly stealth equipment. She scrunches her nose and tugs her boots off.

The gunman says, "What's wrong with this country when patriots like me are painted as bad guys? One man tricked a nation to yield their vote. He's the bad guy."

None and MJ crawl one table closer to the broken window. They drag Shen behind them.

"I'm a patriot. We can't tolerate stolen elections. I want the world to see justice enforced."

None and MJ aren't far from the broken window.

Corella lays her microphone on the window sill and climbs in.

MJ stops in her tracks.

She's climbing in? What is Corella doing? None can't believe his eyes.

Her cameraman sneaks Corella the camera, then runs off.

Corella takes the camera and mounts her mic on it. She scoots closer to the gunman.

MJ looks at Corella, then whispers to None. "We can't leave these people. We have to stop him."

Escape is so close None can taste it. He sighs, then nods to MJ.

The gunman says, "I want the world to know I'm not the bad guy, None of the Above is. Where's that reporter?"

Corella stands. "I'm a reporter." She smirks. "First on the scene, it's Corella Fox."

"Find me that charlatan president or I up the body count." The gunman crouches towards a woman wearing a blouse with clock cog buttons and goggles in her hair. He yanks the woman to her feet by the arm.

The hostage screams, her mouth open wide.

The gunman shoves the barrel of the AK-47 into her open mouth.

The hostage chokes uncomfortably on the gun barrel.

MJ sneaks to an adjoining table in her socks. She rolls across the aisle, closer to the gunman.

None moves to follow MJ.

Shen extends his arm to block None. "You can't go out there. He'll shoot you."

None whispers. "MJ needs a distraction. That's me unless you volunteer."

Shen shakes his head and points at his leg.

None waves his hand above the table. "I'm over here."

"Show yourself." The gunman leads the hostage towards None's voice, the rifle in her mouth, her dance partner.

None shouts from under the table. "Patriot, I value your word. Can we talk first?"

"You have two minutes to sell me on keeping you alive." The gunman looks at his watch.

None emerges behind his table. "You can let her go."

"I was a sniper. I know better than that." He aims the weapon at None, through the back of her head, a mere obstacle.

His hostage quivers, crying.

"When you voted None of the Above, you didn't expect me. This country needs the unexpected. I'll give average Americans a real voice. Our leaders neglect those not rich enough to bankroll their campaigns."

The gunman backs up his hostage one step at a time. None is three tables away. "You're just a rich fat cat like they are. I looked you up."

"Then you also know I'm not a lawyer or politician. I'm a problem solver, an engineer. I build things. You think you voted for me by accident. No, you did not. I *am* None of the Above. I am exactly what you asked for."

"I didn't ask for an Asian man who wears funny clothes," the gunman says. The victim continues her backward death march, two tables away.

"I will help fix America. Give me a chance. I will prove it, with deeds, not words. Our country's in some serious shit. I won't mince words, or sugar coat it. We can't ignore the truth and kick our problems down the road. We're running out of road."

The gunman stops one table away. He glances at his watch. "One minute." He grins at the victim. "Just what I need."

She looks back with hopeful eyes.

His crazed eyes meet hers. "Blood splatter," the gunman says. He yanks the victim's goggles off and slides them on to protect his eyes.

She understands. Horrified.

Her eyes close.

Silent tears.

Silent goodbyes.

None scans for MJ. Nowhere. Perspiration glistens on his forehead. None glances at the clockwork buttons the hostage wears, a reminder of time running out for both of them.

None points at Corella. "You can have a one on one with the president, broadcast live on TV. You want to be heard? I promise to listen. Sell me on your problem. I can help."

"You can't take back what I've seen, or done, for this country. You don't know me."

None says, "Then *let* me know you. Do you want change to be more than a slogan? Patriot, I'm change."

MJ peers from behind a table on the gunman's left flank.

None calculates the physics in his head. It's too far to traverse unseen from her vector. If MJ tries to save him, she'll die. Line of sight, the difference between life and death.

The gunman says, "Twenty seconds, you better pray. You aren't the first animal I put down to save this country."

Deep breath. His life for hers. None steps from behind the table. Only bits of skull and brain of a scared woman remains between him and a bullet. Wide open. Vulnerable.

Exhale. None circles the gunman's right flank with his hands up. "If you kill me, they win."

The gunman and victim reposition to keep aim at None. The gunman's back now faces MJ.

She creeps up behind the gunman. MJ doesn't let his bodybuilder physique, or his, at least double, weight advantage intimidate her.

"The fat cats you meant to vote against will win." None says.

The words bother the gunman. He looks up, pondering.

Could the assassin be reasoned with?

MJ takes advantage of his lapse in awareness. She grabs the gun and kicks the hostage, who stumbles backward.

The gunman fires, as MJ deflects the rifle.

A projectile grazes the hostage. Red streaks across the right side of her face. Blood trickles, from where the missing chunk of ear used to be, down her neck. The hostage flails to the ground with a scream.

Full auto. Bullets spray up the wall and across the ceiling, as MJ and the gunman struggle for the rifle.

Pandemonium.

Hundreds funnel towards the two exits. A stampede tramples those too slow to ride the landslide of humanity out the doors.

The fallen cry out, crushed by the pounding feet of the masses.

Those too terrified to move lie still on the floor.

The gunman's grip on his weapon is too strong. MJ has no leverage. She could grab the end of the rifle, but recoil would rip up her hand, and the barrel is already hot.

She attacks.

The gunman blocks her blows and counters with his own.

Tougher, stronger, and he has boots.

None puts his arm around the hostage. He helps her scurry behind an overturned table.

Smoke wisps off the hot barrel.

Stray bullets slice through a group as they flee.

Glass beakers and flasks shatter onto the floor.

A trampled woman with a head injury crawls towards the door.

The 75-round drum, spent.

The gunman rummages in his backpack for another magazine.

No rounds, no recoil. Still, smoke warns of the barrel's scalding heat.

MJ can't let this madman reload.

She grabs the smoldering rifle barrel. The heat of the metal brands her hands.

MJ uses her new leverage to loosen his grip.

Searing pain. A barely contained wail.

She kicks off his chest with enough force to dislodge the rifle.

Her flesh sizzles. The scent of cooked meat wafts through the bar. "I'm not BBQ, mother fucker."

MJ flings the weapon toward None's hiding place.

"I should have killed you first." The gunman throws the ammunition at MJ.

She ducks.

He pulls out a pistol.

MJ kicks the gun from his hands, but he grabs her foot.

The gunman lands several brutal attacks before she frees her leg.

Off balance, he shoves her to the ground.

He scans the area and rushes for the pistol.

She's closer. She reaches.

He stomps.

She retracts her hand too late. The steel toed boots crash down on her pinkie before she can reach the pistol.

MJ slips her finger from under the boot. "None, get the gun!"

"You're closer!" None says.

"No, the rifle!" MJ backs up.

Shen hobbles off to search.

None sniffs. Smoke. He finds the rifle.

None pulls gloves from his utility belt and grabs the gun.

Empty.

The gunman looks at MJ's socks and grins. He stomps at her feet repeatedly.

Attack. Attack. Attack.

He's relentless. He drives her further from the pistol.

She's no match for his power and skill. At least, not without boots.

MJ gives up more ground. "What kind of pansy killer plays footsie?"

Another stomp.

She curls her toes. His boot pins the edge of her sock.

He goes to stomp with his other foot.

She yanks her left leg back. The sock comes off. Barefoot.

Freed, she runs backward. Glass shards dig into her bare foot. "If I'm going to die, it's with boots on. This is not happening."

Shen finds the ammo magazine. He tosses it to None.

MJ limps back.

The gunman quickly catches her and knocks her down.

She kicks his face with her bare foot. The glass embedded in the bottom of her foot slices across his face.

The gunman pummels MJ.

None fires into the ceiling. "Get off her."

The gunman raises his hands.

Corella moves in for a close-up.

MJ says, "Shen, give me your tie." She tips the gunman over and hogties him with Shen's tie.

Shen points at his leg. "MJ, use your words next time."

"Cut her some slack. She saved the day," None says.

"Tell that to my ankle," Shen says.

"It could be worse." MJ upturns her burnt hands towards Shen.

Shen cringes at her injuries.

The gunman looks up at None from the ground. "You lied. You said you'd listen."

None keeps the gun trained on him. "I'll give you ten minutes. That's more than you gave me."

Corella films the interview.

The gunman speaks passionately about better treatment for veterans and improvements to VA hospitals.

Shen helps MJ to the emergency chemical shower in the corner of the bar. It's not just for ambiance, it's fully operational. Shen turns on the overhead shower. Water drenches them.

MJ runs water over her burns. Rolled up strings of the burnt skin dangle from her hands.

Shen grabs a bar stool for MJ to sit on. He kneels down to pluck glass from her foot.

Twelve minutes into the gunman's rant on veterans issues, dozens of the police and Secret Service enter the bar.

None hushes the gunman. "Your time's up. I agree veterans should be treated better. You tried to murder me, so you're still a double ass. That's what assassin starts with. The 'in' part, you'll learn about in jail."

Separate security details surround None and MJ. Suits are everywhere. They install bulletproof plastic over the windows.

"Mr. President, I'm Agent Vincent. I've been approved to lead your security detail."

"What took you so long to get here?"

Agent Vincent says, "I waited for paperwork. Department of Homeland Security guidelines didn't account for an independent write-in campaign like yours. An emergency meeting of the Congressional Advisory Committee just convened to amend those requirements. I apologize."

None is flabbergasted. "I almost died, due to bureaucracy?"

"Mr. President, welcome to the U.S. government." Agent Vincent waits to be dismissed, then resumes his work.

Medical teams triage the wounded. An EMT applies cream on MJ's second-degree burns and bandages her foot. They check out Corella's injuries from the car accident and her minor cuts from the broken window.

None hands off his prisoner and weapon to Secret Service agents. He heads over to MJ and Shen.

Agent Reynolds introduces himself to MJ. He leads MJ's protective detail.

Shen says, "They're projecting a landslide, with over 300 electoral votes. That's a pretty strong mandate. My brother, the first Asian American president." Shen can't contain his emotions. He gives None a heartfelt hug.

Shen steps back. None and Shen exchange proud smiles.

Agent Vincent and two agents approach None. "Mr. President, we've secured the perimeter. You survived your first assassination attempt."

"Aiya! Did you have to say first? That implies there will be others," None says.

Agent Reynolds says, "Most presidents since Kennedy have had at least one assassination attempt. You haven't even been sworn in yet. We can't rule out further attempts. That's why you have us."

21. Questions

None, MJ, and Shen sit in a row of chairs in front of the emergency shower.

None addresses Agent Vincent. "LDR News coverage alerted me to the gunman. Give them the first interview."

Agent Vincent leaves to make arrangements.

Medical personnel remove the dead and injured. Detectives and lab techs collect evidence systematically, like worker ants.

An hour later, Agent Vincent returns. "We vetted the reporters and secured a press area." He gives None an earpiece.

Agent Reynolds hands MJ her earpiece.

Agent Vincent makes arrangements and returns with LDR News anchor, Bruce Cannon, and a news crew. "LDR sent their best anchor."

None cusses silently and crosses his arms. He expected Corella.

Bruce skips pleasantries. His voice has a hostile edge. "I'm Bruce Cannon, LDR News. None of the Above, what's your real name?"

None and MJ pull out their California and Texas drivers licenses.

"None of the Above is my real name."

"I'm More Jobs. You'd know that if you had seen our videos."

Bruce says, "Come now. Those weren't always your names. Did you change names to deceive voters?"

None stands up. "No!" None clenches his fist and sits down.

"Then why did you?"

None says, "Not long ago, we were the most powerful civilization in human history, an unopposed superpower with global reach. Yet, for average Americans, the American dream gets harder to achieve every year. We're frogs in a pot. The heat turns up bit by bit, so we barely notice. How long before we're boiled alive?"

MJ cringes. She stares at her hands and shivers as she visualizes boiling alive. The image is all too real to her now.

Bruce says, "That seems overly dramatic. America is still number one."

"You don't realize you're in the pot yet." None smiles. "Can't you feel growing frustration? We need to solve our problems, not fight each other. My message resonated with Americans and many made it their own. The problem is, you can't hear a thousand voices. It becomes noise. So, MJ and I took back our message, by becoming the message, with our very identities. That's the first reason."

Bruce says, "I'm probably going to regret asking, but what's your other reason?"

"It's like data compression," None says.

"What now?" Bruce looks confused.

"Condense information into a smaller space." None crunches his hands together. "We changed our names to what we stand for. Instead of remembering Maria Cortez *and* her focus on more jobs, just remember More Jobs, the person. Instead of remembering James Wong *and* his unconventional thinking, remember None of the Above, the person."

Bruce scowls. "That's crazy. You could change names again."

"Not crazy. Unconventional." None grins. "We're not politicians. We're not going to stand for something different next week. If you forget what we stand for, remember our names."

Bruce says, "Uh, OK. Is data compression like when I cram a conversation into a sound bite?"

None nods. "We're the protest candidates for the protest vote."

"This is only the fifth assassination attempt on a president-elect. That already puts you on a short list with Lincoln, Kennedy, FDR, and Hoover."

None says, "With another possible depression looming, I don't like being on any list with Hoover."

Bruce chuckles. The insincerity is obvious. "You already had the killer cornered. You let him talk. You even agreed with him, while his victims lay dead and dying. Don't you think that's disrespectful to the victims and their families?"

"I keep my promises, even to those who try to kill me," None says.

"What was the key to your survival in there?"

None points to MJ. "I'm lucky I picked Rambo as my vice president."

"Who's Rambo? Is she changing her name again?"

None waves away the comment. "Before your time. Check the Internet later."

"Eight people are dead because you ran for president. How do you live with yourself?"

Agent Vincent speaks into the earpiece. "Should I shut down the interview, Mr. President?"

MJ slumps. Drained.

None waves Agent Vincent off, None narrows his eyes, and breathes deep. "I lost friends and countrymen who celebrated with me. I left safety and stared down death. MJ probably has permanent scars. We sacrificed and stopped him. That's how I live with myself. The death toll could have been a lot higher. My thoughts are with the families who lost their loved ones tonight." None stands up.

MJ follows his lead.

Bruce opens his mouth for the next question, but None cuts him off.

"Bruce, I've got a question for you, and then we're done here. You must hate me pretty badly to leave Julie alone in the middle of a news broadcast. What were you doing?"

Bruce says, "I won't answer your question since you stopped answering mine."

22. Invincible

MJ broods. She ignores agents sweeping through her house. She stares at the hole, where her water serpent katana penetrated the bathroom door. Streaks of dried blood stain the wood. MJ never got her katana back from Officer Monroe. She couldn't even find proof Officer Monroe exists.

None whispers into MJ's ear. "Are you OK? You barely said a word. That Rambo comment didn't even draw you out."

MJ says, "Today, I got old."

None sighs. "I guess I got gray hairs today too."

MJ scoffs. "Recognize today for what it was. Escalation."

None opens his hands and shrugs.

MJ says, "Today's a taste of what's to come. We're on a global scale now. We'll face bigger threats than Renquist."

"Aiya! You think I didn't notice the new guy, pointing a gun at my head? I'm on a kill list, with Lincoln and Kennedy! Neither survived their presidency. That's half the list. That's what happens to great men when the world isn't ready for them."

"A great man?" Maria rolls her eyes. "That's how you see yourself? You're so full of it. You weren't the only one in danger."

"I didn't say I was great. I made an observation. You got old? Can you skip the emotional Riddle of the Sphinx and tell me what that means?"

"You're just going to laugh." MJ sighs. "OK. OK. It's just that I'm not invincible anymore." MJ holds up her burnt palms.

None chuckles. "You were invincible before? Now, who's the one who's full of it?"

MJ says, "What separates young from old? It's that ignorance, that optimism, that you can survive anything. Today, I got hurt. I got beat. I got old."

"We won," None says.

"Bro, only because we worked together. He had me. It was over."

None gives a thumbs up. "Together is what counts."

89

MJ sighs. She goes for a bro hug but recoils as she remembers the burns. None and MJ make several awkward attempts at a hug that avoids her injuries. They acknowledge each other with a nod instead.

23. Twelve Courses

Today is the Super Bowl and Christmas, rolled into one. Pizza and snack foods cover folding tables in MJ's living room. The celebration and excitement eclipse the trials of yesterday.

Write-in votes delay official election results into the afternoon. MJ and None wait for final tallies, their unopened presents.

LDR News, with anchors Julie Reed and Bruce Cannon, displays on MJ's 90-inch TV. Julie says, "Every state finalized results, except California. California Secretary of State, Ivan Pushkin, scheduled a press conference in 30 minutes to announce the official results. We're projecting None of the Above will earn 326 electoral votes in a landslide victory."

Bruce says, "It's not over until it's over. Lawyers are challenging the results, citing potential charges of fraud."

"Do the lawsuits worry you?" MJ asks.

"No. Losing California does," None says.

MJ says, "Without California, that's still 271. 270 wins, right?"

None says, "It's the difference between a mandate and narrow victory. Given skepticism over how we were elected, they'll take us more seriously with a mandate."

"Presidents without mandates act like they have one anyway," MJ smirks. "Just act confident."

Secret Service agents escort the Wong family in. Shen and his wife follow None's mother, carrying trays of food and grocery bags. Shen and Lan move the food onto folding tables and set up trays of Chinese food.

None gets up to greet his family. "Ma, they're bringing in catering."

His mom says, "Nonsense. There's no good Asian food in this part of town. I brought all your favorites from San Gabriel Valley. I got Dim Sum from Empress Harbor and barbecue pork from Sam Woo. Eat. Eat." She disappears into the kitchen with groceries.

Lan whispers to Shen. "Why does your mom think they'll get married after all these years?"

Shen hushes her, as None approaches.

Lan hands None a "Congratulations on Winning President" greeting card, with two crisscrossed American flags on the cover. Inside, the card has clip art of the White House. The caption reads, "We're very proud of you. When you come back to L.A., buy a house. I'll paint it white, so you feel at home."

None laughs at the card. "Thanks."

Shen reaches for a donut.

Lan swats Shen's hand. "You still have weight to lose after that month with your brother. No egg tarts either."

Shen frowns. He heaps steamed rice and broccoli onto his plate.

"I heard you and MJ have some big life changes coming," Lan says.

None lights up. "I'm excited. I'm counting off the days until our big date. Then we'll move into our new place."

Lan's eyes widen when he says "big date." His excitement feeds her own.

Shen backs up a step, while Lan is distracted. He sneaks a piece of donut into his mouth. He holds broccoli above his plate with chopsticks and smiles.

"Is MJ going to be picking out decorations?"

"She has better taste than I do, but we'll probably have the staff do it," None says.

Shen opens wide and gobbles down an egg tart. There's barely enough room in his mouth to chew.

Lan glances back, as Shen chomps awkwardly. "Staff? So, it'll be a big event. Will there be a big banquet? I'm guessing ten courses."

None says, "Who knows? Maybe the White House has 12 courses."

"Twelve courses!" Lan gasps. She's so excited that she has to catch her breath. "You're going to have it at the White House?"

"What better place?"

"I've never seen that many courses in any celebration." Lan smiles at Shen.

Shen swallows the last bit of egg tart, leaving no evidence. He smiles back at her.

Lan says, "Our wedding banquet was only eight courses, and we didn't have money for a proper honeymoon either. I had to work two jobs to help put Shen through law school. I was a real estate agent and worked as a hairstylist when I didn't have any showings. Our neighbors knew about four days for free in Las Vegas. All we had to do was listen to a timeshare presentation. So, the four of us went up together."

Shen leans forward as if imparting secret wisdom. "Four is an unlucky number, but we didn't have enough gas money to go alone."

Lan smiles, reminiscing. "The Jacksons, it was the Jacksons. They scheduled their presentation for the first day. That night, Mr. Jackson told us about their four-hour nightmare. They wouldn't take no for an answer."

"Four hours! That's the third four. This is really unlucky. I was terrified. We were doomed." Shen wipes his brow.

"The Jacksons gave in, just to get out of there." Lan shakes her head. "They've been stuck going to Vegas every year, ever since. They haven't enjoyed going for a long time."

"Yes, the Jacksons were very unlucky. I looked at Lan, and I said, 'What are we going to do?'"

Lan says, "We had no options, so we went. We sat there for ten minutes. Then the presenter started the high-pressure sales. I told him we had no money. He went over payment plans. Then I pulled out home listings from my bag.

"I didn't want to spend our whole honeymoon there, so I gave them a taste of their own medicine. I tried selling them houses in L.A."

"I loved watching them squirm." Shen smiles at Lan. "It took everything I had not to bust up laughing. Lan wouldn't take no for an answer. They upgraded our hotel and threw in an eight-day cruise, just to get us out the door. We were out of there in 15 minutes."

Shen and Lan crack up together.

"Eight is a lucky number, so Shen finally calmed down," Lan says.

None laughs. "I can see you doing that. It's so you! Why haven't you told me that story before?"

"Timeshares on your honeymoon is not something you want to brag about," Shen says.

None looks confused. "What made you think of your honeymoon?"

Shen winks. "Right, it's a secret. I understand. Just don't forget us when you make up invitations."

None loads a plate of food and grabs chopsticks.

Lan frowns at Shen. "If you don't take care of yourself, you're going to die and leave me alone. That's very selfish."

Shen stops laughing. "How did you know?"

"There are foods you stuff your mouth with. Broccoli is not one of them." Lan kisses him and licks her lips. "Donut!"

Caught, Shen acknowledges how well she knows him with a smirk. Shen and Lan take food to the far couch.

None sits next to MJ on the closer couch. "I brought you something to eat, so you don't hurt your hands. Let me feed you."

MJ says, "Gross! You know how I feel about sharing saliva."

"I'll get a separate plate after," None says.

MJ nods.

None crisscrosses the chopsticks to cut a barbecue pork bun into pieces. He feeds MJ one bite at a time.

His mother watches, with a knowing smile. "I can take over. Eat. Eat." His mother shoos None off and takes over feeding MJ.

None returns with food. He divides his attention between the newscast and his mother.

"Maria, where's your family?" Mrs. Wong asks.

It's jarring to hear the old name, from an old life. If it were anyone else, None would correct them.

"They still live in Texas. They're celebrating there," MJ says.

"I heard you and James are living together now," Mrs. Wong says.

MJ says, "We were. I had him stay with my parents for a while."

Mrs. Wong lights up. "Family is very important in our culture. When two people come together, so do their families."

MJ says, "It almost feels like we're family."

Yeah, a brother, but nothing more. None has a lifetime membership to MJ's friend zone.

"We'll be spending even more time together in the White House," MJ says.

Mrs. Wong pulls out a small jewelry box. "I wanted to give you this, to show how much I appreciate your place in his life. Today seems like the right day." She pulls out a gold chain bracelet from the box.

MJ says, "It's beautiful. Thanks, Mrs. Wong."

"Should I put it on you?"

MJ nods.

Mrs. Wong circles the chain around MJ's wrist and closes the clasp.

None checks his watch. It's been almost 30 minutes. He turns up the TV volume. "It should be on any minute."

Bruce says, "There's an hour delay for the press conference. Speculation on the delay centers on California Proposition 23, the 'None of the Above Act.' It placed 'none of the above' as an option for most elected offices. Back in 2000, the measure was defeated 64% versus 36%."

Julie says, "Fast forward many elections later when Prop. 23 was reintroduced verbatim in the last midterm election. It won 68% to 32%, a complete turnaround."

Bruce says, "Let's recap the last results with 98% of precincts reporting."

The tallies for California display on TV:

52% None of the Above/More Jobs (Independent)

19% Lucas Garfield/Norman Kiln (Tea Party)

13% Ed Bilsky/Felipe Ramirez (Occupy)

7% John Robertson/Anthony Denver (Democrat)

5% none of the above (N/A)

4% Steve Danvers/George Rotterman (Republican)

Julie talks over the tallies. "Prop. 23 has a provision that doesn't count the 'none of the above' votes towards who wins. In cases like misspelled write-in candidate names, the courts have usually upheld votes, whenever the voter's intent is clear. Should the votes for 'none of the above' count towards the None of the Above write-in campaign?"

Bruce says "52%, or 57%, which should it be? Stay tuned to LDR News extended election coverage."

"We should have California," None says. "Why haven't they announced it yet? Something is up."

"They just explained about what was wrong on TV," MJ says.

None says, "They're guessing. I bet the Democrats and Republicans know more about what's going on with our candidacy than we do. Without a party apparatus, we're left in the dark."

"Hmm, how could you pass an hour?" Mrs. Wong clasps her hands together and smiles at None.

Always busy. How can None relax, when moments of peace are so rare, so alien? None mutes the TV and turns to face his mother. "Ma, how's work?"

"They fired the entire HR department. Marjorie was with the hospital 30 years. How could they treat her like that? They replaced her with something on my phone."

"They're called apps, Ma."

"Can you help me figure out how to use it?"

"Free tech support, my favorite pastime." None forces a smile. "Show me which app."

Mrs. Wong swipes to the third page of apps and points at Slam Dunk Jobs.

None freaks out. "I wrote the original app. MJ, look! Ma is using Slam Dunk Jobs."

MJ says, "That's great. The company is taking off."

"Son, did you get Marjorie fired? She was my best friend at work."

"That's not what I expected to happen," None says. "That's part of the company I lost control of."

"Well, you get it back. Marjorie's still out of work."

None sighs. "I'm sorry. All I can do is show you how it works."

His mother calms down.

"It has biometrics, place your thumb there," None says.

She follows his instructions. He shows her how to fill in her information. The next screen shows her hourly rate. $31.75 is prefilled.

"You've been a nurse a long time. When was your last raise?"

"They stopped yearly raises five years ago," Mrs. Wong says.

None says, "Give yourself a raise. The worst they can do is say no."

Mrs. Wong hesitates. She updates the value to $33.75, and presses send. Rate accepted. "You just got me a two dollar raise! Thank you, son! I'm going to start on dinner." Mrs. Wong rushes off to the kitchen.

None grins ear to ear. Slam Dunk Jobs works just like he hoped, better wages for better skills. This could help a lot of people, at least those not in HR.

He shifts focus to MJ. "Ma just used Slam Dunk Jobs. Wow! Life is good, a little strange though. I had the weirdest talk with my brother earlier. I was talking about my inauguration, and then he changes the subject to his honeymoon. What made him think of that today?" None shrugs.

"Your mother gave me this gold bracelet." MJ holds out her wrist. "That was strange too. Do you like it?"

None's eyes widen at the gold chain. "Ma gave that to you?"

MJ nods.

"Aiya! They're talking about how many courses I'm having for a wedding banquet! Now I know why they were so excited about a White House dinner... You didn't tell her we were dating, did you?"

"We're not dating." MJ scoffs. "I'm not your girlfriend. I'm your—"

"Bro." None stares at the ceiling. "If Ma understood that she wouldn't have given you that gold chain."

None hears his ringtone coming from between the couch seat cushions. His phone must have dropped out of his pocket. He grabs it and answers. "This is None, go ahead."

A man speaks with a strong Russian accent. "Today, your dream dies."

"Do you know who you're prank calling?" None asks.

The Russian says, "You're not safe. You'll never be safe."

"You can't intimidate me. I'm president."

"You'll never be president. It's taken care of."

None remembers Trench Coat Man. "It's you, from the police station."

Dial tone.

None pushes the phone into his pocket. His phone is already there. The couch phone isn't his.

24. Propositioned

The threatening phone call makes MJ's house feel less safe.

"What was that about a police station?" After her recent visit, the mere mention of police stations unsettles MJ.

"I got prank called by one of Renquist's henchmen. He said I'd never be president. I have enough delegates to be president! I guess he doesn't know how to add."

"Agent Vincent!" None flags down his lead agent. "I was threatened on this phone. Find out where it came from."

Agent Reynolds pulls out gloves and collects the evidence in a baggie. "We sweep for electronics. The last sweep was two hours ago. So, the phone was placed recently."

None says, "Renquist owns cops. Maybe he owns Secret Service too. One of your agents wasn't sweeping for electronics. He was placing them."

Agent Vincent says, "That seems unlikely, but I'll investigate."

"Remove anyone on the last sweep from my security detail," None says.

Agent Reynolds calls in the reassignments on his earpiece. Three agents are escorted out the front door.

"MJ, let's play detective," None says. "You're the first question."

"How did they match your phone model, ringtone, and case?" MJ asks.

None says, "They're watching us. If agents sweep for electronics, they'd need to be mobile, like a drone,"

Agent Vincent nods. "We'll turn signal jamming back on. Cell phones won't work anymore." Two agents leave the room to configure the equipment.

"My neighbors are going to love that," MJ says.

None says, "My turn. How did they predict where I'd sit?"

"With more than one phone, they wouldn't have to," MJ says.

None flips up a couch cushion. Another phone.

Shen and Lan get up, as agents from both security details converge on the couches. They find eight phones, all identical to None's.

None says, "He sent a message, that I'm not safe. This was no prank."

"Why is that jerk news anchor, Bruce Cannon, smiling?" MJ asks.

Shen rushes over. "We lost California. It's all over the Internet. I need to borrow your laptop."

"Aiya!" None unlocks the laptop and unmutes the TV. He stares at the screen, stunned.

Bruce grins wide. "We have breaking news. California Secretary of State, Ivan Pushkin, announced finalized election results. We go to Sacramento live."

Pushkin stands behind a podium emblazoned with the state seal of California. "Voting percentages remain unchanged from our last update. Prop. 23 prevents votes for 'none of the above' from counting for who wins the election. We chose to also apply this law to all write-in votes for "None of the Above," including votes specifically meant for a candidate with that name. The winner for California's 55 electoral votes, with 19% of the vote, is Senator Garfield, of the Tea Party." The crowd erupts into boos, with isolated applause.

Bruce says, "Just voting for conservatives in California is a protest vote. It's been decades since a right-leaning candidate took the state. Large celebrations have been scheduled tonight in Orange County."

Julie says, "LAPD is on high alert. There are reports of protests and isolated rioting statewide. Occupy Oakland switched from protesting against None of the Above to protesting against Garfield."

None turns his back to the TV, unable to watch anymore.

MJ says, "271, we still win. Just strut a bit more."

"I'm the only main candidate who didn't take their home state," None says.

"Don't get all mopey-mopey on me. I don't have the energy to jolt you out of a funk. It's amazing we did this well," MJ says.

Mrs. Wong comes from the kitchen. "Son, can you help me?"

"Now is not a good time, Ma."

"Where's your rice cooker?" she asks.

None says, "It's still in the garage. I never finished unpacking. We're moving again soon, anyway."

"How does Maria make your rice?"

"It's not her job to cook my rice." His eyes narrow. "Can I talk to you in the kitchen for a minute?" None grits his teeth and follows his mother into the kitchen.

"Ma, don't plan any wedding banquets. I'm glad you approve of her, but MJ and I are not getting married."

"You moved in together. How can you not get married?"

None sighs. "We're roommates."

"I saw you feeding her." Mrs. Wong holds up a tray of freshly made potstickers. "Food is an expression of love."

"She hurt her hands. I'm just helping my best friend."

She slams the food tray onto the counter. A few pot stickers flop to the ground. "A wife! That's who should be your best friend! There's something more with you two. A mother knows."

None says, "You jump to conclusions. That's why I can't introduce you to any of my friends. I only let you meet MJ because I didn't know any better when I was six."

"What mother doesn't want grandchildren?"

"Ma, 37 and single doesn't mean I'm broken. I promise. If I'm going to propose to MJ, I'll tell you first."

She flashes a guilty smile. "My friend has a lovely daughter—"

None holds out his hand. "I don't have any trouble meeting women. It's my choice who's right for me." None storms out of the kitchen.

MJ and Shen watch the newscast.

None approaches from behind. "You're right MJ. A win is a win."

MJ turns around, shocked. "While you were away..." She points at the TV.

A caption reads, "Mutiny! 43 electors defect." A reporter interviews a Chinese man in a suit. "Why are you breaking your oath to cast your vote for None of the Above?"

"My family and I have gotten death threats. After yesterday's attack, we no longer feel safe. I've decided to abstain."

At another location, a white male wearing coveralls and suspenders talks to a reporter. "I'm Reginald Nesbitt, attorney at law. It's come to my attention that not all votes for None of the Above were meant for my candidate. Just as I fight for my clients, I'm fighting for the voters in the great state of Colorado. I'm supporting Tea Party candidate, Senator Garfield, instead. None of the Above may be The Internet President, but he's no American president!"

"Shen, how did this guy become one of our electors?" MJ asks.

"He's a friend from college. He always wears suits. He must be pandering to potential clients. It's a rural part of the state, mostly farmers."

"Do you trust him?" None asks.

Shen cringes. "Well, now that you mention it, no. I knew he'd appreciate the prestige of being a presidential elector. I didn't know he'd turn it into a photo op."

"Can they switch teams like that?" MJ asks.

Shen says, "They're called faithless electors. It's happened over 150 times but never swung an election. It's not against federal law, but many states have laws against it. Those laws aren't usually enforced."

"Shen, do something," None says. "Make them enforce those laws."

"Do you know how hard it is, just to convince 12 people on a jury? It's impossible to get 41 out of 43 people to unchange their minds. Anything more than two holdouts and you're wasting your time," Shen says.

"What if I ask them to come back? Or scare them?"

Shen says, "Either you look weak, or like a jerk. It's my job to be the jerk, and I wouldn't even do it. Guaranteed backlash."

None says, "I went to bed with a landslide. I wake up facing a re-vote by the vested interests I called out. I'm a cow walking into a slaughterhouse. I'm hamburger!"

"You can still win the House of Representatives over," MJ says.

None sneers. "I'm sure we'll be warmly greeted after all the smack I talked about politicians and lawyers. Take Back America is on the warpath too. Even the anti-establishment forces are against me. This one-two punch is what Trench Coat Man warned me about."

"We've made it this far," MJ says. "We'll just figure out how to win in the House."

None says, "The people voted for us, but will politicians? To them, we're a threat."

"I'm not worried about MJ," Shen says. "No politician wants to be on record as literally voting against More Jobs. I can see the attack ads now."

"It's so lame that 43 people can steal millions of votes." None throws his hands up in frustration. "How can I win the popular vote by such a large margin and still have to go through all this?"

25. **All Your Law Are Belong To Us**

Shen files an injunction to stop certification of the California election. Lawsuits wind their way through the court system for three weeks. The California Supreme Court oral arguments begin tomorrow.

Stock markets drop almost every day since the election. Markets hate uncertainty. At least, that's the Wall Street opinion.

Voters fire so many Congress members, that they decide to cancel winter recess. Politicians must earn favors, before entering the private sector. They will work through the holidays to complete those favors.

One obstacle stands in their way. Democrats and Republicans cannot overcome their bitter mutual hatred. The two-party system locks itself in a death grip. Every bill adds a poison pill amendment, a rider clause so unacceptable to their opponents, that it kills the bill.

Republicans introduce a bill. Democrats add an amendment to raise the minimum wage. The bill dies.

Democrats introduce a bill. Republicans add an amendment to cut Medicare funding. The bill dies.

As a test, a junior senator enters a bill to reaffirm there are seven days in a week. Both sides add their poison pills. The bill dies. It's become a habit.

Not even a looming government shutdown next week jolts Congress to work together. It won't be their problem soon. Will those disgruntled former employees let the government burn?

None wears jeans and a black T-shirt with a cyborg and the text "ALL YOUR BASE ARE BELONG TO US." None slouches behind a black podium, flanked by Shen and MJ. Two steps lead down to rows of reporters in folding chairs.

MJ whispers in None's ear. "Confident. You're president. I believe in you."

None stands up straight. Who needs electors, when MJ has his back? They fought off an assassin. A press conference should be no sweat. None points at a raised hand.

"You're registered to vote in Texas, yet you're here in California. Do you pretend to live in Texas to escape the 12th Amendment?"

None says, "I promised the Internet to spend only $25,000 on campaigning. Moving to Texas allowed me to keep that promise. I voted in Texas. I'm in California to fight for votes that are rightfully mine."

None points at a reporter next to Bruce Cannon, but Bruce stands up instead. None disguises a sigh as a deep breath.

Bruce says, "Your candidacy caused stock market turmoil. Will you concede your legal case to stop the meltdown in the global economy?"

"The markets had no problems dropping before I got involved," None says "The global debt crisis has built for decades. I won't concede as the scapegoat."

None quickly points at another reporter far from Bruce, before he can ask another question.

The reporter asks, "As president, how would you stop a government shutdown?"

None shakes his head. He pauses for a moment of silence, surveying his audience. "Politicians get caught up with their opponents, their agenda, their lobbyists. They forget about the rest of us, about the repercussions of their actions."

He nods and grins. "America belongs to 'We the People,' not 'We the Special Interests.' I want to remind voters we can take back America. We." None pounds on the podium, like a gavel.

"The entertainment industry pushed anti-piracy laws that went too far, SOPA and PIPA. One page with infringing content would block entire websites. Imagine YouTube taken down because Prince played on the radio during a dancing baby video. Over 10 million voters protested the bill. Sites like Wikipedia went dark. The Internet united and defeated it. We did it before. We can do it again." None squeezes past MJ to leave the podium.

"Lines from a mistranslated video game became one of the first memes on the web." None points at his shirt. "'All your base are belong to us' was left, like graffiti, all over the Internet and even places in the real world."

"Congress does the same thing. They leave graffiti, unrelated poison pill amendments, on every bill."

None struts down the two steps. Secret Service agents close in. None waves the agents back. "If Congress can't work together, we face another government shutdown. We have a week to stop it. Are you with me?"

A few reporters nod. Silence.

None addresses the crowd. "Do you want to devastate entire swaths of our economy? Do you want a government shutdown?"

A single meek "no," comes from the back.

None strides down an aisle. He points at Bruce Cannon. "Do you want a government shutdown?"

Bruce glares back in silence.

None throws his hands up in mock shock. "Bruce Cannon from LDR News wants a government shutdown."

"I didn't say that."

None says, "We lose control by doing nothing. When you say nothing, do nothing, you tell me and everyone in this room that you support the government shutdown."

Bruce stands up. "I don't support the government shutdown."

None works the aisle. He points at each reporter along the aisle and asks the same question.

No. No. No. Most of the crowd stands and chants along.

The fourth aisle reporter smirks. "Yes. Turn off the government and leave it off."

None claps. "Thank you! America was founded on dissent. I applaud your participation. Democracy shouldn't be a spectator sport."

None delivers a sermon with fire and brimstone. "I'm calling on the American people. I'm calling on the Internet. Remind Congress who they work for. Send a message. Yell out your window. Post on your news feed. Leave a sign on your lawn. Make them listen."

The crowd claps.

None shakes his hand dismissively. "Even as President, I'm just leader of the free world. I don't have the strongest power. That strength, that power, is yours, if you're willing to use it. Leave no doubts how you feel, online, or off. Make your message a funny one. Winter is coming. We're going to need the laughs."

None fist pumps and imitates a robotic voice. "All your law are belong to us! For great justice!" He swings both arms up in the air. "Unleash the Internet!"

The newscasters laugh and give None a standing ovation.

None, MJ, and Shen leave the press conference.

Within minutes, None's words spread across the Internet on message boards, blogs, and social media.

Corporations also tire of the deadlocked Congress. They send forth their armies of lobbyists.

Social pressure on Congress mounts by the hour.

26. **Rock Star**

Backstage, None spins and fist pumps. "Did you see that?" His jaw gapes open. "Wow! I want to do it again."

Shen and MJ don't share his enthusiasm. MJ purses her lips. Shen stands with his arms crossed.

MJ says, "I felt like a second wheel on a unicycle. Next time you want me by your side, don't expect me to step aside."

"What happened to me discussing our legal strategy?" Shen asks.

None shrugs. "I got swept away in the moment. It got results." He smiles at MJ. "Confidence. You were right."

She doesn't smile back. "Maybe too confident."

None feels like a rock star, but the two buzzkills dampen his mood. He looks for a distraction. "Agent Vincent, how's it going?"

"Mr. President, it's more difficult to protect you when you don't stick to the plan."

Three buzzkills. "Aiya! Everyone's a critic today."

"Corella Fox, LDR News, has a proposition for you," Agent Vincent says.

"We have nothing more to say about Prop. 23 until tomorrow's hearing," Shen says.

"Sir, it's related to a media appearance," Agent Vincent says.

None says, "Let her in. I never got to thank her."

Agent Vincent gives the signal to let Corella in. An agent escorts her to None.

None shakes Corella's hand. "You were my hero on election day. Thank you for all you did."

MJ waits for her accolades. They don't come. She watches as the handshake lasts a few beats too long.

MJ shakes Corella's hand and fishes for compliments. "Yes, *you* were the big hero on election day."

Corella doesn't respond to MJ. She fawns over None, instead. "Mr. President, you were on fire out there. I'm producing a new reality TV show for the Reveal Channel, and I'd like you to star."

None says, "TV is the past. The Internet is the future. It sounds like a step back."

Corella says, "There are still people with landline phones. It's not one or the other. We'll call your show *The Internet President*. If you don't get elected, the name still works."

None zones out in a trance, starry-eyed. TV star and president? Could every night be like tonight?

MJ says, "We'll think about it." She grabs None by the arm and pulls him away.

Corella says, "Don't settle for some five-minute weekly newsreel online. This is real exposure."

None pulls his arm back from MJ. "I'm interested. I have one non-negotiable term. It has to be streamed live on the Internet unless you want to call it *The Landline President*." He grins and holds out his hand to shake.

Corella shakes on it.

"I guess we're in," MJ shakes too.

None says, "This is my lawyer, Shen. Give him the contract."

Shen takes the paperwork. "Now I'm your lawyer? Just not during the press conference." He wanders off, clutching the documents.

Security guards hold back a wall of people, hopeful to see the president. An agent escorts Corella back through the crowds of people.

MJ watches Corella leave. She fixates on a slender French woman just beyond the security perimeter. Most men, including a few agents, fixate on the same woman. Her look belongs in full-color magazines, a canvas of fashion artists. A golden mane flows down to the thigh of this lioness, her most striking feature. Astrid. The Astrid.

Astrid flirts with a seductive smile at the checkpoint. Agents argue over who gets to escort her.

MJ turns to None. Her nostrils flare. "What's Gold Digger doing here?"

"She has a name."

"That doesn't mean I have to say it."

Agent Vincent interjects, "You have another visitor. Astrid Fontaine."

The supermodel sashays towards None, escorted by six agents.

Agent Vincent looks at the size of her escort and shakes his head.

MJ stretches with clenched fists. She yawns like a cat, waking up from a nap, then covers her mouth. "How delightful to see you, Astrid."

"Someone needs their beauty rest," Astrid says.

MJ purses her lips.

Astrid throws her arms around None.

With a mix of concern and jealousy, the agents hesitate.

Astrid says, "It's OK. He's my boyfriend."

"Oh." None is enthralled, more puppy than man.

"I'm so proud of our new president." Astrid French kisses None.

MJ shrugs. "We still might lose."

Astrid retracts her tongue. Her lips stiffen.

"None, you must be looking forward to that first paycheck so that you can pay people back," MJ says with a half-cocked smile.

Astrid loosens her arms and presents her cheek when None returns for another kiss.

MJ says, "In another three months, you'll have spending money."

None can feel the supermodel slipping away. He glares at MJ for embarrassing him.

Astrid releases him and inches away from None like he's contagious.

MJ's smiles aren't fooling anyone. MJ looks Astrid eye to eye. "Astrid, None doesn't really have his own place anymore. He's living with me. Am I going to see you more often, or is he moving in with you?"

None stares down MJ, his breathing heavy, his face red. It wasn't her place to share those secrets, but everything MJ said is true. How can an honest man fight the truth?

Astrid pulls her phone from her pocket. "I have to take this." She steps out of hearing range.

None whisper yells at MJ, careful not to let Astrid hear. "Bro, you are messing with my mojo."

MJ whisper yells back. "I just saved you from four years of embarrassment. You don't make a party girl your First Lady. We're lucky she'd rather fly off to Ibiza, than entertain world leaders. Otherwise, she'd still have her hooks—"

Astrid strides back and points at her phone. "It was work. Can you believe I forgot about a photo shoot tomorrow?" She makes an airhead laugh. "Duh, I'm so stupid sometimes. I have to run to the airport." Astrid blows None a kiss, and she's off, like Cinderella without the glass slipper.

None watches her leave. The buzz of tonight disappears, leaving disappointment. "I don't appreciate your meddling any more than my mother's."

"Nobody watches out for you more than I do," MJ says.

None says, "Who asked you to? My life, my choices. I know it's your house, but you really deserve to be sleeping on a couch tonight."

27. **Day In Court**

A curved, raised desk extends along the back of the courtroom for the California Supreme Court. Court staff work at the center of the main floor with a podium.

A wooden partition separates a row of seats for attorneys inside from the gallery for spectators. None and MJ occupy seats in the first row of the gallery, near Senator Garfield and his running mate.

Shen leans towards None. "We need to talk," Shen whispers. "The way you handled that press conference bothers me. It calls into question how you value me, as your lawyer, and your brother."

None whispers back. "I'm sorry. It was selfish. I won't do it again." He hopes his apology is enough.

The seven California Supreme Court justices take their seats on the raised platform. Shen returns to his seat.

A presidency hangs in the balance, but the court does not skip even the most mundane procedure. Staff attorneys welcome notable guests, go over the court calendar, and read the attorney roll call. Shen stands alone against eight opposing counsel.

A staff attorney reads off the lawsuit information. "Court Case S305471: None of the Above, et al. Petitioner versus Ivan Pushkin."

None broods. Shen alone represents him in a high-profile case. There is no more public way to embrace Shen as his lawyer. How could he still be hurt by one little press conference?

Shen steps up to the podium. Poised and ready, he looks the justices in the eyes. "May it please the Court, I'm Shen Wong, appearing on behalf of None of the Above and More Jobs. This case raises the issue of denying votes meant for my clients based on an unconstitutional interpretation of California Proposition 23. We contend the misapplication of this law, despite winning a majority of the votes, has handed an undeserved and unwarranted victory to our opponents."

Justice Tozen says, "The exact text reads, 'Votes for "none of the above" shall be tallied and listed in official election results, but will not count for the purposes of determining who wins elections.' Your client's name is None of the Above. How was the law misapplied? I don't see the ambiguity here."

Shen says, "Justice Tozen, Proposition 23 pertains to 'none of the above,' the concept, and is thus in lower case. This is distinct and separate from None of the Above, the person, a proper noun and thus capitalized. That law has been interpreted broadly and incorrectly to include my client."

Justice Tozen narrows his eyes. "We're talking about one clearly written sentence from a one-page law. Your client caused the only ambiguity when he changed his name to match the concept. The two are conflated and not disentangled by mere capitalization. Secretary of State Pushkin used a sound interpretive theory."

Losing his pivotal argument unsettles Shen. His eyes drift down to his notes. His hands shake. "California Election Code, Section 15342, states that the intent of the voter will be used to determine valid votes, in spite of not following directions."

Justice Garcia says, "Mr. Wong, how is the intent of the voter relevant? I see no dispute on how the votes were counted, only whether those votes count toward selecting a winner."

Shen's voice cracks. "The votes were stolen."

"The votes were properly counted, just not considered for victory, as clearly stated in the law," Justice Garcia says.

Sneers and dismissive looks greet Shen, from opposing counsel.

None's pulse races. Is it over already? Is Shen mad enough to throw the case, or did he fail his brother once again?

Shen looks back at MJ. Recognition spreads across his face. He smiles at her.

MJ smiles back, confused.

Justice Garcia looks over his glasses at Shen. "Surely, you must have better legal arguments than these?" The words sound damning, but the tone was more that of a mentor, prodding a student to do better.

Shen clears his throat, his hand still shaking. "The crux of this case is not None of the Above, but my other client, More Jobs. Voters for my client had a very high bar, a write-in vote. Each voter had the presence of mind to write in More Jobs, in spite of a much easier ballot option with 'none of the above' by itself. Voters themselves differentiated between these distinct choices."

"Proposition 23 has no mention or relationship with More Jobs. That statute does not apply to her. All of her votes should count toward victory in California." Shen steadies his hands.

None exchanges smiles with MJ, pleased with the change of strategies.

Shen says, "As a response to the 1800 election, the 12th Amendment reworked the electoral process. Each elector votes once for president and once for vice president. His electors are her electors,"

Justice Tozen says, "She shares a ticket with None of the Above. The two are conflated. Consequently, couldn't she also be affected by a statute that affects him?"

Shen looks the justices in the eye again. "I concur that None of the Above and More Jobs are conflated. However, I would draw the opposite conclusion. Proposition 23 has nothing to do with her, yet it is being applied to her. This is a violation of the Equal Protection Clause of the 14th Amendment. The only constitutionally sound interpretation of Proposition 23 is to apply it only to ballot votes, but not to write-in votes."

"That determination is for this court to decide," Justice Tozen says.

"I would like to reserve the balance of my time," Shen says.

"Very well, Mr. Wong." Justice Garcia looks down at his notes. "Mr. Kirchner, we'll hear from you."

The lead attorney, representing the Secretary of State, strides to the podium. He nods at the justice who summoned him. "Justice Garcia, and may it please the Court. The Constitution specifically vests the authority in state legislatures for choosing how to appoint electors. States should have primary jurisdiction, including voter submitted petitions, such as Prop 23. This case is purely a matter of state law."

"The Constitution is the supreme law of the land. Couldn't federal law countermand state law?" Justice Garcia asks.

Mr. Kirchner says, "The plenary power to appoint electors is granted in the constitution. It's a state right that should not be taken away. We have a clearly written one-page law that precisely states that votes for 'none of the above' will not be counted toward winning. None of the Above *and* his ticket are stricken from their eligibility to earn electoral votes."

"Are you saying this man, and anyone who runs with him, is unelectable?" Justice Garcia asks.

Mr. Kirchner says, "There are other states outside California. I would never presume to know what is or is not possible.

"This is a simple case that the plaintiff attempts to distract us from. There are no overvotes, no undervotes, no hanging chads. In spite of unprecedented numbers of write-ins, there is no controversy that votes were miscounted. This law is clear, valid, and binding.

"The Constitution is built on the separation of powers. Taking away the clearly granted right to determine electors would be a blow to the sovereignty of states. If such clear and concise laws can be overridden, what hope do we have of maintaining justice in the face of ambiguous and complex legislation?"

Mr. Kirchner steps away from the podium.

"Thank you, Mr. Kirchner," Justice Garcia says.

"Thank you."

"Mr. Wong, you have 20 minutes," Justice Garcia says.

Shen says, "As opposing counsel states, this is a simple case. More Jobs has been unfairly and unconstitutionally deprived of her rights as a citizen to be elected to public office. Californians have had the value of their votes stripped by this unjust interpretation of a well-meaning law. Only votes using the added ballot option should be taken away.

"The spirit of Proposition 23 was to encourage alternative candidates, like my clients. It's ironic this very law is being used against them.

"If a single sentence can invalidate the votes of the majority of citizens in this state, it will set a dangerous precedent indeed. Our very democracy would be imperiled." Shen collects his notes and pulls them from the podium.

The court adjourns. The justices retreat to deliberate in their chambers.

None and MJ greet Shen to congratulate him.

"Shen, way to use your words," MJ smiles.

28. **Twelve Things**

Two Secret Service agents guard a cramped dressing room. Shen covers the legal strategy with his brother.

None says, "Why did I have to become a Texan? What happened to the 12th amendment? They never even mentioned it."

Shen says, "They argued the law was clear and unambiguous. A second cause of action muddies their argument. Justices would pick up on that. While we wait for the verdict, I'll prepare the appeal."

None says, "You kicked butt. You think we're going to lose?"

"Me, or them, the loser will appeal. The legal steps aren't over until the U.S. Supreme Court weighs in."

"Thanks for getting a head start."

"I didn't do it for you." Shen's words linger in the air.

None pauses for an explanation that never comes. None replays current events in his mind. What is Shen mad about now? "You seemed pretty upset at the oral arguments. Thanks for accepting my apology."

"I didn't," Shen says. "I bottled it up."

"It's not healthy to bury emotions. I hope you're not still mad I didn't highlight you as my lawyer. You were the only one I asked to stand up for me in the Supreme Court."

Shen mock laughs. "The only one you asked, or the only one you could afford? I looked up 'All your base are belong to us.' That Internet meme was classified as Engrish. Were you making fun of my accent? I can't even say it without getting mad."

None stifles a laugh. "Aiya! That's what you're mad about? It had nothing to do with you. It was harmless fun. It was just a bad translation. Besides, it was a Japanese game, not Chinese."

Shen exhales loudly through his nose. He scolds like a disappointed parent. "American Born, you give these people too much credit. If they can't tell the difference between Hawaii and Kenya, how can they tell China and Japan apart? It's just a matter of time before they ask for your birth certificate too. Chinese. Japanese. Korean. Dozens of cultures. They paint us with one brush. Asian."

Ever the precocious child, None smirks. "I'm a dozen things first before I'm Asian."

"Wrong, that's the first thing they see. We bring China with us." Shen sighs and shakes his finger at None. "It's a terrible thing to ignore your heritage. It's a gift, thousands of years in the making. You barely speak Mandarin."

None settles down. He tries to thaw his brother with a guilty smile. Shen ignores it.

Shen says, "A part of me wonders if you changed your name to be less Chinese. Are you a dozen things first, before you're my brother? Family comes first, not MJ, not unless she's family."

None asks, "If you're so angry with me, why are you still helping?"

"Your actions reflect on us." Shen clears his throat. "I won't be going on stage with you." Shen approaches the doorway. He glances back with hurt eyes and departs.

29. United Only In Dissent

Lights, rigging, and catwalks extend above a stage. Backstage, None, MJ, and Corella prepare for the premiere of their new show, *The Internet President.* None and MJ wear their steampunk clothes to begin their TV careers. MJ wears a new addition to her outfit, Gothic arm warmer gloves, with seven buckles on each arm.

"Are you dating anyone?" Corella asks None.

Is she flirting? None blushes and smiles at her. "Why do you ask?"

MJ intrudes, her tone almost defensive. "Yes, he is."

"I knew it." Corella points at None and MJ. "How long have you two been dating?"

MJ points at None and herself, palms down. "There's no us. He's dating Astrid."

Corella says, "Perfect! She's very popular. Viewers love celebrity dating. Who's a bigger celebrity than the president? We'll be trending."

MJ fidgets with her buckles, sizing up Corella. "You seem different. I can't put my finger on it."

"I probably sound dumber." Corella hisses air. "Now I care about ratings and the lowest common denominator. I use tabloids and paparazzi." Corella shrugs. "I run my own show."

None says, "During the decline of the Roman Empire, they used bread and circuses to distract citizens from their problems. I'm the circus. The line I won't cross is becoming meatspace clickbait."

"You'll be a fool, but not a charlatan. Noted. Come out when you get your cue." Corella walks on stage to cheers.

MJ says, "You wanted modern fireside chats, like FDR gave during the Great Depression. That's not the vibe I'm getting."

None asks, "Would FDR give up dignity to get his message across if he faced a 24-hour news cycle and short attention spans?"

"You're on." A stagehand pulls open a curtain.

High energy dance music booms over the sound system. "Welcome, The Internet President: None of the Above, and VP: More Jobs."

The crowd claps in unison, awaiting their rock stars.

MJ bro hugs None, then pulls away. MJ stomps her boots in time with the claps. "You ready?"

The clink of MJ's spurs. The intensity of the crowd. They're waiting for them, for him. None's spine tingles with excitement and dread. Though nervous, None pauses to savor the moment. It feels like a big moment, one he doesn't want to lose. None says, "Unleash the Internet."

Two best friends charge onstage. In front of the audience, they alternate jumping up and down like game show contestants and high five.

A large oak desk takes center stage, with adjoining black couches on either side. Including a second-floor balcony, 235 viewers stand and clap for their new leaders. Shen stands in the aisle of the first row, almost the only person not excited to be there.

Corella claps with the audience. "Wow! Feel that energy. We're also live on the Internet."

None and MJ settle down. Corella stops clapping. The audience follows her lead and sits.

Corella says, "You both wore these clothes at the…shootout, the bar. Are these part of your brand, as unconventional leaders?"

None swings around a wrench from his utility belt. "It's more than branding. We role play steampunk adventures on the airship Nemato. I'm the engineer."

MJ curtsies and tips her augmented fedora. "I'm Nemato's captain."

"So, it's like unpaid acting," Corella says.

None says, "Role playing is exercise for your imagination. Most people only see what is, not what could be."

"Speaking of what could be, any word from the California Supreme Court?" Corella asks.

"Right now, the Tea Party is telling Californians, 'All your vote are belong…'" None looks at Shen, who scowls back. "They're saying, 'All your votes belong to us.' We will fight all the way to the U.S. Supreme Court. I'm not giving up, and neither should you. We're going to storm the castle."

"You're not even in office, and you've already reduced gridlock," Corella says.

None nods at MJ.

MJ chimes in. "Both parties stopped adding poison pills. Congress agreed there are seven days in a week." She rolls her eyes. "Great bipartisanship." MJ claps slowly.

"Aren't you at least proud that you were able to pressure Congress to avoid a shutdown?" Corella asks.

MJ throws up her hands. "It bought us two months: just enough time for the new Congress to take over. The country is led by spoiled children who act out when they don't get their way. They needed a spanking. They just got one."

Corella says, "In opinion polls, head lice are more popular than Congress. Care to comment?"

MJ grins at Corella. "Congress is an embarrassment to head lice. This isn't a government. It's a civil war, and there's nothing civil about it. I hope the incoming members of Congress don't repeat the same mistakes."

"When voters were asked to describe Congress in one word, common choices were 'dysfunctional,' 'liar,' 'gridlock,' even 'clogged,'" Corella says.

MJ says, "Congress is a clogged sewer line. It's backed up and full of crap."

None pulls a wrench from his utility belt. "Did somebody call a plumber?"

The audience laughs and cheers.

None says, "The United States, the U.S., us. There's no us anymore. It's every American for themselves. That lack of cooperation is a toxin that poisons this country. We may not agree on what we want, but we can agree this is not it. We're united only in dissent.

"When I grew up, everyone watched the same three TV networks. Republicans and Democrats talked to each other. The Middle was dominant.

"The middle of the road was ripped up, leaving us on separate highways. Everyone has their own personalized channels on separate screens. We've forgotten how to talk to each other."

"What's the cause?" Corella asks.

None says, "Since they killed the Fairness Doctrine in 1987, one person's crackpot is another's source of truth. If facts aren't common ground, then how can we agree?

"Celebrity relationships get more attention than the wars we wage. How could the voice of reason compete with that?"

MJ says, "It can't. The voice of reason is boring."

None shrugs. "You're right, MJ. We can't put crazy back in the bottle. So, we'll have to live in our separate reality distortion bubbles. Moving on. Let's solve gridlock for real."

Corella asks, "You sure you want to fix gridlock? If they can't pass laws, they can't screw anything up."

None says, "Corella, America can't survive the next crisis without a functioning government. Gridlock has to be fixed. In software engineering, we solve big problems by breaking them into smaller components."

"This is politics, not software," Corella says.

"Decomposition is a first principle," None says. "You can apply it anywhere. Let's try it. MJ, what makes gridlock a big problem?"

MJ says, "Congress represents the entire nation. Most members of Congress have to agree to pass a bill."

None nods. "It's easier to get a smaller group to agree. What if only one state had to agree, or one political party? Wouldn't that be easier?"

"That's absurd, "Corella says. "One party can't make decisions."

None says. "I know an approach that lets everyone be heard, template laws. Who says we have to agree at the national level? A law could have different templates for how the law functions. I assume these versions will coalesce around party lines. Most of the time there'll be two or three versions. States choose which template to use."

"There are already laws at the state level. Why do we need that?" Corella asks.

None says, "It keeps laws consistent across states and avoids conflicts between state and federal laws. Template laws would effectively be both state and federal. I'll run you through an example for marijuana. There would be three templates: marijuana is legal, only medical marijuana is legal, and marijuana is completely illegal. The members of Congress would get involved with the version they prefer. Templates include a framework of how the different types of states work together. As a bonus, you fix conflicts. Imagine marijuana in Colorado is legal by state law, but illegal nationally."

Some members of the audience tune out. They want their problems solved but listening to the details of how to do it is beyond their limited attention spans.

"Even with five choices that might not be enough control for states," Corella says.

None says, "These are templates. States fill in the blanks. For example, Congress could set the minimum age for marijuana to 18. Colorado picks 19, Washington picks 21. The states use boilerplate for the rest. States can experiment. If one template works better, states can switch, on their own time."

None says, "If this works well enough, we could add local customizations. My vision for America is different levels of government working together. National infrastructure, local choices.

"The party system divided us on social issues, like abortion. The sacred cows of each party keep us from working together. All sacred cows will be slaughtered, made into hamburger, and eaten with fries."

The audience stands and claps. Shen remains seated. He plays with his phone.

None leaves the stage and strolls down the center aisle. "Why can't we have laws for blue states, red states, purple states? If my state wants to allow people to smoke pot, that shouldn't be overridden by federal laws. We'll customize our states, our way."

"I want my America!" None walks down the aisle and fist pumps. "My America! Say it with me. I want my America! My America!"

The audience joins the chant. "I want my America! My America!"

MJ walks down the right aisle. She fist pumps and chants along.

Shen jolts up. He swings his phone over his head. "The verdict is in!" His shouts can't penetrate the voices of the crowd in unison. He runs down the aisle and around the front. Agents intercept him and call it in.

None stops mid-chant. He clasps his earpiece so he can hear. "We have details on the case." None runs back to the front.

The chants die down, anticipating the news. MJ runs back to the front.

"What can you tell us, Shen?" None asks.

"We won a 5-2 decision in the California Supreme Court," Shen says.

None fist pumps and rallies the crowd. "Now we're off to fight in the U.S. Supreme Court. We will win this fight."

"No need. They're not pulling a Bush v. Gore this time. They reaffirmed the lower court decision. It's over. You're getting your America." Shen hugs None tightly. He nestles his head on None's shoulder, sniffling. The rift between them disappears in the celebration of the moment.

MJ lingers nearby, letting the brothers have their moment.

Applause erupts. Spontaneous hugs in the audience. Tears of joy. Screams of delight. The jubilant crowd places the burden of all their hopes and dreams on None. The weight of expectations has crushed many a noble man. Perhaps this time will be different.

30. **Promotion**

Backstage of *The Internet President,* agents and stage crew swarm around Corella, MJ, and None. With the presidency decided, None shifts focus to taking over from the sitting lame duck president.

"President Hayfield's transition team will debrief your staff next week," Agent Vincent says.

His staff. Right. MJ and Shen, more like. How do you build an army from scratch? None jerks his head from side to side. "Where's Shen? I need him here."

Agent Vincent points to a couple of agents, who break off.

None says, "We move to Washington. Corella, do you have a D.C. studio?"

"It's on standby, Mr. President."

"I'll write up suggestions for your cabinet," MJ says.

None says, "I've got a better idea. Corella, can you upgrade our TV show from weekly to daily?"

"You bring the ratings. I'll bring the time slots."

None says, "People want to be heard and entertained. We do both, and we're set."

Two agents escort Shen. He shuffles into the center of the group with the tempo of a reluctant prisoner.

"I need a moment alone with my brother." None parts his way through the surrounding stage crew, flanked by agents.

Shen dawdles behind. The brothers enter the cramped dressing room where they last spoke.

Shen says, "That hug doesn't change anything. I was excited about the first Asian American president. You were just closest to hug."

None stifles a laugh. Shen wouldn't hug a stranger. He barely hugs family. "I want you to be my White House Chief of Staff."

Shen eyes None with suspicion. "Are you trying to buy me off? That's how this feels."

118

None says, "I need someone I can trust." It's better not to say first choice, which might stroke Shen's ego. "There is no time to vet anyone else."

A cloud of doubt covers Shen's face. "I didn't earn it."

"So? You'll work that much harder."

Shen says, "Voters will see nepotism. That's not how to start your administration."

"You're on video storming the Garfield campaign with us," None says. "Don't pretend you're not a big part of this campaign."

Shen nods. "Will you listen to what I have to say?"

"I'll listen, but I make the decisions."

"Let me think about it," Shen says.

None says, "I can't wait. The global economy could collapse anytime. When the big moment comes, and I need someone to stand by my side, I know you'll be there. That's why it has to be you."

Shen breathes deeply. "I'm scared to imagine how much trouble you'll get into without me." He absorbs the weight of the responsibility and trust None places on him.

"Yes, Mr. President." He nods and salutes the closest president, who happens to be his brother.

31. The Robots Are Coming

None and MJ lounge on the left couch in the TV studio, for the second and last TV episode in L.A. Corella sits at the center desk.

Corella grins at None. "Which campaign issue is the highest priority?"

None strokes his chin. "If only there were a way to remember." He turns to MJ. "Do you remember?"

MJ perks up. A man cries out, "More jobs."

None cups his hands around his ear. "I can't hear you."

The man repeats his answer. A few more chime in. MJ jumps up and points to herself with mock surprise. "More Jobs," the crowd shouts. The audience rises to its feet.

"Say my name," MJ says. She sprints down the center aisle of the audience. The crowd chants her name, "More Jobs. More Jobs." MJ beams. The intensity of the crowd energizes her. She slaps high fives as she runs down the aisle.

None stands up and walks to the edge of the stage. "That's right. My highest priority is more jobs. My VP changed her name to More Jobs so that you would remember. Let's come together and make that our top national goal."

The spotlight shifts to None. It's like a sudden winter chill during summertime. No more attention. No more high fives. MJ sputters to a stop. She looks at her hands self-consciously and makes fists. She marches back onstage and slumps onto the couch.

None says, "We learn by asking questions. A toddler asks hundreds of questions a day. From the peak age of three and four, children ask fewer questions every year. Creativity and engaged learning hide in those questions. By adulthood, Americans don't ask enough questions. I hope you all remember the six W's from school."

Corella scoffs. "Of course. Who? What? When? Where? Why? How? I was a reporter. I know a thing or two about questions. Here's one. Wasn't today's topic the economy?"

The interruption catches None off guard. "Those questions will explore our plan for more jobs. First, we'll ask 'why'? Why do we need more jobs? Labor has struggled against capital for centuries. That battle is over. Labor lost, permanently."

Corella gets up from her desk and approaches None. "Woah! You can't drop a bomb like that without explanation. How did business owners get the upper hand over their workers?"

None says, "The balance of power has shifted against American workers for decades. With the end of the Cold War and globalization, Americans had two billion more workers to compete with. Act Two is the progression of the Internet, the Internet of Things, augmented reality, and beyond. Billions more joined the competition. Wait until you see Act Three.

"When companies grow faster than the economy, over time, wealth concentrates in fewer hands. The result is income inequality. Those vested interests protect and compound their wealth further using lobbyists to control the government and avoid paying their share. Corporations push for externalities, hidden costs that someone else pays."

None steps down into the aisle. "They want to use technology from our universities, but they don't want to pay for it. They want workers trained with skills they need, but they don't want to pay for it."

He speaks with a rhythm, more song than speech. He repeats words like a chorus and signals the audience to join in. "They want to use our infrastructure, but they don't want to pay for it. It takes taxes to provide these things, but they don't want to pay for it."

The crowd shouts along with his chorus. None speaks louder with each line. "American corporations hate taxes so much. They'd do anything to avoid them. They have American profits that should have American taxes, but they don't want to pay for it. They hide trillions in profits offshore, but they don't want to pay for it. With inversions, they buy foreign companies to pretend they're not American. They pollute our environment."

The crowd finishes his sentence. "But they don't want to pay for it."

None throws his hands up in frustration. "They don't even admit to it. To build a factory and bring jobs, they want us to pay for it. When incentives run out, they leave to find another deal. So-called job creators don't grant jobs out of benevolence. If they can outsource, automate, or downsize your job, they will. Less for you is more for them.

"Rising productivity no longer means higher wages. Management sees employees as cost centers to be slashed. Consumer spending drives the vast majority of the economy. Without those cost centers to buy products, there is no economy."

None addresses the crowd. "Are you just a cost center?"

"No!" reverberates through the crowd.

"I work my butt off for my boss, and he doesn't notice," a woman says.

None nods. "We all know how that feels."

MJ crosses her arms on the couch. "Like you notice," she says under her breath.

"Are you ready for the third act, that will impact workers more than globalization and the Internet combined?" None looks around like he's sharing a secret. "The robots are coming. The robots are coming. Cheap robots. They're not coming to kill you, unless you live in the wrong country. No. They're coming to take your jobs. Robots don't have to pick up their kids from school. They don't call in sick. They don't ask for a raise. Unions won't save you. Robots don't have to cross the picket lines. They're already at work. It doesn't matter how cheap you are. Robots will be cheaper. Workers in China and India will be next to you in the unemployment line. Sure, robots exist now, like Predator drones and autonomous vacuums. I'm talking about cheap multi-purpose robots you'll buy off Amazon. If you are not highly skilled, your future is low wages and long bouts of unemployment. For many, that's not the future. It's now."

The room is silent. Does the audience think None is crazy, or are they scared?

"Robots, huh?" Corella laughs haltingly.

None reverses course toward the stage, "The next wave of competition will be brutal. Telemarketers don't have the decency to be human anymore. The technology improved enough to fool me on occasion. Wait until they pair that with some quality artificial intelligence and millions of jobs will disappear. Plan for it now. it's closer than you think."

Some in the audience squirm in their seats. They grow increasingly uncomfortable.

"On par with the transition from hunter/gatherers to farmers or the Industrial Revolution, the next great epoch awaits. We live in the pre-sci-fi era. Technologies imagined in books and movies will arrive. Precursors are already here: genetic engineering, self-driving cars, nanotechnology, and artificial intelligence. If we don't destroy ourselves, a new golden age could arrive in our lifetime.

"Technology will transform faster than we can adapt. Entire professions will turn obsolete. Lifelong learning will replace college. Our future is a struggle between hope and fear. If productivity rises so high that workers are unnecessary and unemployment hits 30%, then fear wins. We'd relive another French Revolution, like in a *Tale of Two Cities*. The poor would introduce the kings of Wall Street to their guillotines."

Corella says, "Guillotines paint a pretty bleak picture. What would it take for hope to win?"

None reaches the stage. He walks toward Corella. "Socially conscious business leaders make a difference. Benefits corporations consider the impact of their decisions on society, not just shareholders. A pledge to do no evil is not enough. America should think long term and focus on the big picture, instead of being distracted by quarterly earnings. Place the highest value on the good of our planet and the health of our people, not just the dollar. Aspire to the ideals of *Star Trek*, where people work for their own fulfillment and the good of humanity.

"Our destiny is not assured. Will the future look more like *A Tale of Two Cities* or *Star Trek*? Income inequality will be the deciding factor. We must push the future in the right direction before wealth and power concentration becomes unstoppable. If the 99% don't stand together, it will be global elites who make that decision."

MJ says, "Come on, *Star Trek* future. Voyager's my favorite. I love me some Captain Janeway."

"Next show, I'll talk about what we can do about it," None says.

Corella claps. The audience murmurs, contemplating his speech. "Have you picked your cabinet yet?" Corella asks.

None slides next to MJ on the couch. He's all grins, but she doesn't reflect his enthusiasm. None says, "I'm announcing my White House Chief of Staff, Shen Wong, who joined our campaign early on."

The audience applauds. Corella says, "Roll it, Marty."

Shen campaign highlights display on a screen behind the couches. Blurred faces of the Garfield and Robertson campaigners chase Shen to campy music. The audience cackles as Shen drives the getaway vehicle in a narrow escape. The escapades play like a prank on the entrenched political parties.

None says, "Politicians and lawyers run this country, but I'll keep them to a minimum in my cabinet. We need problem solvers. I want outliers, original thinkers."

Corella asks, "Isn't your brother a lawyer?"

"Yes, but I chose him because I can trust him," None says.

MJ says, "We don't care what party you're from, just bring big new ideas."

Corella asks, "How did you pick your cabinet?"

"I haven't yet." None stands. "We'll choose them right here, on *The Internet President*." He outstretches his arms towards the audience. "I want you to find them."

32. No

A long white corridor leads to rows of dressing rooms backstage at the Washington D.C. studio. It widens to an open area with whiteboards and craft tables. None, Corella, and MJ plan the next episode of *The Internet President*. Agents Vincent and Reynolds analyze the security implications of every detail. A ring of security and studio staff eat while listening.

"Mr. President, you have a visitor off your open access list," Agent Vincent says.

"Good, Shen is late," None says.

"It's Astrid Fontaine," Agent Vincent says.

MJ says, "Your supermodel is sniffing around again? I thought you dumped her."

"I did." None sighs. "Take Astrid off the list, but I'll see her now."

"Yes, Mr. President." Agent Vincent nods.

"Dump her again," MJ says. "I want to see."

"Not gonna happen," None says.

Astrid prances like a cat, flanked by a handful of agents.

MJ counts the agents accompanying her. "You're slipping. Only five agents this time?"

Astrid says, "If I had a face like yours I'd slip and fall, so I had an excuse to hide under bandages."

MJ seethes. "I'm glad you're back. Next assassination, I nominate you for human shield." MJ flaunts burns on her arms. "Your turn to burn." MJ clasps her fingers like a snake and snaps Astrid hard on the forehead.

None jumps between them. "Don't say that!"

"The assassination part, or human shield?" MJ asks.

None gives MJ a stern look and turns to Astrid. "Let's go talk, alone."

"You can use my dressing room," Corella says.

Two agents follow Astrid and None into Corella's dressing room. None crunches numbers. Her dressing room is 30% larger than his. These rooms become a box for uncomfortable moments of near privacy. Thinking outside the box won't make his room any larger.

Astrid moves to embrace None.

He backs away. "What happened? Were you checking with the Supreme Court before returning my calls?"

She says, "I told you, I forgot about a photo shoot. There was no reception on the island."

"I could feel you pulling away."

"Don't you want to see me again?" Astrid pouts.

None burns every detail about Astrid into his memory. Porcelain skin. Flowing golden hair down to her thigh. Two-hour hair: the very definition of high-maintenance. How did None love every vain shriek and indulgence? He dreamed of raising a litter of cubs with his lioness. What beautiful children they would be. Did he truly love her, or the dream of her? MJ is real but forbidden. Astrid is fake but available, at least as an illusion. Can you marry an illusion? Embrace the power of no. His vocabulary includes no. He merely has to use it. No is the only answer of self-respect. Any other answer would shrivel him to the size of nothing, smaller than an ant.

"No."

She gasps. Arrogance seeps out. "How could *you* not want *me*? I'm Astrid. No, is *my* word."

"You will not break my heart again, vile temptress." He flicks his fingers at her like casting a spell. "I dispel illusion. I see the wretched creature you truly are. You are without merits as a person. I hope someday the hair that fuels your vanity is scalped from your head. You deserve any suffering fate can imagine. Begone! I cast you out and exorcise you from my life."

"You'll beg me to come back. You always do." A sneer, a witch's cackle, and she's gone.

33. **American Dividends**

Corella stands in front of a large oak desk in the center of the D.C. studio. MJ and None sit on one of the two brown leather couches adjoining the desk. The audience fills all 235 seats of the amphitheater. If None switched the brown couches for black ones, it would be as if he never left home. None shakes off the déjà vu, reminding himself that he's on the opposite side of the country.

Astrid groupies, her Glitter Girls, line the front rows on the faint hope they might see their goddess, a hope None dashed in the dressing room.

Corella says, "Social media fired off millions of recommendations for cabinet positions. Our first position is Administrator for the Small Business Administration, the SBA. We vetted the top five, and the President-Elect picked his top two. Your votes tonight select who wins. Let's meet them now. Deputy Administrator Mike Fitz worked his way up the ranks in the SBA for 15 years. He's ready for a promotion."

Mike, a pasty man in his forties, saunters onstage to applause. He sits on the empty couch.

Corella says, "Judy Goddard wrote 'Stuck in the Queue,' a book where she recounts how she nearly lost her business waiting a year for an SBA loan approval."

Judy struts out, punching the air like a boxer coming to a grudge match. The audience claps, egging her on. She eyes Mike with disdain and sits at the opposite end of the guest couch.

None walks to the front of the stage. "Debt strangles our future." He outstretches choking hands around an imaginary neck. "Bad promises… Unrealistic expectations…" His hands clench tighter. "Tax loopholes… More debt… One of the most powerful forces in the universe will destroy us: compound interest."

None squeezes his hands together. He gasps for air as if he were being strangled. "If debts outgrow us, the America we know is gone."

A few suggestible audience members fall for the theatrics, gasping themselves.

None breathes an exaggerated deep breath and wipes his brow. "Want to hear a crazy dream?" Lackluster applause filters from the crowd. He screams like a maniac. "It's really crazy, but it might put a smile on your face." Clapping and hollering emerge as None steps down from the stage into the aisle.

"Today's dream is sponsored by the letter W and the question what? What if we make government more efficient and improve the economy? What if we create surpluses by lower spending and higher government revenue?" None extends his left hand towards the audience. "What if we keep politicians' hands off those surpluses?" He slaps his own hand.

An involuntary laugh shudders through the audience.

"I told you it was a crazy dream." None smirks and motions for the crowd to join in.

"What if...a miracle happens, and we pay off our debts? What if the government builds up savings? What if compound interest works for us, instead of against us?"

A trickle builds into a bellowing shout. The crowd joins in. "What if?"

None nods encouragement to his followers. "What if we save enough money to pay off our debts, achieve escape velocity, and run the government like an endowment? Save enough, and we don't need corporate or personal income taxes. Imagine American dividends to every citizen, the shareholders of America."

Corella scoffs. "That impossible."

None smiles back at Corella. "It's been done. Alaska made so much money from oil that it has no income tax. Every year, Alaskans get a payday. It's better than Christmas. We could get there in one generation if we dream together and fight the special interests. What if? Ask yourself that question, whether for this dream, or your own. Don't settle for what is. See what could be."

His vision moves the crowd, in spite of more than a few skeptics. What if? It's a question many in the audience ask themselves.

None rushes back onstage. "Last show, we talked about why jobs disappear."

MJ raises her hand. "No, I'm still here."

A few isolated claps.

None ignores the interruption and turns to the audience. "The jobs that disappear aren't coming back."

MJ jumps up each time he mentions the word jobs.

"New jobs and new professions must replace them. Most net job creation comes from new businesses less than five years old. The SBA should help create more jobs."

MJ goes nuts behind None when he says her name.

The audience cracks up at her antics.

Their laughs irritate None. What's so funny about jobs? He turns around.

MJ plays it straight.

None turns back to the audience, none the wiser. "Commercial real estate loans are the main impact of the SBA now. Big banks guarantee their profits and stick SBA with the losses. Do you want to fund big bank profits?"

A heckler shouts, "No! We want more jobs."

None takes the heckler seriously, "Exactly. We need more jobs."

MJ raises her fists over her head like she just won a prize fight.

The audience cracks up.

None turns around too late.

MJ puts her hands behind her back.

None says, "Let's revamp the program. We should have an SBA and infrastructure bank, like Fannie Mae. We'll lend directly to businesses at low interest. We're not here to make banks money. Let's turn the SBA into a profit center for us. We're America. Let's make money."

The idea of profitable government draws an uncomfortable silence.

MJ slumps onto the couch and yawns.

None says, "Government needs to become smart money, not a piggy bank for political donors. To compete with China, America needs competition and collaboration. Don't just seed new companies, cross-pollinate. Combine multiple ideas into breakthroughs, those epiphanies that every entrepreneur chases. Places like Silicon Valley, or Silicon Beach in L.A., thrive with startups because incubators spread ideas across companies. Set up incubators in major cities with mentor entrepreneurs. Invest in the best ideas and innovations.

"I don't trust politicians and their pork barrel spending hands near this. They devour everything in the public trough. Every year, those pigs get bigger and bigger and bigger." None imitates pig noises. "You have to slaughter pigs before they get too big, or they can knock you down and devour you. Pigs are one of the few animals that can eat their way through bone. Gigantic hogs might win a prize at the county fair, but they're eating us." None mimes chomping on his arm. "We're the bacon. I don't want to wind up on a plate next to scrambled eggs."

Laughs and hollers follow the absurd imagery of being eaten.

"There can be no better stewards for taxpayer dollars than the American people. We'll build a crowdfunding platform called the Invest in America Program. Like Kickstarter, crowds of American consumers will determine which companies to invest in, or loan money to. Instead of pork barrel horse trading, we find what consumers want. Most startups fail because they don't have products someone needs. Show me the customers. The crowdfunding model builds in an initial customer base, to better select successful companies.

"SBA needs a bigger impact than loans and incubators. Government is distributed across local, county, state, and federal levels, without coordination. I propose a government concierge system, to give small businesses a single point of contact, across all government. A business portal will track all needed steps to complete projects and the current status. Find the bottlenecks and overlaps between different government agencies. Find the government workers who don't do their jobs and fire them. Find the understaffed agencies and hire more workers to speed turnaround. Inter-agency cooperation improved law enforcement. The move to national crime databases helped catch serial killers. Coordination needs to happen for government at large. The entire government ecosystem needs to be transparent and responsive."

MJ imitates non-stop talking with her hand and fake yawns.

None says, "Make life easy for startups and small business. With advances in 3D printing, we can be the USA that makes things again. New Deal 2.0 will put America back to work on infrastructure projects. The last long-term national infrastructure plan was the Interstate Highway System in 1956. We'll investigate new high-speed transit, like a potential interstate Hyperloop system." None pauses for applause, then returns to the couch.

After a few seconds of hesitation, Corella and MJ clap. The audience bursts into applause. Do they approve of his ideas, or are they relieved his long speech is finally over?

Corella says, "Let's interview our candidates. Mike, what's your big idea for SBA?"

"I'd track loan performance." Mike smiles, confident in his answer.

None lurches forward. "You don't track loans?"

Mike says, "Banks do loans, we do insurance. We find out loans fail when banks file a claim."

None says, "Screw the banks. Loan direct to customers. That's easy to track."

Corella asks, "Judy, what's your big idea?"

"The SBA wastes tax dollars. Shut it down."

MJ laughs. "There's an outlier for you."

None glares at MJ, then refocuses on Judy. "A profitable SBA wouldn't waste tax dollars."

Judy says, "Government shouldn't pick winners and losers."

"The American people pick winners and losers, via crowdfunding," None says.

"Crowdfunding is unproven technology," Mike says with a smirk.

None says, "Kickstarter crowdfunded $2.4 million for Oculus Rift, a virtual reality prototype. The company was later sold to Facebook for $2 billion. I'd call that proven."

Mike says, "What if…you come back to reality? The SBA isn't a venture capital firm. That's not how the SBA works."

"That's my point," None says. "We'll change it."

Mike stands up and points at None. "Unions will stop you."

None rises and walks inches away from his finger. "Technology eats inefficient industries for lunch. How long can government be inefficient before taxpayers revolt? The unions can play nice and help shape how it plays out or wait until they become lunch."

Mike slinks back to the couch. Judy grins at None.

"Judy, have I sold you on not closing down the SBA?"

Judy shakes her head. "My business nearly failed because I didn't get help when I needed it. No one else should be strung along like that."

Mike cringes. He scoots to the edge of his couch.

None says, "Slow banks with slow government, that's a deadly mix. There was an awesome three-wheeled car called the Aptera. It was so sleek, like driving an airplane. They waited for a Department of Energy loan, but it only covered four-wheel vehicles. Aptera sunk money into a four-wheel design, and it destroyed the company. With crowdfunding, it'd be clear whether enough customers were interested in a three-wheeled car, instead of keeping the company in limbo."

"Mr. President, you're talking about socialism," Judy says.

"Call it socialism, but I call it value. I want *more value*." None turns to MJ for encouragement.

"Don't look at me. I'm More Jobs, not more value."

The audience laughs.

None grins and turns away. "We have socialized institutions, public schools, police, and fire departments. Corporate socialism, like bank bailouts, is the bad socialism. We socialize losses but privatize profits. Banks keep their casino winnings. We eat the losses. That's not fair. How about socialized profits for a change? What's more capitalist than wanting a return on my investment?"

"Mr. President, if you were a true capitalist, you'd let the SBA fail on its own," Judy says.

"If the SBA doesn't provide enough value after a few years, then yes, I would shut it down." None says.

"With reservations, I'm willing to try your approach," Judy says.

"How about you, Mike?" Corella says.

Mike crosses his arms and stares off into the distance.

34. **Powered by Failure**

Backstage at *The Internet President* D.C. studio, None paces between agents of his protective detail. His breath uneven, he shakes his hands along his sides.

MJ stands amidst a circle of his entourage. She smiles encouragement. "I believe in you."

"Live demos make me nervous," None says.

MJ advances to him and grasps his hand. Her touch instantly calms him. "Sell your vision. Don't worry. You're ready."

Corella clears her throat.

MJ yanks her hands back, self-conscious.

Corella says, "Viewers think I'm too easy on you." Corella casts a stern gaze at None. "That's over. Don't get speechy. This is TV, not college. You're not here to give lectures."

None asks Corella, "How good is the Wi-Fi here?"

Corella shrugs. "It's OK, as long as too many people aren't using it. Why?"

None says, "I didn't finish offline mode. If Internet access goes down, my demo won't work."

Corella flings open the curtains. She smiles back at None and struts onstage to applause.

None summons his brother. He scribbles down instructions and hands them to Shen. "Adaptonium is the most important thing I've ever done. If anything goes wrong, I need this done."

Shen says, "I'll take care of it."

None and MJ come onstage and approach the host couch.

Glitter Girls dressed in black with gold glitter tears fill the first few rows. Word that None dumped Astrid make the studio a place of mourning. They sulk in disbelief that anyone could turn down their goddess. None senses Astrid watching him through the eyes of her superfans. He does his best to tune them out.

Corella announces None and MJ, before moving on to the guests. "Welcome your two candidates for Secretary of Education. Jake Faraday, CEO of Rainbow Unicorn Education Centers."

Faraday shamelessly hypes his schools to the crowd. "With over 3,000 schools to serve you, pick Rainbow Unicorn Education Centers. Dreams of a better future can be yours today." He sits and spreads out, to engulf as much of the guest couch as possible. One arm extends all the way to Corella's oak desk.

Corella says, "I'm very pleased to introduce Anwar Motindy, a Nobel Prize winner for his innovative teaching program in Somalia."

A fit Indian man wears sunglasses and a sports coat with rolled up sleeves. A standing ovation greets his entrance.

None intercepts him to shake hands, but instead, they embrace. "I'm a huge fan of your work," None says.

Anwar says, "You're the first politician to give me hope. Anything I can do to help, just ask."

Politician? He's no politician. None ignores the unintentional insult and leads Anwar to join him on the host couch.

MJ squeezes onto the host couch's far corner to make space. She leans forward to look past Anwar at None. "You about done with your bromance?"

The crowd's laughter jolts None to attention. He grows serious and heads to the front stage. "Today's question is how. How do we power more jobs? With new skills, innovation, and knowledge. Education is the key, a better education than our schools provide."

None speaks louder. "The privatization movement takes any opportunity to replace public schools with chain stores to sell our kids out for a profit. Politicians cannibalize public school budgets to fund charters with little oversight, which makes matters worse. They take down public schools, not fix them."

Faraday fake coughs, "Bullshit."

Faraday's comment gets a few laughs.

None pretends he didn't hear anything. He steps down into the aisle and addresses the audience. "Do you want Wall Street controlling your schools?" After a few scattered replies, he asks again.

The crowd shouts with raised fists, "No!"

None says, "In major cities, like L.A., when you look for highly rated schools, what do you find?"

Corella says, "Let's take a look. Search Redfin, or another real estate site with school ratings."

Most of the audience pull out their phones.

None says, "Expensive houses surround the best schools. Now find the worst schools. What do you find?"

The audience cries out with competing answers. The most popular response is cheap houses.

"Those houses are much cheaper. Poor kids are priced out of a better future in the best schools. How? Rich neighborhoods have higher home values, which leads to higher taxes and better-funded schools. Wealth correlates with higher test scores. How do we make a huge difference? Spread out school funding at the state, or county level. Of course, the rich have lobbyists and lawyers. They'd never settle for getting the same education as the rest of us."

Corella says, "Us? Aren't you a billionaire?"

"Only on paper." None sighs. Will he look incompetent when people learn he was tricked out of his company? He sulks, ready to end the charade.

MJ laughs and slaps the arm of the couch. "Right, because dollars are just paper." She stares straight through None.

He smiles, thankful for the distraction. None resumes before Corella asks follow up questions.

"Every president claim to be an education president, but I see punishments, not solutions. Our schools are drowning, but we throw them an anchor."

Corella asks, "Don't we have to punish bad teachers?"

None says, "Why not improve teachers? During the summer, teach the teachers. It's the perfect time to mentor, or train for hard to fill specialties like math, science, and special education."

"In school, have you ever suffered through painfully boring lectures?" None asks.

"I sense I'm about to," MJ mutters under her breath.

None says, "Or, did your teacher not understand their topic? Why be stuck with the teacher you have, when you could have the best? Have you ever seen a TED talk and been inspired? Use teachers as mentors. Move lectures from teachers to the computer. We'll find rock star teachers on the Internet."

None fawns over Anwar. The audience applauds the rock star teacher in their midst.

"Is a teacher boring, or not getting through to you?" None asks. "Change the channel."

MJ mimics a remote control. She points it at None and pushes the button. The audience laughs.

None turns around to see what's funny.

She grins at None and points her imaginary remote control at the audience and presses again.

"That's right, MJ." None turns back to the audience. "Change the channel. Try another lecture on the same topic. Favorite a lecture, or teacher's feed, to help us find those rock stars. We don't have to teach to the lowest common denominator. Some of the brightest students underperform because they're not challenged. With computerized lectures, everyone goes at their own pace. A student can master a topic, before moving to the next one. The smartest and most self-motivated can whiz through. We won't hold anyone back.

"To push everyone forward, I present the future of education, my new platform, Adaptonium." None spreads his arms with an implied 'tada.'

Corella says, "Roll it, Marty." The text, "Network timeout," scrawls across the back wall.

Sweat drips from None's forehead. "Aiya! It was working." None looks at all the people in the audience on phones. Corella overloaded the network when she encouraged them to go online. He warned Corella his demo might fail without Internet access. Was it deliberate? Was she getting back at him for being long-winded?

Corella laughs. "I hope you're a better president than you are tech support."

MJ cringes, unsure how to help.

None says, "All it took to scuttle the launch was spotty Wi-Fi. Tada." None throws his arms up. "That, ladies and gentlemen, is how not to do a software launch."

The crowd heckles the President.

Corella expresses disbelief. "You expect us to believe you planned to fail?"

None says, "Failure is part of the demo. Failure powers Adaptonium."

"Just what our schools need, another failed software rollout, like Obamacare," Faraday says.

None says, "Too late. The L.A. Times reported, in 2014, about the catastrophic launch, five times over budget, of MiSiS, a student records system for the L.A. Unified School District. At Jefferson High School, students waited weeks in an auditorium because they weren't assigned classes yet. Transcripts for colleges had the wrong GPA, and graduation requirements were missing. Imagine losing out on your dream college because of a software glitch. LAUSD is no small-time outfit, with about a $6.7 billion budget and 640,000 students.

"It took about two years to correct all the problems. Just like the Obamacare launch, attempting to deliver everything at once without enough testing courts disaster."

Faraday says, "Just another example of public schools that are too big to succeed."

"Funny, coming from the guy with 3,000 schools," MJ says.

None says, "It's not about size. The best software evolves with feedback from its users. Look at the Internet. Don't do it all at once, do it in pieces. Build a platform, then build on top of that.

"Custom software comes with higher costs and lower quality. When schools go it alone, the nation pays over and over for the same features. When even the second largest school district in the nation can be brought to its knees, we need a national education infrastructure.

"National infrastructure, local choices. Say it with me." None repeats the words with the audience like a mantra.

"Devote enough resources at the national level, and we get the most advanced platform possible. When software deploys to thousands of schools, it's well tested. Add a feature to the platform, we all benefit. Remember how standardizing on Windows pushed the PC revolution? Remember how the Internet connected the world like never before? Remember how the Apple App Store provided a single market that drove smartphone adoption?

"Once we have a shared platform, any business or educator can offer their apps and content. Schools can vote with their feet on various curriculum and teaching approaches. If Common Core falls out of favor, a school can publish their own curriculum. Why not Anwar's Core?" None smiles at Anwar.

Faraday chimes in. "Or Faraday's Core."

"A shared platform will drive innovation and cost savings better than privatization," None says.

Faraday goads None. "If only it worked."

None accepts the challenge to prove himself. "Shen, pull the plug."

Audience members using their phones, signal irritation, as the Wi-Fi cuts out.

None scans the audience. He uses their irritation as a cue to outstretch his arms in victory. "Behold. Adaptonium, the future of education."

Adaptonium still doesn't work.

Corella shakes her head. "Just admit you didn't fail on purpose."

MJ and Anwar flash concern.

Mortified, None blushes. All eyes, all phones are on him. An entire audience of paparazzi document his failure. Months of work led to this moment, this embarrassment. Everything rests on his credibility. None panics.

"Maybe you should fix your software before you fix America," Faraday says.

Corella enjoys Faraday's insults with a snicker.

None's face distorts into a constipated scowl. He tunes out Corella and Faraday. It's harder to tune out an entire audience. He ignores the tall man that sounds like a hyena and the fat woman with the snorting laugh.

Surrounded by jeers and flashing phones, None clutches his head in his hands and kneels on the ground.

He stares at the one friendly face, MJ.

She mouths the words, "I believe in you."

None crawls behind the couch towards Corella's desk

Corella's lips curl into a jeer. "Are you hiding?"

The audience laughs louder.

Anwar bows his head in disappointment.

None crawls behind Corella to the other side of her desk.

Corella looks behind her. "Mr. President, I saw you face an assassin. I never expected such cowardice."

None examines the Adaptonium demo laptop under Corella's desk. The network cable is gone.

Faraday peers behind the guest couch at None and smirks.

None crawls halfway down the guest couch until he finds the missing cable. It disappears under Faraday's left hand.

None lifts Faraday's left arm. The missing network cable dangles into the air.

The room grows silent as they realize Faraday sabotaged the demo and hid the cable.

Faraday moves his right arm away from Corella's desk. He holds up the end of the cable. "Oh, is this important?"

None plucks the cable from Faraday with unspoken rage.

Faraday feeds off the anger. "Too bad your precious Anwar wasn't sitting over here, where guests belong."

None plugs the cable into the laptop. Adaptonium comes to life on the screen behind Corella.

He returns to the front of the stage to begin the demo. "What should be in the Adaptonium platform? No more one size fits all instruction. Adaptive learning and big data are the future of education. You already use big data. Apps aggregate GPS signals to generate traffic maps on your phone. More data means a quicker ride home. More data means better Netflix recommendations. More data means a better education.

None says, "Adaptonium records how students do on each question. Who needs a report card, when the parents and teachers have real-time updates?"

"Fine, your demo works, but you never explained how failure powers your platform," Corella says.

"Each math topic builds on the one before," None says. "When a kid gets lost on a math concept, they're lost forever. Math becomes a hated subject. When I tutored, I saw kids fail on concepts as basic as fractions. Powered by failure, Adaptonium analyzes wrong answers to locate students lost on fundamentals. Adaptive learning software drills all the way down to the missing concepts and repairs a student's math knowledge. Once they have a solid foundation, math won't be a problem subject.

"Powered by failure, Adaptonium uses patterns of incorrect answers to diagnose potential learning disabilities, like dyslexia and adapt appropriately.

For non-native English speakers, Adaptonium can teach content in both languages.

"Powered by failure, Adaptonium detects common mistakes and misconceptions, which lead to tailored instruction on how to improve. Educational researchers and artificial intelligence can analyze the data for new insights. Integrating insights into Adaptonium allows for quick improvements."

Faraday says, "Your speech, powered by failure."

None turns his back to Faraday and continues. "Different types of content will be available to different learning types, whether text, video, audio, or even games and songs."

"I wish I had that in Somalia," Anwar says. "They're auditory learners. Somalis have oral traditions, so listening is more natural than textbooks."

None nods to Anwar. "Another key feature, Adaptonium integrates current knowledge about the brain and memory. Not just facts are important but learning strategies. What's the point of learning if you don't remember? Don't teach to the test, teach to the retention.

"Everything you learn is connected to something you know. Your brain is all about connections. Synapses, neurons, all the interconnections, they get rewired depending on how you use your brain. Past knowledge isn't lost, just unreachable. Over time, information is forgotten. A memory strengthens if accessed again before it's forgotten. Each time the memory lasts longer. Wait too long, and information has to be relearned.

"Based on how well you remember, Adaptonium can decide how often to remind you about old information. Knowledge: use it or lose it. Every day is a pop quiz, drilling on a variety of topics to help you remember. We'll have contests to find the best spaced memory algorithms to improve Adaptonium." Algorithms. The word may as well be a sleeping pill. Deep, meaningful ideas are difficult to convey in a world based on impatience.

"I thought it was better to test on one subject at a time," Corella says.

Anwar gives None an approving glance. "I was taught that growing up too. The research shows it's better to mix topics. It's also good to study in different locations. It's more potential associations in the brain."

Faraday says, "Adaptonium sounds more advanced than my current software. My online schools handle up to 270 students per teacher. How many students can Adaptonium handle?"

None facepalms. "270? That explains why only 35% of your online students pass state assessments. Children still need social development. It's hard to keep motivated on your own. That's where online-only schools fall short. Leave no child behind? We leave entire schools behind."

Corella says, "You speak so poorly of online charters, but aren't you pushing mostly computer instruction?"

None says, "While most students learn on their computer, the teacher can run collaborative learning exercises. Sports aren't the only way to teach teamwork. By focusing on a subset of students at any given time, teachers get a smaller effective class size. Studies prove small class sizes work. Tutoring is so effective because it's a class size of one. Adaptonium isn't meant to replace teachers. It's so they can give smaller groups of students more individual attention."

Anwar says, "There are ways besides computers to get students more individual attention. One of the best ways to learn something in depth is to teach it. Teachers aren't the only ones who can teach. Students can teach to learn."

None fist pumps with a smile. "What good is knowledge, if you can't communicate it? Great point. I can take your insight, and I'll integrate it into the software. When a child flags they're stuck on a concept, another student can accept a knowledge share challenge. One student gets assistance. The other cements their knowledge, by testing their own understanding. Exercise that brain. It's similar to pair programming, where two software engineers work together on a single problem, even sharing a keyboard. Pair programming is great for knowledge shares or tackling difficult problems.

"Schools haven't changed much in a century. Once we plug schools into an adaptive system, innovation will be part of the system. That changes everything.

"As Adaptonium gets more efficient at teaching, those who hate school can graduate earlier, or get free vocational training. What if we could finish most college coursework in high school? In the new economy, education isn't something you finish, you learn for life. We cannot stand for an expensive college system with massive student debt. Adults will endure multiple career shifts, as automation devours entire professions.

"Tech moves so quickly that if you're stuck somewhere for four years, the world has moved on. That's why the wealthiest tech billionaires are college dropouts like Bill Gates, Larry Ellison, and Mark Zuckerberg.

"Adaptonium can extend into college, lowering costs and accelerating degrees. With adaptive learning and memory retention techniques, can the genius of today become common tomorrow?" None marches back to the stage.

Corella asks, "Why did you create Adaptonium?"

None says, "In kindergarten, I hid in the back at school because I was shy. At that age, I assumed I was normal. I didn't know I was supposed to see the chalkboard up front. My teacher called me retarded. Now they say intellectual disability."

Corella gulps and shifts in her seat. "I'm sorry. I didn't know."

None shrugs. "My mom believed I was smart. She took me out of school and tested me. They discovered my terrible eyesight and gave me glasses. I also had trouble learning to read. With extra help, I became an avid reader. When students were tested years later, I was accepted into the Mentally Gifted Minds Program. Without my mother, my reading troubles would be seen as confirmation of my intellectual disability. I might have lived a very different life. I had to redo kindergarten, but it worked out." None smiles at MJ. "I got to share classes with my new best friend."

"I created Adaptonium so no other kid would get mislabeled like I was." None focuses on Anwar. "With the right educational system, we'll discover how bright our kids really are."

The story connects with Anwar. He embraces None. "I'll champion your vision."

The audience takes to its feet. Some concepts elude the audience, but None has concrete solutions to fix the education system. The potential of the platform drives optimism. What if it works? Every parent listening wants Adaptonium for their children.

Corella says, "Amazing! Would the standard curriculum teach evolution?"

None throws up his arms. "Local school boards pick what content and curriculum they want. Let's stop rehashing old fights and solve education."

Faraday clears his throat, unheard over the audience clapping. After the noise subsides, Faraday slow claps.

Corella says, "Mr. Faraday, we haven't forgotten you. In fact, we're going to show the world premiere of your newest commercial."

On the back screen, a rainbow unicorn jumps over the moon. The image flattens into the charter chain logo. Children in the video grin with the authenticity of a sugar substitute. The hard sell video leaves None with an aftertaste.

Faraday goes into sales mode. "Rainbow Unicorn Education Centers provide cost savings because we aren't held back by government regulation."

None says, "You cannibalize public school funding to provide your profits. Taxpayers bought your schools computers and other property. Why should they have to pay again?"

"Our fee schedules were clearly listed in the contracts," Faraday says.

None says, "You force school districts to repurchase all equipment when the contracts end. When your charters failed academically, you forced them to double pay."

Faraday leaves the couch to confront None. "They didn't have to sign."

None says, "Just because you beat the school district in court, doesn't make it right. You must have excellent lawyers."

"I do." Faraday puffs his chest out, nearly bumping the president. "Briggs, Hultz, and Cantaker. Continue this witch hunt and they'll sue you into financial oblivion."

Renquist's lawyers? Should None be angry, or intimidated? "I have a confession. I picked you, not to be in my cabinet, but to call you out. There are good charter schools, but yours aren't among them. You stand for everything wrong with school privatization. I'd rather claw my eyes out with razor blades than have you in my cabinet."

Faraday points at Anwar. "I see where he's sitting. It doesn't matter. All I need is votes."

"This isn't an election," None says.

"Isn't it? You told viewers their votes would choose your cabinet." Faraday slumps onto his couch with a satisfied grin.

Could Faraday win? None could veto. No. A veto with two choices makes a vote meaningless. "I'll respect the vote."

Faraday says, "Now that that's settled let's talk about Adaptonium. I'd like to invest."

"Not everything has a price tag." None crosses his arms and looks away from Faraday.

Corella says, "It's time to announce the winner for Administrator of the Small Business Administration. Will it be Deputy Administrator Mike Fitz, or Judy Goddard, well-known author and SBA critic?"

Mike and Judy step onstage.

Corella opens an envelope. "The winner is…" Corella draws out the moment. Cameras capture reaction shots. "Going to be announced after this commercial."

Corella muffles her microphone and whispers in None's ear. "I said no speeches, professor."

None retakes his seat and hides behind a smile.

Corella toys with the envelope. A stagehand counts down the end of the commercial break. "The winner is…" Corella draws out the moment. "Judy Goddard."

Judy screams and jumps up and down with excitement. She turns to Mike. "That makes me your boss."

"Judy, what's your first act when you head the SBA?" Corella asks.

"I'll talk to Mike about improvements he's going to make to keep his job," Judy says.

35. Academic Probation

None takes center stage during the next episode of *The Internet President.* "Of the six W's, I'm going to answer the two shortest ones today. Where and when? Right here, right now. Tomorrow is built today. That's why we can't wait anymore to improve the infrastructure the nation depends on. Civil engineers issue report cards on how the nation is doing. Our current grade? It's a D+. As a nation, our infrastructure is on academic probation. The report card lists concrete suggestions. Why aren't we following them?"

The president covers the entire report to the smallest minutiae. Both candidates for Secretary of Transportation record notes.

Corella sensationalizes the nation's decaying infrastructure and turns the back screen into a disaster movie.

None lists high priority bridges and roads to upgrade while a symphony of destruction plays out behind him. Bridges collapse. Trains derail. Airplanes crash. A hundred cars pile up on a foggy Texas highway.

MJ abstains from the duel between information and entertainment. She stares at the back screen, transfixed by the escalating violence of endless disasters.

None modulates his voice into a monotone stream of data, information in its purest form, without emotion. Corella selects more graphic images from her tablet with ominous music. Fatal car crashes flash faster and faster. An unstoppable dystopia of broken glass and twisted metal wears down the crowd. The drone of the president's voice transports the audience past despair and futility.

The hyper destruction of every transportation mishap washes over MJ.

The interplay between Corella and None intensifies until it loses shock value.

The speech and video end simultaneously. MJ goes numb, as None and Corella climax together.

The audience reaches catharsis with a collective sigh.

Abruptly weaned from the strong graphic images, all but the most astute observers sit dumbfounded in a haze. Most seem unable to pay attention to the transportation candidate interviews. A dozen unremarkable questions later, it's time to announce the Secretary of Education.

Corella asks, "Who will win Secretary of Education? Will it be Jake Faraday, CEO of Rainbow Unicorn Education Centers?"

The audience boos. None thrusts his thumb down.

Corella continues. "Or Nobel Prize winner and rock star teacher, Anwar Motindy?"

Applause rings through the amphitheater, as the candidates enter.

Corella pulls out an envelope and immediately reads the name. "The winner is Jake Faraday."

None steps over to Corella's desk. "Aiya! Worst Internet prank ever."

Corella shoves the card in the president's face.

Impossible. His knees wobble. None slumps onto the couch, speechless. His mind wanders over every plausible explanation.

Faraday spreads out on the guest couch. "It's a triumph for markets. America's schools will be open for business."

36. **Voting Records**

None walks into Corella's office flanked by MJ and Shen.

Corella watches the Astrid breakup from the show on a wall mounted monitor. She rushes to minimize the window. "You may be president, but this is my office."

None pretends he didn't notice her screen. "I need to see the voting records."

"And I need higher ratings." Corella scowls at None. "You read that infrastructure report like a damn phone book. Another stunt like that and the show will get canceled."

None says, "I'm working to fix America. That has to bring in ratings."

Corella shrugs. "Right now, people would rather grab a snack from the fridge than hear your boring solutions. So, why did you need the voting records?"

"We think Faraday cheated," MJ says.

"The votes were independently verified," Corella says.

"I scoured through property titles for Faraday's charter schools and found business connections to Alvin Renquist," Shen says.

Corella shrugs. "He's one of the richest men in the world. He does business with everyone."

None crosses his arms. "He's my sworn enemy,"

"Mine too," MJ says.

"Isn't that a little melodramatic." Corella scoffs. "What? Did he swear vengeance and threaten to kill you?"

None and MJ nod.

Corella looks at Shen. "You too?"

"I just loathe him," Shen says.

None says, "The mega-rich subvert everything to make even more money. What wouldn't Renquist do to get a golden ticket to an education overhaul? Hundreds of billions are at stake that he could siphon."

Corella says, "Voting was secure. Online safeguards prevent hacking. See for yourself." Corella grants None read access to the voting database.

None writes SQL scripts to parse the data. "Most votes are text messages."

He groups the text message votes by area code and prefix. The president exports and charts the data.

The results look uneven. "The votes cluster around certain area code prefix combinations."

He scans through the highest concentrations of votes. "This number looks familiar."

None searches his phone address book. He enters the partial phone number. Only one contact matches. "Renquist. He's sending us a message."

"Where did the other numbers come from?" Corella asks.

MJ says, "The charter schools."

None goes to the Rainbow Unicorn website and parses out contact phone numbers for all 3,049 schools. He inserts them into the databases and compares. "They match."

"That doesn't mean they cheated," Corella says.

None says, "Let's play detective. I'll go first. There are only 10,000 phone numbers in a prefix. Each phone number gets one vote. How do you explain 115,000 votes?"

Shen says, "Technically, the rules state one vote per device, not per phone number. If those are unique devices, it doesn't break the rules."

MJ asks, "How many devices are there per phone number?"

None pulls upvotes for one number. "How do 42 devices share a phone number? You'd need a cell phone carrier to pull that off. Impossible." None says.

Corella looks worried. "A cell phone carrier, huh?"

MJ says, "Maybe a phone carrier is involved. What about season 8 of *American Idol?*" MJ searches online. "In the final episode, it was down to Kris Allen and Adam Lambert. There were rumors that 38 million votes out of 100 million came from Kris Allen's home state of Arkansas, a state with less than 3 million people. According to the New York Times, AT&T reps at two Arkansas viewing parties handed out phones that could power text. The power text app could send out 10 or more votes with a single button."

None says, "That 38 million number was retracted. It's unproven."

"The only truth I know is Adam Lambert had my vote," MJ says. "True, or not, what if a phone carrier *is* involved this time?"

None squints at his phone. "When Trench Coat Man threatened me, those phones had my settings. What if…"

None searches online, "My carrier is Galaxor Telecom. Renquist is on the board of directors."

Corella covers her mouth with her hand. "Galaxor Telecom is a major sponsor. We can't let this get out. That kind of controversy would destroy the Reveal Channel. We're just a fledgling media company."

None says, "You're asking me to let Renquist's spy in. I can't exactly exclude Faraday from my cabinet."

"Viewers saw your bias against Faraday," Corella says. "If you dispute the vote when it goes against you, how will that look?"

MJ says, "They don't know we know. It could work to our advantage. After everyone forgets about the vote, find an excuse to fire Faraday."

None nods. "Fine, we run damage control on Faraday. None of Renquist's people will sneak past us again."

Agent Vincent steps from the background. "Mr. President, I recommend we put security and control of the voting process under government control."

MJ says, "Agent Vincent is right. This isn't just a TV show anymore. These votes affect real lives."

None looks to Corella. "I don't care what the contract says. My new portal will control voting. Otherwise, we go public."

"I'll put Shen in touch with corporate legal right away," Corella says.

"What portal?" MJ asks.

None says, "I guilted Rick Slater into building a web portal I designed. He returns my calls now that I'm president. Rick said I was the Titanic. I'm the iceberg. I'll sink Renquist."

37. **Spellbound**

The next episode of *The Internet President* goes on as planned, ignoring Faraday's cheats like it never happened.

None tunes out Astrid's groupies, her Glitter Girls, who permanently camp out in the front rows. They can paint on all the glitter tears they want. None won't feel guilty for dumping her.

MJ hides a book in the couch cushions.

Candidates for Secretary of Treasury wait on the guest couch.

None paces across center stage. "One question remains. Who? I'll save the answer for my inauguration. I'm going to ask a different question. How can we make government better? What if we ran government like a software company using Agile? What steps would we take? First, decide goals and add them to a giant to-do list, called the backlog.

"Every goal is critically important to someone. If we can't decide which is most important between 50 goals, it's doubtful we'll accomplish anything. I'm not asking what's important to Democrats, or Republicans, or Occupy, or Tea Party. What do we want, as a unified nation? Do we need lobbyists and political parties to tell us what we want? If we unleash the power of the Internet, we decide our own national priorities. Most people put more effort into a grocery list than America does on prioritizing issues.

"Split those goals into smaller achievable pieces. Always work on the most important goals first. That's definitely not what we do now. The country would be better off if Congress acted more like project managers." None lectures on every detail of how Agile companies schedule and complete work.

The dry material overshadows his passionate delivery. Boredom spreads. Phones come out. Eyes glaze over.

MJ pretends to read her novel to earn laughs. The audience rewards her with giggles.

None drones on about the importance of customer feedback to create better products. Adopting Agile could generate huge benefits, but the audience didn't come for an efficiency lecture.

He switches back to politics. That's what they came for. "Who should be Secretary of Treasury? What are your priorities for America? You can answer these questions and more on the new IWantMyAmerica portal. I'm calling on every American to log in and participate. Rank your priorities, and I'll present the results to Congress. I handed you a megaphone. Make your voice heard. I want my America. Go online today." IWantMyAmerica.com displays onscreen.

None says, "We're shareholders in America. We deserve our money's worth. We deserve efficient government. Let's take back government and transform it. Do you have ideas to improve government? Let's hear it. The IWantMyAmerica suggestion box is open. No idea is too small, or too big. Upvote or downvote your favorites. I'm calling on every government agency to do a retrospective to identify possible improvements. I want innovation. I want government 3.0."

Corella says, "What is it with tech people and their versions? Why government 3.0? What happened to government 2.0?"

"It often takes three versions to reveal the best design," None says. "We don't have time to waste, so let's skip to where we need to be."

"You're talking about fixing an entrenched bureaucracy," Corella says. "You can't solve people."

"Don't accept the status quo. Don't give up. Even without big wins, constant small improvements add up over time. Optimize the world."

Corella looks back at the candidates. "The focus today is the Treasury Department. Let's talk about our dysfunctional tax system."

None nods. "The tax code is over four million words. Taxpayers waste billions of hours filing taxes. The same income is treated differently based on how it was earned. Should working families pay higher tax rates on their income, than investors do on capital gains?"

The crowd yells, "No!"

"How would you fix it?" Corella asks.

Nothing riles up voters like taxes.

None says, "Income is income. Tax it at the same rate. Simplify."

"Tax breaks hide government subsidies. Tax breaks that expire keep getting extended. Special interests pay campaign contributions for each renewal. Tax extenders work like annuities for lobbyists and Congress. Government spending is larger than it appears. Reduce government intervention through the tax code. No more distortions. Remove the exceptions and simplify." Celebration echo through the amphitheater. None waits for the din of the crowd to settle.

A politician would end on the high note, but None delves into an exhaustive list of tax code simplifications.

Taxes bore MJ. She escapes into her novel for real.

The panic of imaginary users changing the channel contorts Corella's lip into a half snarl. Corella flips through her tablet and nods with a widening grin.

None grinds to the end of his monologue on taxes. He notices MJ reading the book. He realizes he's lost his audience and returns to the host couch.

Corella's eyes light up. "I have a special guest who will—"

MJ sighs relief. "No tax questions?"

"I have something more interesting." Corella cracks her knuckles.

The screen behind Corella shows hidden video from Corella's dressing room. None's break up with Astrid is cut to make Astrid more sympathetic.

None narrows his eyes at Corella. Use my dressing room. That bitch!

Onscreen, None growls at Astrid. "I see the wretched creature you truly are. You are without merits as a person." The camera focuses on his mouth. He spews venom with such enthusiasm that he nearly spits. "I hope someday the hair that fuels your vanity is scalped from your head. You deserve any suffering fate can imagine. Begone! I cast you out and exorcise you from my life."

MJ drops her book.

His words send a chill through the audience. Jeers follow shocked silence. Without context, he is no longer a man standing up to an arrogant, heartless woman, but a villain meting out undeserved punishment. How could he speak that way to a woman he loved? He had to push her away, to protect himself. None blushes with shame and guilt.

She did this. Is he mad at Corella, or Astrid, or himself? All of the above.

He plots his escape. He sprints towards the exit, towards privacy. Horrified Secret Service agents close in.

His path is blocked. It's Astrid.

A star field shawl drapes down to her feet. Twinkling stars, sparkling hair, glistening nails: glitter walks on two legs. Where is the evil witch he remembers? He sees only his Rapunzel with her long blond hair. She was his fairy tale, now just a Grimm tale.

Astrid wears sadness like poor fitting clothes, accessorized with crocodile tears and melodrama. Astrid casts her spell. "I love you more than life. Your words shattered my soul and skewered my heart. I've never been hurt so deeply."

Astrid lives performance art, every day an outsized adventure. He misses seeing the world through her eyes, where the straight lines of math turn into curved lines at preposterous angles. Every over the top word casts him deeper under her spell. The glimmer from her regal hair reminds None of better days. Wait. She said love, present tense. A hint of true feelings from this shallow creature bewitches him. He takes her by one hand. "My lioness, can you forgive me?"

"Was I evil for wanting you back?" Astrid slips her hand away. "You tried to play exorcist on me." She slides her fingernails down her dress seductively. "I'm not evil. I just dress that way."

He swishes his hand with a playful smile. "If I meant it, I'd use Latin." None cringes. "I lash out when I'm scared."

Astrid delivers her lines as an incantation. "You lit my heart on fire with dragon's breath, then tried to extinguish the embers. My love is an inferno, which consumes me. How dare you leave me alone in hellfire!"

Can anyone possibly believe that exaggerated delivery? He does. None is oblivious to everyone else in the room. Her siren call lures None to crash upon her shore. He grasps her hand and draws her close. "I summon you back into my life. We burn together."

None cradles Astrid with a deep kiss.

MJ puts both hands on her throat and pretends to choke. She sticks her tongue out and makes a goofy face.

Astrid pulls back from None. She spins and twirls her star field shawl. It extends outward in a spiral. Underneath, a wire harness covers a flame embroidered dress with red ruffles. A flame embossed scepter provides the centerpiece of the outfit, affixed to a belt across her midriff.

The crowd grows tense with anticipation.

None steps towards her.

Astrid flings her shawl to him and steps back. She yanks the scepter from her belt. It's a release, like pulling a pin out of a grenade. The belt drops. The spring-loaded dress explodes with the sound of a hundred scissors, turning into a five-feet wide bell-shaped ballroom gown. Astrid twirls to let the flame ruffles flutter in the breeze.

The shift from taxes, to Astrid's theatrics jars the spectators. The audience turns on like a light bulb ending a decade of darkness. Several Glitter Girls faint.

Jealousy shields MJ from the excitement. She mock gags and mutters to herself. "Next time she pulls a pin, I hope she holds a real grenade."

None presses closer, her expansive dress holding him back.

Astrid presses her lips towards plastic flames at the end of her scepter. It's a microphone. "I'm the Queen of Flame. I've come here to light the flame in your heart like you did mine."

None stares back. Googly-eyed in a trance.

Corella improvises flames and fireballs on the back screen.

She serenades him with an original song, "Queen of Flame." She's never sung to him before. A wild spirit, Astrid rarely chains herself in commitment.

She cherishes each new commitment in music. None transcends from boy toy to something more.

Every man she's ever sung to is a legend to her fans. None joins the immortals of her universe.

Glitter Girls wail with high pitched screams of exhilaration at hearing their idol sing. Anything that brings their goddess closer to them makes their lives seem less small.

Astrid clasps None's hand tightly. She peers deep into his eyes. Her gaze freezes him like Medusa.

MJ covers her ears. She watches sadly as her best friend slips away from her.

The melodious creature with her siren's wail casts deep love magic, entrancing None. He is spellbound even deeper with each note. Astrid spins a tale of the Queen of Flame and her faithful knight. She sings of two souls chained together with molten bonds, forged to last an eternity.

Corella marvels at crowd reactions. The haunting melody of Astrid's voice leaves no one untouched. She casts her spell on every listener. Corella fights back her own emotions.

The crowd applauds her theatrics. She's won them over.

"Kneel," Astrid says.

None consumes her with his eyes. Astrid. Her name is a power word, melting any remaining resolve. Self-respect is a small price to pay for a return to myth and legend. In Astrid's universe, he can climb her hair into lofty towers and vanquish ancient dragons with his +5 battle ax. Which does he love more, Astrid, or her universe? Does it matter? His knee is ready.

MJ motions with her hands as if attempting to lift up None using telekinesis. She watches helplessly, as None kneels. MJ can't look anymore. She hides her eyes behind her novel.

He melts like a cheap candle.

"Henceforth, I'm your queen." Astrid wields the flame scepter. She touches one shoulder, then the other. "I knight you, Sir…Excalibur."

None appears blissful, while his inner geek demons rage. Excalibur is a sword, not a knight. Is that all she remembers from Arthurian legend? Does it matter? He refocuses on what's important, her.

Sir Excalibur, yet another moniker. At last, he can update his relationship status from 'It's Complicated.' He rises, bewitched by his Astrid. None sees the world through her eyes and joins her spectacle. He twirls her in a short dance. Her dress flutters outward.

MJ facepalms, ignored on the host couch. Alone. She muffles her microphone but speaks too loudly. "I thought cats only had nine lives. Damn Pet Semetery up in here." She sighs and returns to her book.

Astrid strides towards MJ, with the President-Elect in tow. "What are you reading?" She pulls the book from MJ and flips the cover up. A wench in shackles kisses a pirate on the neck on the cover of "Unfettered Hearts." Astrid tilts back her head with a sinister laugh.

MJ yanks her romance novel back.

Astrid says, "Don't worry. You'll find a man, someday."

"It's research, on how to be a *royal* slut like you." MJ pitches the book at her head.

Astrid dodges. The romance novel swooshes through Astrid's hair. She rushes at MJ.

The audience cheers their favorites. They chant, "Fight! Fight! Fight!"

Security restrains the Glitter Girls from rushing the stage.

MJ sweeps Astrid's feet and shoves her to the ground.

The bell-shaped dress flops up when Astrid capsizes. She should have worn underwear.

None dives onto her dress. He's held aloft for a few seconds before the undercarriage supports of the dress buckle. He drops onto Astrid with a thud.

MJ stands over the couple with fists drawn.

None rolls to face MJ. "Haven't you done enough?"

Corella delights. She brought a president down to her level. Entertainment won. The show will be bigger than ever.

MJ glances at the audience and gives None a stern look. "They got what they wanted." She shakes her head and storms off.

Astrid points at MJ from the ground. "That's right, run." Astrid revels in the drama.

None spring to his feet. He pulls Astrid up, careful to avoid metal supports jutting out of her crumpled dress. None swings her around and dips her for a kiss.

Corella says, "Let's talk taxes."

None gestures for her to wait, continuing his kiss.

The candidates for treasury wait in irritation.

"Or, we could continue this episode with a live Christmas special with your family."

None gives a thumbs up, lip-locked with his siren.

Corella says, "After the holiday break, join us for a special L.A. Christmas with the Wong family."

38. **Separate Planes**

He'd rather fly with his precious Astrid. Separate planes. None insisted.

Agents dedicated to MJ's protection surround her, but she feels vulnerable. MJ had been separated from None before but never felt apart. Since kindergarten, her world revolved around the two of them. She changed her name for this man, only to be abandoned. A little push from Astrid could turn separate planes into separate lives.

MJ wishes the plane was flying to Texas. Her family would watch old home movies and film new ones. They would cheer her up in no time and forgive her for being too busy to visit. Instead, she'll be stuck in L.A. for a TV Christmas with the Wong family.

MJ yearns for a Texas Christmas. Dad would hug her. Mom would cook her favorite carne asada enchiladas and make fresh guacamole with avocados from the backyard. Instead, strangers will eat her favorites at her parent's family restaurant in Texas, while she's stuck in L.A.

MJ goes to L.A. to fight Astrid to stay in None's life. Astrid can't force None to get rid of her that easily. She's Vice President. MJ knows how to fight. It's what got her into this mess.

39. **Dark Omen**

MJ enters Beer Beakers for the first time since the shootings. Online rumors tell of paranormal episodes in the bar: cold spots and strange knocking sounds. It's a strange contrast, the supernatural in a science-themed bar. She doesn't need ghost stories. Memories haunt her. MJ rubs the burn scars on her hands. A shiver travels down her spine. Vengeful spirits might blame her, but there's nothing more she could have done about the people who died.

The bar seems less whimsical, more somber. Plaques, pictures, and candles on several tables commemorate the dead with mini-shrines. Goths and people with creepy auras fill all of the memorial tables. Fewer steampunks remain. MJ sees a familiar face. Agents intercept The Blacksmith as he approaches MJ, but she waves him through. He wears eight black armbands on his left arm.

The Blacksmith says, "You're a dark omen. If I didn't need the business, I'd kick you out."

MJ asks, "Where are all the regulars?"

The Blacksmith says, "People drink to forget their problems, not be reminded of grisly murders. The only thing keeping the bar afloat are dark tourists. Every time the lights glitch, we get more popular. As long as the ghosts pay their tab, it's fine with me. It just isn't Beer Beakers anymore. You and your partner did that." The Blacksmith leads MJ and her agents to a table with Astrid, None, and his own circle of agents.

MJ points at Astrid. "Why is she here?"

"You said you wanted to apologize," None says.

"To you, not this trollop," MJ says.

Astrid lunges from her table.

None pulls Astrid back and wraps his arms around her. "Apologies start with Astrid. Your cat fight jeopardizes my credibility as president. Corella wants you two to ruin my Christmas and embarrass my family, all for ratings. When you're ready to apologize, she's right here."

MJ says, "Why are you taking drama queen's side? Bros before super hoes."

None winces. "Truce? Insult each other all you want. But, no fights."

"I could just finish this here," MJ says. "There are no cameras."

None points at the table behind MJ. Two Goths record with their phones.

"If you touch Astrid again, you and me, we're done." None put Astrid first. She took MJ's place.

Astrid extends her hand. MJ slaps Astrid's hand hard, to make the handshake painful. Astrid winces with a smirk.

MJ says, "I wish we were just James and Maria again. Everything changed, along with our names."

40. **Second Christmas**

None gave MJ a green Yoda lightsaber toothbrush for Christmas. A toothbrush. Slapping Star Wars onto everything only makes great gifts for geeks. MJ is only geek-adjacent. The least he could have done was get one in her favorite color, not green. Give her money in a red envelope, and she could choose her own toothbrush.

None had less money in grade school and still gave her better gifts. Much to the chagrin of their families, growing up, they wanted every aspect of their lives to be closer. MJ trained her family to take off their shoes at the door, a custom they took with them to Texas. None got Mrs. Wong to replace the floor sitting table with an American style dining room table and chairs. The mahogany table and matching chairs still dominate half the living room. A plastic covered couch and TV fill in the other half. The couch brings back childhood memories for None and MJ. Mrs. Wong replaced the springs at least five times, but still calls it her "new couch." If MJ's family hadn't moved to Texas, she might have more childhood memories next door.

With the entire Wong family, Astrid, MJ, and assorted agents, Mrs. Wong's cramped two-bedroom house turns claustrophobic.

Mrs. Wong says, "I'm glad you finally brought someone home, but this one? She showed the whole world her lady parts."

"I can hear you," Astrid says.

None throws up his hands. "Astrid didn't expect to be spread eagle."

Mrs. Wong points behind her, at MJ. "Why do you bring her back? You can't keep two women. You're president, not emperor."

"She's my vice president. We're not dating."

"I can hear you too," MJ says.

Mrs. Wong says, "You bring another woman with cameras to our house. I have Internet. I watch your show. We're private people. You sold out your family for a kiss."

Astrid says, "Don't treat us like we're not here."

"I know who's in my house." Mrs. Wong points at Astrid, then MJ. "I'm speaking to my son. He chose for none of us to have privacy." She shifts focus back to None. "I want you to clear out your room. With all these women, I'm sure you have many places to put your things."

Shen cringes. Lan enjoys watching the president get scolded by his mother.

MJ mutters under her breath. "At least Corella doesn't have to hear this."

None says, "You're disappointed enough to kick me out?"

Mrs. Wong says, "I need the space. These men in suits just stand around. Put them to work. I've already boxed up your things."

"The Secret Service is here to protect me. They might save my life someday."

"Until then, they can carry a box." Mrs. Wong retreats into her kitchen, sparing None further interrogation.

Corella, her crew, and the candidates arrive.

Mrs. Wong greets them at the door.

Corella grabs the money tree with flourishing green leaves from the Southeast corner. She places it on the dining table as a centerpiece.

Mrs. Wong haggles with Corella over arrangements.

"I don't remember agreeing to sing Christmas carols," None says.

"What kind of Christmas special doesn't have songs?" Corella asks.

The candidates for Secretary of Treasury sport dour faces at the indignity of singing on TV.

Corella leads caroling rehearsals, while dinner finishes cooking.

The songs and smells of a home cooked meal infuse holiday spirit. Even the candidates turn jolly, swaying at the dining room table.

Corella leads the carols, as the show starts.

Mrs. Wong and Lan bring out trays of rice and home cooked Chinese food. Mrs. Wong leads a short prayer. Then everyone digs in.

Corella says, "None of the Above, can you introduce your family?"

None introduces the Wong family and his girlfriend Astrid.

Corella says, "We're pleased to present, once again, our candidates for Secretary of Treasury." She pauses, watching both candidates.

A pretentious British gentleman shoves a forkful of BBQ pork fried rice into his mouth.

Corella talks fast. "Nigel Kruger, what's your big idea?"

Caught off guard, Nigel chokes on his food.

Corella looks like she's opening a Christmas present.

He triple-coughs and regains his composure.

Corella frowns. Would a Heimlich maneuver be good for ratings?

Nigel clears his throat. "You may address me as Chief Economist Nigel Kruger of Kruger Torus Investments. We need tax cuts for the job creators and businesses, a return to Reaganomics."

None says, "Almost tripling the national debt as Reagan did, might pay for another period of prosperity if we weren't already so deep in debt. A 2015 report from the IMF discredits trickle-down economics. Those policies jump-started the trend of growing income inequality devastating our economy today. Prosperity needs to come from the bottom up."

Nigel says, "Trickle-down works because when the rich have more money, they create more jobs."

"The only thing that trickles down is shit when it rolls downhill," None says. "When businesses automate away jobs, workers get pink slips, not a share of the profits."

"The rich create jobs," Nigel says.

None says, "I was poor when I started Adaptive Unlimited. Anyone can start companies. Thank you, Internet."

MJ fumes. None didn't introduce her. Corella skipped her. Does she have to choke to get attention?

Corella addresses an African American man wearing a conservative suit with a loud tie covered in pie charts. "Fred Jefferson is the founder of Amazing Pie Chart Enterprises. Pie Charts? How are you a finalist? Was the president hungry?"

Jefferson says, "I run a data analytics company. Big data is useless if you can't glean insights from it. Customers can read a histogram but don't ask them what it's called. Everyone knows pie charts." Jefferson straightens his back and glares at Corella. "My big idea, thank you so very much for asking, is to push all things treasury to transparent constant data feeds."

None sticks both thumbs up.

"Please translate your geek speak for the rest of us," Corella says.

Jefferson says, "The economy happens every day. Instead of the Federal Reserve releasing Beige Book reports eight times a year, it'll be more like Twitter,"

"I like Twitter." Corella nods.

MJ stuffs her mouth with food. She shoves in more until her cheeks puff out like a blowfish. She acts out unnoticed.

Astrid says, "My big idea is having a second Christmas."

None and the economists laugh hysterically.

MJ almost chokes on her food.

"The questions aren't for everyone." Nigel can't stop laughing.

Corella narrows her eyes at Nigel. She turns her head towards Astrid and smiles. "Tell me more."

Astrid says, "Isn't shopping good for the economy? Create another holiday that requires gifts."

Corella claps. "Astrid, let me know when you have Second Christmas worked out."

Astrid snuggles up to None. "Land the right boyfriend and every day is Christmas."

None stops laughing and instinctively grabs his wallet.

Corella grins. She takes a deep breath and shifts gears. The joy drains from her face. "I'm sharing a moment of clarity. I'm scared. Every major economy sits on the brink of collapse. We're drowning in debt. We can barely keep our government open. Politicians play chicken with government shutdowns."

Corella points at None. "We elected this guy, just because he's not a politician. Mr. President, you told us a few shows back to ask 'what if.' What if America goes bankrupt? What if cities can't afford to pay their police? What if riots fill the streets? What if it all just falls apart?" Corella gets hysterical.

Corella's words unsettle everyone, Astrid most of all.

None must calm the crowd. He rushes to take Corella's hand. He stares into her eyes and tries to calm her down. "It's going to be OK. That's why I'm here. I'm a problem solver. We'll fix it." When Corella calms down, None pulls back his hands. He sits next to Astrid.

Corella puts her game face back on like her outburst never happened. "I'm told that banking crises turn into sovereign debt crises, which causes inflation. How would you stop inflation?"

Nigel says, "Economist Milton Friedman said 'Inflation is always and everywhere a monetary phenomenon.' All we have to do is pull the right levers at the Fed. If inflation gets too high, we'll stop it with high-interest rates, as Volcker did in the late 70s."

None says, "Volcker did that at the end of the Carter administration when we were the biggest creditor nation. Reagan turned America into the biggest debtor nation. Bush and Obama piled on more debt. Higher rates on so many trillions would sink us. We don't have that option anymore."

Panic streaks across Astrid's features.

MJ delights in her fear. She is so tired of third wheel city. There must be some way to take down this weak creature that dares to take her place.

Jefferson says, "The economy is like musical chairs. America can survive until the music ends. Confidence is that music. Each time the music stops we lose a chair. The first chair we lost was junk bonds, the weakest companies. Municipal bonds for cities and states are failing now. Next are the federal government and strong companies. Companies that rollover debt can't make payroll and go out of business. Once confidence is gone in bonds, the banking system seizes up from counterparty risk."

Corella asks, "Can you explain counterparty risk for the non-economists?"

Jefferson says, "Banks don't trust each other and won't transfer funds. The whole system grinds to a halt. Banks stop withdrawals. Then you can't pay your bills. Even if all the major economies weren't having problems, America is too big to bail out. No one will save us. No one can."

Astrid has a panic attack and flees the room. MJ follows her. Their departure ruffles the other guests. Astrid enters an oblong bathroom crying. A white vanity with sink and drawers is near the entrance. Astrid tries to close the bathroom door.

MJ forces the door open and shuts it behind her. She has Astrid alone. MJ's not sure what to do, but she must do something. There are ways to fight other than fists.

Goose pimples raise on Astrid's skin. She backs away.

MJ steps towards her.

Astrid's heel hit the tub. She slips backward, grabbing the shower curtain.

The curtain rips off of the rod. The shower rod loosens. The whole assembly falls with Astrid into the tub.

Astrid wheezes, too terrified to make a sound. She flings the shower rod at MJ.

MJ jumps back with raised hands. "I'm not here to fight."

Astrid scrambles to her feet. She turns the bath water on hot and straddles the tub to stay dry. "Stand back. This house still has old plumbing." Astrid grabs the spray hose from the shower attachment. She flicks a lever. Warm water streams from the hose.

MJ laughs. "Pretty weird threat. I'm not afraid of water."

Astrid sprays MJ with warm water.

MJ could close the gap and turn the hose on Astrid. It's just three feet of slippery tile away. If she hurts Astrid, she loses None forever. Did he really mean forever? Astrid gives her an impromptu shower. MJ soaks. Her waterproof mascara runs.

Something else holds MJ back, something important. What a weird threat. She can almost remember.

"Step back, or you'll burn again," Astrid says.

James was mad at MJ long before all of this None of the Above crap. A cold shower. Remember, damn it. House rules. Don't turn on the dishwasher. Don't flush the toilet, while someone is taking a—

Astrid reaches towards the toilet next to the shower.

MJ remembers. The toilet sucks out the cold, leaving only scalding water. She backpedals, fumbling and sliding, until she slams into the bathroom door. She yanks a drawer out of the vanity. Personal hygiene items from the drawer flop onto the floor. An ornate glass container cracks on the tile. Liquid seeps out. MJ coughs, overpowered by the scent of perfume.

Astrid flushes.

Steam rises where the water sprays. The bathroom mists up.

A sudden terror grips MJ. The heat reminds her of the burns, the sounds of sizzling flesh, the curls of boiled skin, the smell of her cooked meat, the pain. Most of all, she remembers she's not invincible.

MJ jumps like a cat onto the vanity. She uses the drawer as a shield.

Two women, scared of each other, peer across a drenched bathroom.

Water seeps out under the bathroom door.

Agent Reynolds knocks on the door. "Everything all right in there?"

MJ shouts. "We're OK. I wanted to check on my friend Astrid." She feigns concern.

Agent Reynolds taps his feet on damp carpet. "There's water in the hallway. I need to verify your condition, Madame Vice President."

MJ whispers, "Help me clean this up."

Astrid shuts off the water. She holds onto the shower hose.

Agent Reynolds says, "I need to come in." Water squishes under the agent's shoes as he steps back.

MJ worries Agent Reynolds will knock down the door. She jumps down towards the door and flings it open.

He stops himself coming towards the door. He assesses the disaster of a bathroom. He notes the clothes clinging tightly to MJ. He sniffs the perfume and cocks a smile. "I'll give you some privacy." Agent Reynolds nods and pulls the door shut.

MJ puts the drawer back and dams up water at the door with a towel. "We're frenemies, aren't we?"

Astrid says, "I appreciated the heads up when he lost his company, but I decided I'm keeping him. You didn't want him. The president is mine."

MJ rolls her eyes. "Hit that all you want, but don't force me out." A red lightsaber toothbrush sits in a toothbrush holder wall mounted atop the vanity sink. His toothbrush. He's spending the night. If MJ doesn't clean up this mess, she will never hear the end of it. MJ grabs an eight-ounce paper cup next to the toothbrushes and scoops water into the sink.

Astrid puts down the hose. She sits on the tub wall and kicks water.

MJ says, "Tell me what's wrong. Did I get wet for nothing?"

"There's nothing you could do to make me wet," Astrid says.

MJ splatters Astrid with a cup full of water.

Astrid splashes MJ with her feet. "We look like drowned rats."

The splash fight turns into laughs.

Astrid owns her own cosmetics line. She's a drowned rat with perfect makeup.

"I wish I had battle worthy makeup, like yours," MJ says.

Astrid extends her hand. "Frenemies?"

"Yes." MJ grabs Astrid's hand.

Astrid flinches. A firm handshake, Astrid sighs relief. She takes a conspiratorial tone. "If every country fails at the same time, there's nowhere to run. I won't be pretty forever. I have to monetize my looks and get ready to cash out."

MJ snickers. "Monetize. What a gold digger." MJ calls Astrid a gold digger so often. It's a nickname.

Astrid tunes it out. The economists got to her. "The way he's picking his cabinet is a farce. I hope it's not that British guy. He thinks inflation is only about monetary policy. Milton Friedman didn't account for the velocity of money—"

MJ gasps. "Velocity of money? What are you talking about? Who are you?" She stares at Astrid, shocked.

Astrid says, "The velocity of money is how quickly people spend. It didn't change much from the mid-50s to the late 70s when Friedman did his research. You can lower interest rates all you want, but if Americans are too scared to take out loans, it won't matter. The velocity of money drops, as people hoard their cash in fear."

"Why didn't you say something smart like that, instead of Second Christmas?"

Astrid says, "If they assume I'm a ditz because of the way I look, why not incorporate it into my tactics? You'd be surprised what people say around you when they think you're too stupid to understand. Besides, Second Christmas is trending. My brand is stronger than ever."

MJ says, "You played dumb this whole time?"

Astrid says, "I disguise my intelligence in the language of imagination. Whenever None is about to catch on, I throw in an intentional mistake. Excalibur is a sword, not a knight. I thought for sure he'd call me on it."

"The truth will blow his mind," MJ says.

Astrid says, "I thought you of all people would see beyond this. The president will never believe you. People believe what they want to. Conspiracy theories online say that I have a ghostwriter feeding me lines. Sometimes I wear an earpiece to feed the rumors. Who do they think runs my cosmetic line and music label? It's laughable. The prettier you are, the easier men are to manipulate. So, how is that 'accept me as a brainy woman' thing working out for you?"

MJ scrunches her nose. "None assumes he's the brightest person in the room until proven otherwise. He doesn't take anyone that seriously."

Astrid points at MJ. "It's you he doesn't take seriously. I influence the president. Do you feel him slipping away?"

MJ won't admit it, not to Astrid. "Tell me why you're really with him. Money? Power?" MJ scoffs. "Love?"

Astrid looks away. "Fear. Corella isn't the only one who's afraid. The White House is a good place to ride out a storm."

MJ shoves her burn scars in Astrid's face. "It also makes you a target. Astrid, I know an option with more money, more power, and safer than the White House."

Astrid's ears perk up.

"Alvin Renquist."

Possibilities dart around in Astrid's head. "What's my play? He'll be suspicious, if I dump the president, right after we got back together."

MJ says, "Renquist hates the president. You'll score major points with him if you dump None in the most humiliating way possible. His self-confidence will plummet, but he'll bounce back eventually."

"You want me to hurt the president, just so that you can be his number one again? I realize you're a frenemy, but does he?"

MJ says, "I'm his best friend. None deserves a girlfriend who's there for better reasons."

Astrid asks, "What else do you know about Renquist?"

"I kicked his ass once. He might be wary of aggressive women. Go slower. Make him think it's his idea."

Astrid says, "Quid pro quo. Here's my advice to you. Diamonds are a girl's best friend. It used to be the S&P 500. Don't bother with gold. They might confiscate it. Convert your assets to diamonds and be ready to travel. You don't know which country will fall first. It could even be this one."

"Astrid, I'm vice president. I'll be OK." MJ gives up on the cup and sops up water with a towel. "Hey, you made this mess. Why am I cleaning it up?"

"The president will fight his mom for me. Will he do the same for you?" Astrid points at pooling water behind the toilet. "You missed a spot."

MJ cleans up the bathroom, while Astrid watches. When she finishes, Astrid points at her hair. MJ plucks shower curtain rings from Astrid's hair, then combs her golden mane. "I feel like Cinderella. That would make you my evil stepsister." Cinderella never betrayed her best friend, just to be their number one.

MJ and Astrid take turns with the blow dryer to dry out their clothes. Body heat will eventually remove the remaining dampness.

Applause passes through the walls. Astrid and MJ primp their outfits and return to their seats to see what the clapping is about. Their wet clothes attract a few odd looks.

None gives the speech of his life. He doesn't try to pave over the problems with easy money. The president calls on Americans to unite and make shared sacrifices. He skips his techno-babble for once. He replaces it with Christmas cheer and connects with the country on an emotional level. The fate of the nation, perhaps the world, lies in confidence. The American people must have confidence in their president, or the music stops.

41. **Poisoned**

MJ didn't touch Astrid in the bathroom encounter, but None still holds her responsible. He doesn't exile her from his life, but MJ is on the outs. The ear-full he gets from his mother, he passes on to MJ. Turning the bathroom into a water park ruins MJ's reputation with Mrs. Wong.

None keeps Astrid and MJ apart to keep the peace.

MJ realizes her terrible mistake. She must talk Astrid out of dumping the president. MJ can't tell None why it's urgent she speaks to Astrid. How could she be so selfish? Too much rides on his confidence.

Not being introduced at Christmas is no fluke. None and Corella treat her like a rotting appendix.

Nothing stings harder than the call from her parents asking why she isn't on the show anymore. Corella edited out her comments and ignored her contributions. She may as well sit in the audience. The Glitter Girls get more attention than MJ.

Only one person could poison None and Corella against MJ so fiercely, Astrid. Now Astrid's just showing off.

Astrid uses her online influence to sway votes. She backs Fred Jefferson, the "pie guy." Pie mania infests the Internet. People share how happy they are in pie chart status updates. Everyone discusses their day in terms of good pie or bad pie. It's like a contest to find the craziest way to work pie into a conversation.

Fred Jefferson wins in a landslide. Astrid's lemmings choose pie. He was probably the best fit, but Astrid's endorsement mattered more than substance.

None and Corella process each cabinet decision like clockwork. One week from the inauguration, only two positions remain open.

Big changes wait only days away. Five days before the inauguration, None and MJ will move from their hotel rooms in the Lancelot Court Hotel, into Blair House, the President's Guest House. It's the final transition before the White House.

Tomorrow he celebrates his birthday. The president's confidence and approval ratings peaks at an all-time high.

42. Succubus

Agent Vincent approaches None and MJ backstage at the D.C. studio for the Reveal Channel, where *The Internet President* is filmed. "Mr. President, Corella added a special guest to today's set list that could potentially embarrass you."

None shushes Agent Vincent. "No spoilers. Authentic reactions are part of my deal with Corella."

Agent Vincent says, "There should be operational directives, limits to what can happen on TV."

None says, "No limits. The more shocking, the better. It's difficult to keep voters watching. I compete with other channels and snack food. If I keep them entertained, I keep their attention. Otherwise, I lose them to a ham sandwich in the fridge."

"Understood, Mr. President," Agent Vincent says.

None and MJ walk onstage. Birthday banners and balloons decorate the stage. MJ and None sit at opposite ends of the host couch.

Corella says, "We have a special birthday guest. Give a warm welcome to Astrid Fontaine."

MJ squirms in her seat. She stares at Astrid with wide-eyed horror.

None tunes MJ out and hopes there's not another cat fight.

Corella plays "Queen of Flame," the love theme of None and Astrid. The new hit song premiered on this very stage.

Astrid saunters in with her runway walk to enthusiastic applause and hoots. She dresses like a dark angel, with folded wings and a skin-tight latex catsuit. A black corset crushes her into an hourglass. It's so sexy it could take a man's breath away, and so tight she can barely breathe in it herself. The golden glow of her hair shines brightly against her all black outfit.

"I have a special gift for the president." Astrid moves with undulating sexual thrusts. Her hair sways, as her head bobs. The black feathers in her wings flutter with each stride.

None licks his lips. He sits forward in anticipation. This could be the best birthday ever. This is Astrid. Anything is possible.

Corella says, "Please, nothing X-rated, like last time."

Astrid straddles None and sits on his lap, facing him.

MJ scrambles away and flees to the guest couch.

Corella fiddles with her tablet. Security guards emerge at the edge of the stage. Corella stares down Astrid. "It's so tight. How do you even walk in that?"

Audience members record Astrid with their phones, hopeful of another wardrobe malfunction.

Astrid slinks off. She kneels in front of None, careful not to let her wings touch the ground. Astrid stares into his eyes. They focus on each other and block out the world. It provides an illusion of privacy, a bubble of intimacy.

"Good news, today I give you something you've asked of me many times, the gift of honesty," Astrid says.

Disappointment registers in the audience. They hoped for something more titillating.

Astrid says, "I tried you on so many times. You still don't fit."

None feels like expired food. Is she dumping him on his birthday? "Aiya! Happy fucking birthday."

How could she tell None he doesn't fit? Only a supermodel could see men as wearables, to toss in a hamper. None studies every piece of Astrid's ensemble. "What am I to you? A pair of pants? A scarf? A feather?"

Corella cocks an eyebrow.

Astrid rises. "I don't mean it like that. You never defy gravity and take flight. You only soar into imagination, when I lift you."

None gets up to face her. "I don't need you to fly. I have my own airplane."

Astrid touches the back of her catsuit. The spring-loaded wings extend outward in a six-foot wingspan. "Keep your airplane. I fly on metaphors. I fly on dreams." Astrid raises her chin, aloof.

None says, "I have lots of imagination. What about all those times we role played?"

"My world is not a game to be played," Astrid says. "It's meant to be lived, epically."

"My lioness, I love you."

"I know."

She Han Soloed him.

"If I stay with you much longer, you'll strap a saddle to my back. I'm not your Pegasus," Astrid says.

Corella covers her mouth with her hand.

None says, "Don't you want to get married someday?"

"You and me, we're good ingredients, but we make a bad pie."

Corella realizes Astrid is dumping None. She stops Astrid's song abruptly.

Humpty Dumpty had it easy. He only broke once. Astrid shattered his heart enough times it ought to count as exercise. The black hole inside pulls so strong, not even love can escape the event horizon. He blocks out emotions, an aspiring Vulcan. Yes, he feels nothing, like the deafness and shell shock after an explosion. None yearns to immerse himself in code, to commune with his machine and manifest his will. Computers do what you tell them to. People, not so much. The president speaks with stoic acceptance. "Acknowledged."

Astrid stares back with impatience.

He shares a common truth with every victim of reality TV. In television, there is nowhere to hide, as your sorrows broadcast to the world. His misery is recorded for all time. His enemies can delight in his pain, whenever they want. Is Renquist laughing? Or his fifth-grade gym teacher?

Unlike the fairy tale, his Rapunzel and the witch that owns her are the same person. Her witch cackle and claws will come out any moment. Astrid won't let down her hair so that he can enter her tower. She'll strangle him in a golden noose of hair.

Astrid's a tabloid favorite. Her dating history lists like a graveyard of celebrity flameouts. Some of them still stalk her, willing to do anything to win her back. She's a super drug that fulfills your greatest desires, then annihilates you. She lifts you into space, then crashes you into the crushing depths of the Marianas Trench.

She flits in and out of her boyfriend's lives at her whim. It's only really over when she burns your effigy and sings a song dedicated to how much you loved her and how you suffered for your folly.

An album of Astrid breakup songs went platinum. She performs only for her hapless victims and recording studios. She's not a singer. It's part of her brand, along with the trail of her discarded patrons. Each man thinks he can tame the shrew and end her reign of sensuality and cruelty. MJ warned him every time. He never listened.

Honesty is too core to his being, for him to hide his feelings. He speaks the truth, while Astrid speaks fantasy. If he could lie like her, perhaps she'd truly love him. She's an actress. All her emotions are fabricated, her persona manufactured for cameras. She claimed None was the first to hurt her. What can hurt a soulless simulacrum? All that beauty hides an evil hag.

No, Astrid is something worse. She's a succubus: a demon of seduction. This is her true form with black outstretched wings. All that's missing is horns on her head. Even in the midst of all this pain, his body still yearns for her. How many souls has Astrid stolen, or left in torment?

He turns away from Astrid, towards Corella. This is the second time she's ambushed him. Corella must have planned the whole thing. How long has she schemed against him? Corella brought Astrid back into his life, only to take her away. What was his crime? Long speeches? Lower ratings? She got her ratings, and Corella's going to pay for it someday.

The president glances at MJ.

MJ writhes in her seat like she's sharing his suffering.

What a good friend that she would share his sorrow over losing Astrid. He's been too hard on MJ. He let Astrid turn him against her. She warned him about Astrid. Why didn't he listen? MJ did everything she could to protect him. Astrid rips out his guts and tosses away his heart for all to see. Why didn't he listen this time, or last time, or five times ago? MJ always has his back. She's the one he should trust. MJ would never hurt him. He pushed MJ away. Now he has no one.

Astrid stands on the couch and points down at the president. "Just sit there, lost in your head. Is that all you can do?" She cackles like he knew she would. "You have as much emotion as those computers you love so much. You're dead inside. It's like dating wallpaper."

Astrid says, "Roll it, Marty."

Corella scowls at Astrid. That's Corella's line.

The screen behind Corella displays footage of the Astrid breakup recut to make the president sympathetic.

None looks up at her from the couch and relives the hurtful memory. "I could feel you pulling away." He repeats the words with the playback.

In the video, Astrid says, "Don't you want to see me again?" A close-up reveals her insincerity.

None stands up to his kryptonite onscreen. He quivers and gathers strength. He invokes the power of no, a power he no longer has.

"How could *you* not want *me*? You'll beg me to come back. You always do. I'm Astrid. No, is my word." The playback completes. Astrid took back the power of no, her way to rebuke None for his insolence.

His seductress revels in his suffering with a witch's cackle. She waves two fingers with a haughty sway, as she speaks. "No, is my word. You mess with the lioness, you get the claw." She slams down two fingers and says "tssss!"

Frat boys high five. They talk about how sizzling hot she is.

Astrid groupies imitate her burn sizzle gesture, with a swaying wave as they speak. "You've been Astrified." They slam down two fingers with a sizzle sound and imitate her cruel laugh.

She points at the president. "Pro Tip: You can't dump Astrid. Astrid dumps you." Astrid proves her mean girl cred. She does one final burn sizzle and retracts her wings.

Astrid exits to the cheers of her fans.

The Glitter Girls anticipate adding his suffering to their music playlist.

Astrid's lack of respect for the presidency antagonizes older audience members. None's public loss makes him sympathetic and bolsters his image as a transparent man of the people. Others perceive None as weak, crushed like a worker ant by his queen.

Corella hides her gloating.

MJ rushes over to the president. She gestures for a bro hug. None latches on.

She hesitates, then embraces him into a full hug. "If you chase that kind of girl, you can't expect to be more than a walking ATM." Her words provide no comfort.

None broke the rules when he said no to Astrid. Her love felt real this time. The shift happened after Christmas, after she ensnared him, after she knew it would hurt. She got him back just to be the dumper, to take back her word, 'no.'

MJ helps him off the stage, like an injured player limping off the field: the quarterback everyone depends on whose spirit is broke. The nation needs the president at his best, not a distraught lovesick boy. What a messed-up world when a president's love life becomes a forward-leaning economic indicator.

He pines over Astrid offstage. She hasn't burned his effigy yet. He still has hope, her cruel birthday present.

43. Two Bells and a Timer

Corella hoped a dating angle would keep the country talking about *The Internet President*. Mission accomplished. It distracts from real challenges facing America. The breakup shatters None's confidence and imperils the faith of voters in him. The stock market declines on concerns None has weakened too much to fight a do-nothing Congress.

No Astrid. No energy. No more speeches. No more audience participation. None broods on the host couch, a shadow of misery and grief. He continues a wake for his relationship. How many days will it take to accept Astrid's love is dead? His suffering adds a melancholy tone to the show.

Corella asks None questions about Astrid. He ignores her, catatonic.

MJ finishes what None started. MJ interviews candidates on the episode for Homeland Security. Corella won't be able to edit out MJ now. MJ wanted more input, but not with this steep price.

MJ leads the next show for Secretary of Defense. Civilians control the military at the highest levels. To keep it that way, the second in command below the president, the Secretary of Defense, can't be in the military within the last seven years. MJ picks two beefcakes in their late 40s with distinguished military records as her top picks.

Corella asks questions about their military missions. Neither of them can say anything more than "classified." The answers agitate Corella until she spices things up by asking them to take off their shirts and show their muscles.

Half the audience hollers in approval.

Corella has gone too far, again. MJ has to rescue the decorum of the show. She suggests an arm wrestling match instead. The candidates quickly roll up their sleeves and challenge each other.

The candidates turn the arm wrestling into a fun highlight, before moving on to what they did since they left the military. Results for Secretary of Homeland Security finish the show.

The next episode of *The Internet President* ends the selection of cabinet members.

Since Secretary of Defense is the last cabinet position to fill, Corella runs a clip show of bloopers and highlights until it's time to announce results.

MJ's favorite, Fernando Rivera, won the arm wrestling match and today he wins Secretary of Defense. A former U.S. Navy SEAL, his action hero looks conceal a calculating intelligence. A recent Pentagon audit showed trillions in funding couldn't be accounted for properly. What put Fernando over the top was his pledge to bring transparency to military budgets.

MJ is ecstatic Fernando won. Besides being a SEAL, he taught mixed martial arts after he left the military. She fantasizes about learning new skills from him. If Fernando had a dojo nearby, he'd be her ideal sensei. She waits for guests to file out after the show ends.

The show is over, but None remains slumped on the couch.

MJ hides her guilt about Astrid. Even if he never finds out, she'll know. MJ looks into None's eyes. "Astrid wasn't here to stay, but I am."

He struggles to speak, drained of life, drained of will. None bears MJ's friendship, like a crutch. "I just." He trails off weakly. "I can't go back to the ordinary world without her."

MJ cringes. He'll be playing that Duran Duran song, "Ordinary World," all day. It's DEFCON 1. His love life went nuclear. If MJ can't bring back his confidence, the next thing to go nuclear will be the economy. The dominoes will fall if None can't restore confidence: the bond markets, the banks, the government. He's roadkill. He can't even restore confidence in himself.

MJ did this. She has to fix it.

None lies down on the couch.

MJ approaches Corella. "Why did you have to bring Astrid on the show?"

"I didn't know what she was up to," Corella says.

"Your guy, Marty, ran recut footage of the breakup," MJ says.

"I fired Marty last night. That diva crashed my show with her own agenda and used my people to do it."

"Corella, are you sure you didn't do this for ratings?"

"I heard the economists. I know what's at stake."

MJ decides to believe her. She'll take any help she can get. "Let's brainstorm how to get him out of this funk. I tried everything last night." MJ walks over to Agent Reynolds. "Get Shen. We need a plan."

"We could give him a stimulant." Corella looks around at federal agents and catches herself. "If he has a prescription."

Agent Reynolds calls to have Shen sent in.

Shen enters the stage, moments later.

Desperation tweaks MJ's voice. "Shen, how do you cheer up your brother?"

171

"Why ask me, MJ? You spend more time with him."

MJ turns to Corella. "I've never had a boyfriend. I've done research, but most of what I know is based on movies."

Corella sneers. She looks at MJ from head to shoe. "How could you not have a boyfriend?"

"You seem like someone who's been in a bunch of relationships," MJ says.

Corella glares at her.

MJ asks, "How do you get over guys who dump you?"

"Why do you assume I've been dumped?" Corella's nose flares.

"Do you know the answer, or not?"

Corella says, "Time."

"We have three days before the inauguration to get him back to normal," MJ says.

Corella blushes and talks in a quiet voice. "I date other people until I forget the idiot who left me."

MJ sighs. "He's a lot of work when he's mopey-mopey. Who's going to date the president?"

Corella raises her hand. "I could take him out for a spin."

MJ shakes her head. "He's not a car. Go test drive somebody else."

Shen stares at MJ. "Only dating you can break Astrid's spell. Will you do it?"

MJ looks at None. His body lies on the couch like a corpse, not far from the crime scene where his confidence was slaughtered.

"I can't," MJ says.

Shen says, "MJ, you know what's at stake. He cares for you. This is when he needs you most."

"I'm his bro, not his girlfriend." MJ lurches towards Shen with clenched fists. "Are we clear?"

Shen retreats. "If you're worried about commitment, None would marry you."

MJ sighs. "I have personal reasons. It wouldn't work. Astrid did such a number on him. He'll have to date a hundred women to forget her. I don't want to be on such a long list."

"I do," Corella says.

MJ asks, "Where are we going to find enough women?"

Corella raises her hand. Ignored.

"I've had success at bars," Agent Reynolds says.

"You can't pimp out my brother with a bunch of alcoholics," Shen says.

Corella sulks. "I'd make a great first lady."

None stares at the ceiling with vacant eyes, immobilized by depression.

"Fine, Corella. Do what you want." MJ points at None on the couch. "It's going to take something massive to wake sleeping beauty."

"Great. I'm just asking so I don't get tackled by the Secret Service." Corella approaches the president with glee. She sets down her tablet without bothering to turn it off. Corella kneels next to the couch.

Agents approach Corella, watching every move.

"This kiss is a matter of national security," Corella says.

The agents don't find that amusing.

MJ groans and waves for the agents to back off.

Corella lifts up None's neck. Her lips meet his. Corella sucks out death and breathes life back into him.

None's eyes remain closed, but a smile creeps onto his face. He acts half asleep but pulls Corella on top of him.

They explore each other with a series of sensual touches.

"He's the Internet President. Let's use the Internet." This time, it's MJ's turn to be ignored.

All eyes stare at None and Corella on the couch.

The steamy exchange makes Shen uncomfortable. He retreats offstage.

MJ panics. She remembers Corella editing her off the show, helping to push MJ out of None's life. Corella can't be allowed to take Astrid's place.

None responds like he's dreaming. He sucks Corella's kiss deeper. His tongue probes her mouth.

MJ grabs Corella's tablet. She types frantically on the virtual keyboard. MJ will use the Internet to keep Corella at bay. MJ wonders who None dreams of, but she can't be distracted. MJ pulls photos of None off the web and generates a dating profile. She knows everything about her best friend. It's not hard to fill in all the details. Every moment Corella's lips remain locked with None, MJ gets more agitated.

Sleeping beauty awakens with bulging eyes. None realizes he's kissing Corella. He seems to expect someone else from his dream. None doesn't register disappointment that it's actually Corella.

Corella comes up for air. She gloats at MJ. "The rest of you should leave."

MJ thrusts Corella's tablet in None's face. "I put you on a dating site a few minutes ago. It's already viral. Over a thousand women want to meet you."

"O.M.G." He slides from under Corella and jolts to his feet. He grabs the tablet and eliminates matches based on pictures. He wanders off, excited with his new toy.

"We'll see who edits out who," MJ mutters under her breath.

Corella springs from the couch and grabs her tablet back. "That's mine."

MJ pulls None by the arm. "It's OK. We'll set up our laptops, and you can log back in."

None and MJ rush off with their agents and grab Shen on the way out. Corella stays behind with resentful eyes.

None, MJ, and Shen set up three laptops in the lunchroom and log in.

With each picture, None gains confidence. He has to be selective, with so many choices. None scours photos for geek girls and the drool-worthy.

None addresses MJ and Shen. "Get rid of any deal breakers. No lobbyists. No mean girls."

"No supermodels?" MJ asks.

None pauses clicking to think. "Supermodels are fine, but no dumb girls."

He still hasn't learned. MJ rolls her eyes. No dumb girls. If only he knew. Astrid was awful, but this is worse. MJ sinks to a new low, helping None date other women.

Over ten thousand women answer the casting call to be the first lady. It takes hours to whittle it down to a couple hundred. The media picks up the story, and his message box runs out of space. Only outbound messages work. Three people can't keep up with the deluge. It's a Sisyphean task.

"You already favorited 205 matches," MJ says. "How are you going to date that many women?"

None says, "Speed dating. We'll call it one on ones with the president. Schedule the women in 15-minute slots. Put together a waiting list to fill in for no-shows, or early eliminations. Buy two bells and a timer."

He seems more confident than ever. At least he'll get over Astrid quickly.

None and MJ already have rooms at Blair House, but it's a government-owned building. Hosting speed dating there seems questionable. Instead, None rents a conference room in the Lancelot Court Hotel where they stayed before.

MJ says, "A conference room. How romantic." MJ sets up two bells and the timer on a folding table where None sits. Leave it to None to turn dating into work.

MJ slips into the hallway. Aspiring starlets wait outside the door, None's three next appointments and four women on the waiting list. MJ says, "Time is precious. When you come in, a timer will start your 15 minutes with the president. There are two bells, one for you, and one for the president."

A dancer in cut-off jeans and ballet shoes interrupts. "President-elect. He's not president until the inauguration."

MJ scoffs. "Be sure to mention that when it's your turn," MJ says to the dancer. "One last instruction. Anyone can ring the bell to end the date early."

"You're up." MJ points to the dancer. She leads the dancer in and introduces her to the president.

The dancer snaps her finger at MJ. "He's not the president. It's president-elect. Pay attention."

None rings the bell before she sits down.

MJ says, "A smart aleck who doesn't understand sarcasm, epic fail."

The dancer flicks off MJ and storms out.

MJ brings in Sarah, a yoga instructor off the waiting list.

Sarah veers the conversation into None's inner feelings. Her soothing voice calms him. None joins her in yoga poses, while they talk. The timer goes off in the middle of downward facing dog. None rises to his feet and blows her a goodbye kiss.

MJ leads in the next woman with waist-long golden hair and glitter eyelashes. The Glitter Girl wears a miniskirt and a sweater. MJ introduces them and steps back.

The Glitter Girl fawns over the president, her knees shaking. "I can't believe I get to meet you."

"Yes, it's not every day you get to meet the president." None eats up all the attention.

"I want to be with you so bad." The Glitter Girl says the words intensely and begins to pull up her sweater.

She has None's full attention.

MJ gasps until she sees the shirt under the sweater. Astrid, in her Queen of Flame outfit, sings to None on the Glitter Girl's shirt.

The Glitter Girl points to her bosom. "Would you sign my shirt before I ravage you?"

None fumbles for a pen.

MJ assumed None wouldn't settle for one of Astrid's groupies. Maybe she gave him too much credit.

The Glitter Girl says, "What was it like being with Astrid?"

MJ cringes hearing Astrid's name. She steps closer. The whole point of this unfortunate exercise was to forget Astrid.

The Glitter Girl softly caresses None's hands. "These hands touched Astrid. Now they're touching me."

MJ considers whether she should intervene.

None gets drawn in by the heat of her desire reflected back at him. He kisses the Glitter Girl on the neck.

MJ rings the bell.

The Glitter Girl remains ecstatic for even a lingering touch, enraptured.

None scowls at MJ. He breaks away from his admirer to confront her. "You had no right. I ring the bell."

"Trust me, Bro. Groupies are second level crazy. Look at her."

The Glitter Girl pretends her arms are still around None, outstretched into thin air. She shuts her eyes to help maintain the illusion.

It's tempting to be wanted, but None makes the right choice. He motions for agents to remove her.

Guards lead out the Glitter Girl, smiling with her eyes closed, still clinging to an imaginary president.

MJ collects herself and leads in a woman in a well-tailored suit with a briefcase.

The woman opens her briefcase. "Mr. President, you can create more jobs with oil drilling in the Arctic."

A lobbyist got through.

Ding.

The lobbyist talks over the bell. She tosses a 386-page report about Alaskan jobs onto the table. It lands with a thud. Secret Service agents lift her up to remove her from the room. She struggles and kicks as they carry her away. "You want more jobs? Zentripity Energy can help."

Zelda steps forward from the waiting list. This one did her homework. She wears a steampunk bowler hat with leather goggles on the rim. Her saloon girl dress shows leg in front, but the fabric extends to her ankles in the back. Two empty holster belts crisscross her waist.

MJ stares into her deep-set blue eyes, which make you feel seen. "I see why he picked you," MJ says. Zelda's steampunk fashions remind her of how easily she could be replaced. When None finds the perfect girlfriend, she'll never be number one in his life again. The thought terrifies her. MJ ushers Zelda in.

Why does confidence have to be so important? MJ swears under her breath. "The things I do for my country." MJ hands over pimp duties to Shen. She needs a break.

MJ finds Agent Reynolds. She reads through the move-in arrangements. MJ frantically pages through the paperwork. "I don't see my items on the White House manifest."

"The vice president gets their own residence at the Naval Observatory," Agent Reynolds says.

"I belong with the president at the White House." MJ rushes off to confirm with Agent Vincent.

Agent Vincent doesn't give MJ the answer she wants to hear.

MJ has no other choice than to confront the president about it. She passes steampunk Zelda in the hallway. Has it been 15 minutes already? Time flies when MJ gets pissed. She barges into the conference room.

A blond bombshell talks at None about all her unfaithful boyfriends.

MJ smiles at None from the doorway. He smiles back.

The blond looks at None, then back at MJ. "Don't you dare flirt with her. No wonder Astrid dumped you." The blond slams down her bell.

Ding. Ding. Ding.

None rips the bell out of her hand to stop her.

MJ watches her leave and taps Shen on the shoulder. "Hold the line. I need a few minutes with your brother."

MJ flings her finger past her ear with a swishing sound. "Another crazy jealous blond. You dodged a bullet there. You're welcome."

"What is it MJ? I'm busy."

"They're moving me to the Naval Observatory."

"That's where vice presidents go."

"There's plenty of space for both of us at the White House," MJ says.

None checks the time on his phone. He grows impatient.

MJ says, "You said co-president. I don't want to be shutout again if there's another Astrid."

"Fine. I'll get you a room in the White House. I'll make it so." He points her to the door and resumes speed dating.

44. **Drowned Ants**

The Secret Service reserve rooms for the Wong and Cortez families to stay at the Lancelot Court Hotel, where None and MJ stayed before moving into Blair House. None repurposes the conference room at the hotel from speed dating into a family room. Catered Mexican and Chinese food cover two tables.

Inauguration festivities begin early tomorrow morning. None and MJ celebrate their last night as private citizens with their families. Lan and Mrs. Wong arrive first.

Shen embraces Lan tightly. The two of them may be strangers for a while, as Lan winds down the real estate business in L.A.

MJ's parents and two younger brothers come from Texas. The Wong and Cortez families were neighbors until MJ's family moved to Texas 20 years ago to start a restaurant with her Uncle Diego. Both families reminisce and catch up, after their long separation.

MJ's father, Sergio, delves into challenges opening a third restaurant in Texas. He jokes about opening one in D.C. MJ latches onto the idea and tries to persuade him to open a D.C. restaurant, but her father only makes jokes about the catered food.

None steps closer to his mother to talk. "Without all my stuff you must have a lot more room at home."

Mrs. Wong says, "I almost forgot there was a bed under all your boxes."

"What did you do with the extra space?"

Mrs. Wong looks at the floor. "A roommate."

None whispers, "After you gave me grief for living with MJ?"

"Mothers can learn from their sons. No one will replace your father, but I still get lonely. Perhaps my children could visit more often."

None pretends not to hear her suggestion.

"MJ told me you didn't come to the airport because you were on many dates," Mrs. Wong says.

"Aiya!" None throws up his hands. "You wanted grandchildren. I'm working on it."

Mrs. Wong imitates ringing a bell. "People shouldn't come when you ring your bell. They're not servants."

"The bell is for them to leave."

Mrs. Wong gasps and covers her mouth. "That's even worse."

"The easiest way to get over someone is another girl. I'm going to get way over Astrid."

"That one is not allowed in our home again." Mrs. Wong wiggles her finger at her son. "You act like you have no relatives."

"I'm president. Aren't you proud?"

"Proud and embarrassed. Even your blind grandmother listens to your foolish videos."

"How much more do you expect from me?"

Mrs. Wong says, "I always expect more from you. How did you achieve so much? Would you be here today if I let you be lazy? If I let you pick easy paths? No. I taught you to work hard, be honest, and do well in school. When your teacher told me 'this boy is stupid and cannot learn,' did I listen? No. My son is smart. That teacher is stupid. Your grandmother hisses at me when you embarrass the family. I never gave up on you and never will."

None nods and reflects.

Mrs. Wong says, "I no longer have room for your items. But, you are always welcome home. Is this what troubles you?"

None's chest tightens. "I'm worried about my speech."

"You talk so much on TV. How can you be worried?"

"Being president is the most important thing I will ever do. What if I fail?"

"My son learns from failure. My son pushes forward against setbacks. He will be an A+ president."

None opens his arms for a hug.

She grins and latches onto him tightly. "Do I have to coddle you like a child to know that I love you?"

He chuckles and releases her. She sniffles and wipes her eye.

None moves to his brother next. "Do you remember what I said after the Supreme Court?"

Shen rolls his eyes. "You're 12 things before you're Asian?"

None says, "I was flippant that day. I wrote a list of all my traits to show you. Half my traits reflect the values and culture Ma raised us on."

Shen says, "So, you're six things before your Asian?"

None chuckles. "I'm a hardworking, smart engineer who's good at math. I'm a walking Asian stereotype. I rebel against it, but I'm proud of my heritage. I'm sorry if I ever made you feel otherwise. We're underrepresented in corporate leadership and the halls of power. Tomorrow that changes.

Tomorrow I break through the bamboo ceiling, to the highest office. I'll add one more trait to my list, president."

"That's seven things," Shen says.

"Math," None says. They exchange smiles. "We're about to jump from basic math to calculus. Are you ready for the jump to light speed?"

"Always, Mr. President," Shen says.

Mrs. Wong complains about her boss at the hospital and tells stories about little James and Maria. She dragged a photo album to D.C. to illustrate her stories.

MJ's pulse races. She would rather forget certain childhood memories.

Mrs. Wong flips through page after page of embarrassing childhood photos. Each story gives Mrs. Wong more attention until everyone listens to her. "James and Maria knew each other a few months, and they were already inseparable. When they were five and six years old, Maria followed James around anywhere he went. One day they found an ant hill in a neighbor's yard. James stomped on the ants and walked off. Maria stomped on the ant hill and sat down, right on top of it."

Her parents shift in their seats at the fearful memory of that day. They know the story well.

Decades later, the memory haunts MJ. She grimaces and stares at None as if blaming him again in adulthood. Why did she have to follow him? Why did she sit down there? If only she'd worn boots.

Mrs. Wong says, "James kept walking. He didn't realize anything was wrong until he heard her screams."

MJ felt them crawl up her leg. Hundreds of fire ants swarmed onto her legs and under her skirt. They bit every inch of her. Ant stingers penetrated her to inject their venom. The searing pain overwhelmed her. MJ closed her eyes, tensed up from the surge of pain. Fear immobilized her. She sat on the anthill, helpless. MJ didn't realize what was wrong until she heard her own screams.

Mrs. Wong says, "James ran back to her, but couldn't tell why Maria was crying. Her screams at some unseen danger terrified him, but he had to help Maria. James lifted up her skirt. Hundreds of fire ants attacked Maria, hidden underneath."

Each word from Mrs. Wong makes MJ's memories more vivid. The fire ant venom burned through her body. Tears washed down her cheeks as she wailed on in agony. Her screams were loud enough to scare away a banshee.

Mrs. Wong says, "Maria couldn't move. All she could do was cry. James lifted her up and carried her to our house. He screamed for help."

If only she had boots on. The ants could have bit and stung the leather all they want. MJ was a dead weight, as None carried her. Her shrieks of torment must have left his ears ringing. She was a heavy burden, but he never complained.

Mrs. Wong says, "I came to the door. All James could say is ants. I ran inside to fill the bathtub. By now, the ants attacked him too. James yelled out in misery. I helped them into the bathroom."

MJ remembers None running out of stamina. His adrenaline wore off. His last few steps dragged.

Mrs. Wong says, "Maria held onto James and wouldn't let go. So, James jumped in the bath with Maria in his arms, fully clothed."

To MJ, that was the moment that connected them. Over three decades later, she'll fight Astrid or anyone else who comes between them. When he had nothing left, None spent his last bit of energy heaving them into that bathwater.

Mrs. Wong says, "Ants crawled for higher ground. They climbed towards Maria's mouth. James told her to shut her mouth, but she couldn't stop screaming. He dragged her underwater with him to drown the ants."

MJ gasps for air. She feels like she's drowning all over again. The details haven't faded over time. Some ants never let go and drowned in the tub. Others swam together. They used their bodies and surface tension to form a living raft. Ants that weren't submerged yet converged towards higher ground on her face. A swarm of ants crawled up her face towards her bulging eyes. When MJ thought she couldn't be more terrified, None dunked her. The water muffled her cries. MJ choked down dirty bathwater and a clump of ants. Choking coughs replaced her screams, as she struggled to breathe. Were the ants she swallowed alive, or dead? She imagined them crawling down her throat, their bites and stings assaulting her from the inside. Alive, or not, she still felt the burn of ant venom. Her whole body felt ablaze with hundreds of bites.

Mrs. Wong says, "Maria choked down some water, but she was fine. She was too scared to leave the tub. James stayed with her. They shivered in the tub together."

While underwater, the fire ants abandoned MJ like a sinking ship to join the living raft. New ants interlocked their mandibles and legs to grow the ant raft further.

MJ watched in horror at the growing ant menace she still shared a tub with. She splashed them to move them away from her. She had screamed in terror. "Kill them!"

Mrs. Wong says, "That day, my James was a hero. As president, he will be a hero again."

MJ remembers how None protected her. He attacked the ants with a shampoo bottle, a bar of soap, and a bath towel. He sunk the living raft and drowned the ants.

MJ grew silent and clung to None. She watched for any hint of movement, afraid the ants would revive.

Mrs. Wong coaxed MJ out of the tub and applied ammonia to her bites. The ammonia stunk like cat pee. The smell of ammonia still makes her think of that day. It's the reason she never got a cat.

The ammonia neutralized the pain from the venom that masked her itchiness. Mrs. Wong grabbed MJ's hands and explained why she couldn't scratch. She told MJ to leave the white bumps alone, or pus would ooze from a pustule, and they would bleed, leaving behind scars. The urge to stop the itching drove her mad.

Mrs. Wong says, "When Maria finally stopped crying, she wasn't a little girl anymore. I never heard her cry again. That day, she grew into the brave, strong woman we know today."

MJ's parents nod politely. She still follows None. MJ's parents worry she'll stomp on some new ant hill.

MJ loses herself in the past.

After that day, her parents didn't want MJ to play with None anymore and grounded her. MJ begged them for weeks before she could play with him again. It's hard to avoid a next-door neighbor. If she lived farther away, the ants would have ended their friendship.

That night MJ had screamed herself awake. Her childhood imagination ran wild. She had nightmares about ants crawling down her throat. They carved tunnels inside her and turned her body into their nest. They consumed her from the inside, transforming her corpse into an anthill.

Cooped up in her room, MJ vowed she would never be scared and helpless again. She would be the rescuer, the hero. MJ faced her fears. She sneaked into the garage to grab a gas can and a box of matches. She stormed over to the anthill. MJ stomped on the ground to antagonize the fire ants. When the soldier ants swarmed on top, she poured gasoline and burned them alive. She got grounded for fighting fire ants with fire, but it was worth it. She slept well that night.

MJ convinced her parents to let her take martial arts classes. She used her new skills to defend None against school bullies, just as he had protected her from a horde of fire ants.

MJ kept the scared little girl bottled up for decades, a prisoner inside her.

None associates the ants with his day as a hero, but to MJ they still represent her fear. When None first got an ant habitat, it took MJ a couple of weeks before she'd go in his office.

She kept her fears in check all this time until a gunman showed her how precarious her survival was. MJ yearns to bury her fear once again and become invincible.

MJ isn't sure how well she'll sleep tonight. She alerts None that it's past midnight.

"You ready?" None asks.

MJ nods and gives None a bro hug.

None and MJ say goodbye to their families and leave for Blair House. They're ready to stomp on anthills.

45. **Untapped Resources**

January 20th arrives, Inauguration Day. None sports a tuxedo with a Linux penguin bow tie and a top hat with goggles. MJ also wears a tuxedo, the long-tailed jacket tailored to her narrow waist, and black cowboy boots. She tops the outfit with fedora that has been augmented with clockwork pieces.

None and MJ meet outgoing President Hayfield and his wife for tea. President Hayfield's sweaty brow betrays his anxiety. This is less a transfer of power, more a salesman unloading a used car with bad brakes and a failing transmission.

President Hayfield accompanies None and MJ to the Capitol without words. When they arrive, None and MJ get out of the limo. President Hayfield closes the door behind them and drives off.

None shakes his head at the departing limo. He worries the government he inherits runs on wheels fastened with duct tape and Band-Aids. He must make Americans believe again before those wheels come off. Two million Americans want to believe. They fill the National Mall, back to the Washington Monument.

Loudspeakers introduce None and MJ. Over lush red carpet, they descend the Capitol steps down to the stage for their big moment. None and MJ stand and pledge in unison to uphold the constitution against enemies foreign and domestic. These political marriage vows bind them to the country and each other. None continues alone with the presidential oath.

President None of the Above.

Vice President More Jobs.

It is done.

A battery of Howitzers fires a 21-artillery salute with a flash, boom, and plume of smoke. Cheers, flags, and songs of patriotism greet the end of the Hayfield regime.

The President addresses the nation behind a podium with his presidential seal. A plastic shield protects him from chest level down along the stage. "Too many are out of work or can't find enough hours. Too many lost hope and gave up looking. Too many live paycheck to paycheck, without any savings. Those lucky enough to have jobs, cling to them, afraid of unemployment."

Spectators turn glum. They remember their own predicaments.

"Our economy and government crumble, as we watch. You're frustrated. You're scared. What did you do about it?" None points at the crowd. "You elected me."

His words remind the voters of their own courage. They dismantled the two-party system and made a protest candidate president. The audience cheers in self-congratulations, for him and for themselves.

None pounds on the pulpit. "I won for not being a politician. My presence here is a revolution. The top tweet when I won the election was, 'They can't be worse than the other guys.' Has government trained us to expect so little? The system is broken. Let's build a better system. Let's raise expectations. Expect more from me, more from your Congress, more from your government, more from yourself. Why does it take hours in a line at a social security office? It shouldn't take months to cut a check or approve a loan. Replace the lazy and incompetent. Reward those who deserve it.

"Government employees, remember the historic words of J.F.K. 'Ask not what your country can do for you, but what you can do for your country.' Americans lost faith in their government. Help me restore it."

Applause reverberates all the way back to the Washington Monument.

None says, "The U.S. Government invented moonshots, but we forgot how to think big. The biggest supercomputer is Chinese. The biggest particle accelerator is European. NASA astronauts relied on Russian rockets to reach the International Space Station. Let's remind the world of the America with big ideas. We'll turn education into our next moonshot and begin a technology revolution." None waits for applause to die down.

"What do moonshots have in common?" None tilts his head towards the crowd and cups his hand to his ear.

Onlookers yell out answers like game show contestants.

"I heard more jobs, new technology and…" None glances sideways at some hecklers. "Budget overruns. Those are all valid answers. My answer: computers and software. Any future advance will involve software. Tesla updates functionality of their cars, with software. Smartphones add new functionality with apps. What drives self-driving cars? Software. Nanotechnology? Software. Swarms of drones constructing buildings? Software. Genetic Engineering? Software."

None pats his chest. "I'm software." He points at the audience. "You're software. Your bodies are built on code called DNA. It's not just for catching criminals. DNA is the key to mapping diseases and how all life on Earth connects."

185

None widens his arms towards the audience. "We are software." He pauses to let the words sink in.

"Scientists can cut and paste DNA now. Soon we'll program life itself. How do we improve America? Software!

"For those who watched my TV show, *The Internet President*, I left one question unanswered: Who? America needs more STEM workers, which is science, technology, engineering, and math. We need new jobs, which means new companies. Innovative companies require entrepreneurs with new ideas and engineers to power them. We have a shortage of technical people to power those startups. We waste our brightest people on Wall Street. I don't consider finding new ways to bilk money out of investors innovation.

"Who won World War Two? It wasn't just soldiers. Scientists and engineers constructed atomic weapons that ended that war decisively, beginning the era of American dominance. Even rogue states, like North Korea, can reach out and attack in a digital world. We need to train a new generation of digital warriors. If you combine American creativity with Asian respect for intelligence and learning, we could be unstoppable. It will be scientists and engineers that herald the next age of American dominance. We won the space race and the arms race. America can't afford to lose the tech race."

None asks, "What is America's greatest untapped resource?" He pauses for answers.

Shouts ring out. Freedom. Solar energy. Americans.

None says, "My answer, women in technology. Engineers shape our world. The phone in your hand, the car that you drive, it's a world designed by men. What will the world designed by women engineers look like? If we tapped the potential of scientific women centuries ago, maybe we'd have the Internet 50 years earlier. Ada Lovelace was the first computer programmer, but only about 18% of engineering majors are female today. When I was an engineering student at UCLA, I remember going along the walls in Boelter Hall, looking at graduation pictures. As I walked through time, I saw one, or two female faces a year, then a trickle. Forward-thinking schools, like UCLA, expand their engineering schools and improve their diversity initiatives to address the problem, but it will take a national effort to solve.

"Every American female engineer is a symbol of freedom and rejection of the terrorist worldview. Islamic militants pursue a global campaign of violence against educated girls. In Afghanistan, men on motorbikes throw acid on girls for going to school. They poison wells at all-girls schools or blow them up with hand grenades. In Nigeria, Boko Haram abducted over 2,000 schoolgirls to be used as sex slaves or cooks. The most famous example was 276 girls kidnapped from their school dorms in Chibok. In Pakistan, Malala Yousafzai stood up to the Taliban for the right to be educated. She was shot in the head on her bus ride to school. Malala survived and became the youngest person to win a Nobel Peace Prize, at age 17."

Education as a weapon in the cultural wars strikes a chord with the audience. It can't be worse than current American policy. Bombs rarely bring peace, whether delivered from a drone, or a suicide vest.

Drawing the connection against extremists distracts from his core message. None waits for the multitude to calm. "To bring more women into STEM, we need to start young. Toys for girls used to send a clear message. Play with your dolls, little princess. Stick to pink. The sciences aren't just for boys. Kill that stereotype with girl targeted STEM toys. Once girls imagine themselves as engineers or scientists, the next step is encouragement. 66% of fourth-grade girls like math and science, but so few end up there without that encouragement.

"Maryam Mirzakhani was the first woman to earn the most prestigious math award, the Fields Medal. It almost didn't happen. A dismissive teacher damaged her confidence and interest in math. Only after a supportive teacher, did she go on to greatness. How many beautiful minds have we missed because women weren't encouraged? World War Two brought the necessity of Rosie the Riveter. I call on the best of a new generation to become Katie the Coder.

"Women aren't our only hope to find more STEM recruits." None imitates Yoda's voice. "There is another." He clears his throat. "Once we fix schools, poor neighborhoods are the next opportunity. What would the inner city look like if they looked up to engineers and doctors, instead of rappers and basketball players? Smart kids hide in poor areas, where they're harder to see for cultural and economic reasons. Minority students face all the same hurdles as women, plus worse schools in their neighborhoods.

"Boy, or girl, not everyone has the aptitude and smarts to be an engineer or scientist. To those who do, I challenge you to seek your full potential.

"I propose a national college system geared towards STEM. In return for scholarships, students will take government jobs for four years. The program will work like an ROTC for technical people. I talk about fixing government, but these young people will be the ones that do it.

"What will they fix in government? The social security website takes nights off. Are government computers lazy? Websites don't have normal hours of operation. That's at least one agency that doesn't understand the web." None facepalms. "The world doesn't run nine to five. This is not the 1980s. That's why new blood will revitalize government and bring it into the 21st century. We can run national hack days to drive innovation. Tear down the old ways. Efficiency is more important than tradition.

"What will we do with an army of engineers? Most of that brainpower will filter out into the private sector. America achieved so much with one Silicon Valley. What happens with enough skilled engineers to power 50

Silicon Valleys? We'll power a technology revolution that will make America's past accomplishments seem quaint. That's America's power-up. Don't just make America great." None stretches his arms out. "Make America awesome!" None rips the mic from the podium and slams it to the ground. He swivels and struts out.

Desperate to believe, the crowd latches onto his vision of tech-fueled prosperity. They cheer like their lives depend on it. None restores the American swagger, in spite of the looming economic crisis. America spits in doom's face.

46. **Toothbrush**

The Tea Party and Occupy Party attacked the special interests from the left and the right, but corruption already spreads into the new insurgent parties. It will not be easy to root out.

None's campaign transformed average Americans into an army of grassroots activists. They organized and shared ideas on the I Want My America Portal. An army of thought leaders from this movement follows None into government positions. His staff begins tomorrow. Conservative, or progressive, they all shared two things in common, a focus on solutions and a desire to take back America from the corporate-sponsored parties.

The inaugural festivities last into the night. None dances up a storm. His full dance ticket includes MJ, Corella, his mother, and an endless stream of well-wishers. The magical night fills with the hope of real change from the non-politician they elected as president.

None and MJ explore the White House. They tour the President's Bedroom where None will sleep. They stop by the Lincoln Bedroom nearby, where MJ will sleep.

The Lincoln Bedroom feels meant for royalty. From the gilded, crown-shaped canopy above the rosewood bed to the gilded frames that enclose Lincoln's portrait and the mirror over the fireplace. Golden exteriors suggest a thin veneer of wealth from another time.

MJ strokes the mirror frame. "Lincoln signed the Emancipation Proclamation in this room."

None and MJ stand side by side in front of the fireplace. History surrounds them.

She walks into the bathroom and leaves the door open. "When was the last time we had a moment alone without cameras or agents? The only place with any privacy is the bathroom. Will we get used to it? At least the leash is longer in the White House."

None steps over to the bathroom door. "At least you didn't have an embarrassing TV break-up. I haven't decided whether to forgive Corella for conspiring with Astrid, or not. What do you think?"

If MJ answers, she might say something to give away her guilt. If she ignores the question, he might get suspicious. MJ grabs the red lightsaber toothbrush and brushes her teeth. She mumbles a fake response.

None asks, "What was that?"

Leave a genius alone, and he'll answer his own questions. MJ mumbles again.

He hears what he wants to hear. None responds to her babbles like sage advice. "You're right, MJ. I should forgive Corella, but not trust her."

MJ smiles knowingly. Her ruse complete, she pulls out the brush. The red brush.

The wrong brush.

MJ isn't color blind. How could she not notice the difference between red and green?

Her pulse races. The vile, disgusting toothbrush falls into the sink. Every loathsome fungus and bacteria from None's mouth, they invade her. MJ drops to her knees and retches. Every meal he's eaten since Christmas rots in her mouth. Putrefied food particles continue their decay inside her.

None rushes to her aid. "Are you OK?"

The onslaught of microbes and germs overwhelm MJ. She pukes. The remnants of dinner splatter into toilet water.

None approaches to comfort her. "Is it food poisoning? I can get the doctor."

She pushes him back. Her stomach empty, she dry-heaves. MJ hopes the stomach acid killed the repulsive gunk. Prone over the toilet, she turns towards None. "That's your toothbrush!"

"The movers must have mixed them up. No problem. I must have yours." None picks up the red toothbrush and rinses it.

"What are you doing?" MJ springs up and jerks the toothbrush from his hand. She flings it into the trash.

"That was mine," None says.

MJ says, "Are you crazy? I used it."

None peers in the trash can. A plastic bag covers the interior. "Do you think that's a clean bag?" None mulls over his options.

MJ knows what lengths None will go to for Star Wars. MJ says, "You can't share toothbrushes. I'll buy you a new one."

"Is that what this is about?" None chuckles, relieved that it's nothing important.

"You can spread blood disorders like Hepatitis, or HIV, not to mention the bacteria for strep throat. Then there's all the nauseating muck and the infections I can catch."

None shrugs and grins. "I'll bring your toothbrush over."

MJ says, "No. I want a new one, sealed in plastic."

"Overreact much?" None asks.

"Leave." Just because he doesn't understand her feelings doesn't make them any less valid. MJ nudges him out of the bathroom.

None stomps off to explore his new home.

47. **Law is Awful Software**

None and Shen shepherd legislation through Congress. Congress remembers how None got voters to flood their offices with angry letters and emails during the government shutdown. That was *before* None took office. How many voters could he organize against them as president? Members of Congress perceive *The Internet President* TV show as a weapon. They fear what a fully armed and operational None of the Above could do. Fear keeps them in line. Congress puts aside partisan feuds long enough to deliver their version of his platform.

MJ redesigns the Oval Office. She combines elements from the Clinton, Bush, and Obama era carpets. The centerpiece remains the eagle seal with olive branches in one claw and arrows in the other. Golden sunbeams radiate outward on a deep navy background with inspirational quotes around the periphery.

After three weeks, both projects finish. None and Shen peruse the new furnishings in the Oval Office. MJ leans against the Resolute Desk, made from timbers of the British ship *HMS Resolute*. It was a gift from Queen Victoria in 1880.

None says, "You didn't use the cyborg eagle shooting lasers from its eyes."

MJ shakes her head.

"I only asked for one thing. One thing. Did you think I wasn't serious?"

MJ deadpans. "I knew you were serious."

Agent Vincent enters. "Senate Minority Leader Robertson and Senate Majority Leader Garfield are here to see you."

None asks, "Wasn't Robertson the Democratic candidate for president?"

"Yes, and Garfield was the Tea Party candidate," Shen says.

"Do you think they'll hold a grudge?" None asks.

Shen says, "They're senators. I'm sure they're refined gentlemen."

The two senators carry hefty stacks of papers.

"If it isn't the graffiti artist and his henchmen," Senator Robertson says. MJ clears her throat.

Senator Robertson says, "And henchwoman. Senator Garfield and I are not fans of your handiwork. You turned your supporters into hoodlums."

"At least I engaged voters," None says.

"Screw you and the Internet you rode in on," Senator Robertson says. "I paid my dues. You can't skip to the front of the line like this. It's my turn."

None says, "I can, and I did. Innovation and communication. Try it sometime."

"In a show of bipartisanship, we're delivering the rope you asked for," Senator Garfield says.

"Rope?" None stretches his hands out in confusion.

"So, you can hang yourself." Senator Garfield plops a 1,183-page document in his hands.

Senator Robertson hands MJ a copy of The More Jobs Act. Her eyes light up when she sees the final title for the bill.

"Government is the problem, not the solution," Senator Garfield says. "Your proposals have the stench of big government. You have the poll numbers, for now. We'll let you pursue this fallacy of efficient government. When it fails, we'll blame you. If by some miracle it succeeds, we'll claim credit."

"Heads we win, tails you lose," Senator Robertson says.

Senator Garfield says, "The Internet made you. It can unmake you. You'll slip up. I'll be in the front row watching."

"Then we'll crucify you, and you'll never pass another law," Senator Robertson says. "You might do what you want to our faces on a sign, but there's nothing you can do to them on the floor of the Senate. We'll be here after you're gone."

None drops the bill on his desk, then claps. "Great, so you've learned to work together. I'm glad I could help." None points to the door.

The senators expected a different reaction. They leave in a huff.

Shen types on his laptop. "We got almost everything we asked for. Adaptonium, your education platform, had overwhelming support."

"How much did they cave on election finance?" None asks.

"Campaign ads will show corporate logos of who's sponsoring them," Shen says.

None fist pumps. "Political ads will look like NASCAR races with all those logos. How much were we able to raise the minimum wage?"

Shen says, "Your political capital has limits. That's the one major area where I couldn't get traction."

None throws his hands up. "Aiya! That was a critical part of the plan."

MJ says, "We got the jobs package. We'll be OK. More jobs mean employers will have to raise wages to compete for workers."

None nods.

MJ attempts to read the impenetrable document named after her.

Shen says, "I can't believe they caved. Just sign it, and it becomes law."

None scowls at Shen. "You expect me to sign something I don't completely understand, after what happened last time?"

"Like sausages, you're better off not seeing how laws are made," Shen says.

None reads the bill. "What is this crap?" He flips through and reads a different spot.

Shen takes on an air of superiority. "You're not a lawyer. It's complicated. Don't expect to understand everything."

"It's just code. This part changes some other law." None reads aloud. "The Social Security Act 42 U.S.C 1396a...is amended in subsection a by striking 'and' at the end of paragraph 75, by striking the period at the end of paragraph 76 and inserting a semicolon. That's crazy! This is the worst example of turn by turn instructions I've ever seen. You have to grab a copy of the Social Security Act and put on a magic decoder ring, just to make sense of it. Imagine if novelists didn't give you their book but gave you instructions on how to modify Moby Dick into their book, by hand. Awful!"

None chuckles. "I know what this is. It's a software patch using diffs. Aiya! Don't they even have version control? No wonder Congress doesn't read their own code."

Shen asks, "Version control?"

"In software, we solved this problem decades ago. Create a file, version one. Make a change, version two. Version control keeps track of all your changes. We call these differences a diff."

"I'm not a coder. How does that help?" Shen says.

None says, "Let's imagine Robertson used version control. Instead of giving us lame instructions, he makes his changes and saves a new version of the Social Security Act. Boom. Now I just look at the before and after and see the changes in context. You don't say change this semicolon and remove 'and.' You say look at version three. Check out my latest changes. You don't tell them what changes to make. They see it."

"Version control sounds fantastic." Shen shakes his head. "It'll never happen. There's power in controlling information. If you make the system easy and transparent, it's not in their vested interests."

None says, "We've only been here a few weeks, and I'm already sick of vested interests. I should vomit on a few lobbyists and see if that chases them away."

MJ says, "I doubt it. I warned you day one. The government will change you, not the other way around."

"I came here a dreamer. I'll leave here a dreamer," None says.

"Stay awake, or this place will mangle you," MJ says,

None ignores her attitude and resumes reading. "Shen, what does educational software have to do with a factory in New Hampshire or a Naval base in Florida?"

Shen says, "The easiest way to pass something is to amend an unstoppable bill, like an AMBER Alerts bill. The More Jobs Act is too big to fail, so everyone piled on."

MJ flips through pages. "I guess that explains why it's so long."

None says, "Laws only focus on the deal. They never consider the rest of us stuck with their convoluted crap. Well-designed software uses separations of concerns, to split a program into modular components. This is the opposite, like steak in a blender. Calling this crap sausage would be an insult to sausage. In software, the more complex the code, the more bugs. I can't imagine the law is any different."

MJ chuckles to herself and fiddles with her phone. "Check this out." MJ underlines a section of the bill. "There's a tax exemption for existing charter school chains with more than 3,000 schools based in Delaware."

"That seems strangely specific," None says. "How many companies could that be?"

Shen says, "Almost half of corporate America incorporated in Delaware. With tax loopholes and business-friendly laws, it's the Cayman Islands of America. You can create shell companies with little information. I'll find out."

While Shen brags about his skills, MJ researches on her phone.

Shen says, "Businesses hide behind shell companies, but I find their footprints in real estate transactions. They can't hide assets from me."

"Tell me it's not Rainbow Unicorn," None says.

MJ looks up a list of largest charter schools. "Rainbow Unicorn Education Centers is the only match."

"Who makes a law that only exempts one company?" None asks.

Shen says, "Sometimes lobbyists write bills and lawmakers put in their wording verbatim. When lobbyists write a law, they leave themselves presents."

None says, "That's like letting murderers write the police reports on their victims. Conflict of interest, anyone? So, Renquist and his spy, Faraday, built themselves a competitive advantage with my legislation?" None pounds his fist on the Resolute Desk.

"Corporate sponsors pay Congress to sabotage laws they hate," Shen says.

None says, "Intentional bugs, unbelievable. In tech, we have QA teams to find bugs. In Bizarro World, Washington tries to hide their bugs in laws too long to read."

Shen waits for None to calm down. "It's the best we can do. Take the deal."

"This sausage is rotten. This law is against everything I stand for, even as they give me everything I asked for. Is this a test? Am I selling out if I sign my name?"

"We have to be practical," MJ says.

"The system rewards bugs," None says.

Shen says, "You can't change the system."

None smolders with anger. "Yes, we can. We start with version control. Shen, maybe if you used version control to track changes in my contract, we could have checked out a diff to see his changes." None narrows his eyes at Shen. "Oh look, Renquist put in a new clause that lets him fuck me out of my own company!"

Shen slumps onto the couch. He runs his fingers through his hair and bites his lips. "I'm your brother, not your chew toy. I need you to get over Slam Dunk Jobs and Adaptive Unlimited."

"I'm not sure I can." None downshifts his anger. "I'll try. I won't give up until I get my company back."

MJ says, "Focus. Focus. Fear got you elected. If we don't stop the money Apocalypse, we'll be first in line at the guillotines. You can't solve every problem."

None sighs. He writes notes on his phone. "Fine. I'll push version control later, but I have to do something. I gobbled the power-up. I'm president. When does it get easy?"

48. TL;DR

None slurps down a slice of chocolate cream pie in the Oval Office. He chews on the flaky crust with perfect texture. It's easily the second-best pie he's enjoyed. The chocolate explosion in his mouth should erase his sadness, but it doesn't. It's Pi Day, one of the biggest geek holidays. On March 14 (3/14), lovers of math celebrate the number pi (3.14...), by eating pies. Astrid created her own pie Internet memes. She should be celebrating with him. He was good pie with Astrid. How could she say otherwise? None would trade the daily treadmill of lovers to get her back. On his runaway train of a love life, 30 potential matches remain. Any competition has losers. The only certainty about his future is 29 more disappointed women.

The runaway train of his presidency races alongside his dating dilemma. None takes immediate steps to implement his new programs and calm a desperate nation. He musters government and private resources to move quickly. National colleges rent cheap office space to train a new generation of engineering talent.

None open-sources Adaptonium, the education platform. Thousands of programmers contribute to the new software. Student tests show impressive results. Educational companies ready their software to run on the new platform. Nationwide, schools scramble to upgrade their network infrastructure.

The Invest in America Program opens hundreds of business incubators across the country. Private and public money fund thousands of new companies selected on the IWantMyAmerica website. New companies create new jobs. Those jobs power the economy better than qualitative easing, or any other Fed financial engineering. Unemployment drops four percent by the end of July.

The new budget buys the president another year before the next shutdown battle.

Stock indexes plummet in emerging markets, China, and Europe. The United States, as a rare pocket of strength, attracts flight capital, scared foreign money looking for somewhere safe to invest. When everyone is scared, it's a battle of the uglies and America is least ugly. American stocks shoot to new highs, as other world economies falter. Global markets talk about the miracle of the American oasis, the only thriving economy. After the unprecedented success of his programs, None's ego grows large enough to cross state lines.

Secretary of Education Faraday reclines with his arms spread out, owning one of the two couches in the Oval Office. Faraday shows his palms. "Don't worry. My hands are empty this time."

None, Shen, and MJ crowd into the opposing couch, separated from Faraday by ideology and a coffee table.

None obsesses about Faraday's black cowboy hat with a jewel-encrusted belt along the rim. In software, black hats mean the thieves and troublemakers that attack computer systems, the pirates of a new age. It galls None that Faraday would be Renquist's spy and flaunt such a symbol. How dumb does he think they are? None fumes, as Faraday drones on about the first school trials of Adaptonium.

Faraday says, "There's not much content yet, but I'm impressed. I'm advertising it for my schools."

"You probably leave no dollar behind," None says.

Faraday takes it as a compliment. "That's about right. It's how you run a good business."

None looks Faraday up and down. "Nothing about you screams Delaware."

Faraday sits up. "So, you know." Faraday smiles and tips his hat.

"I read the law, the whole thing." None leans forward and glares at Faraday. "The rules don't apply to your schools. I'll cancel your advantage."

Faraday stares right back. "No, no you won't. Do you really want to highlight to the American people that a member of your administration gets special treatment? Who's to say you didn't give it to me? I might confess."

None digs his fingers into the couch armrest. MJ winces.

"You gave us free markets. That's what we wanted. If I don't have to take a sledgehammer to every public school to get it, I can live with that. You can label the privatization movement evil, but we're not." Faraday stands.

None jolts up and lurches forward, bumping his shin on the coffee table. "When you lobby an advantage, that's not a free market."

"We don't have to like each other, it's just business. I'm your champion, whether you like it, or not." Faraday tips his hat. "I know the way out."

Faraday leaves as the next appointment, Senator Garfield, arrives. They exchange knowing glances as they pass.

Senator Garfield marches towards the couches. "Who are you to summon me? You're not in the Tea Party."

"I'm your president. If you're going to claim credit for my ideas, you at least need to do some of the work." None hands a ten-page printout to MJ, who hands it to Senator Garfield.

Senator Garfield flips through the document. "What's this?"

"It's a prioritized backlog of topics to address, based on votes at the IWantMyAmerica Portal," None says.

"I didn't ask for this," Senator Garfield says.

None says, "We can't just ignore what they voted most important. It sets expectations that changes will be made."

Senator Garfield balks. "We aren't making any changes. Expectations you set are your problem. I only care about priorities for my donors." Senator Garfield steps towards the door.

None says, "I emailed you some legislation I want introduced."

"You're overdrawn on favors," Senator Garfield says. "When you see the numbers I have, you'll know the American oasis is just another patch of desert." Senator Garfield rushes out.

His ominous words disturb None. What does the senator know that he doesn't? Is someone withholding data from him?

"If Garfield really had something, he would gloat. I'll see what I can find out." Shen leaves to consult his staff outside.

MJ addresses None. "Everything about our lives changed, but you're still acting and talking the same. We should all call you Mr. President."

"Sure, normal people can, but I don't expect you to, MJ."

MJ says, "As president, you have to be more diplomatic. You shouldn't even insult Robertson and Garfield, at least to their face."

None asks, "Is that the same speech you gave me when I became a real CEO?"

"Basically, but you're president now. Everything I said is truer."

None chuckles. "If I didn't listen last time, why do you think I'll listen this time?"

MJ says, "Your actions reflect on an entire country. This isn't tech. There aren't free passes for being a genius."

"I made an oath to myself as a child, MJ. I'll never grow up. I've seen how these 'adults' lose their imagination. I'm nothing without my imagination."

MJ tries to argue that adults can keep their imagination, but it's a wasted effort.

"MJ, it sounds like you're trying to turn me into a politician, but I'm not selling out," None says.

"No, I'm asking you to be an adult. A presidential adult."

None ignores her comment, then gets lost in thought. None mulls what Senator Garfield meant by his 'patch of desert' comment, while a camera crew sets up for the presidential address.

MJ steps up to None, but he remains lost in his head. She waves. Nothing. MJ clears her throat. "Do you ever miss our old life?"

The interruption startles him. "I hope the idea I just lost wasn't important. What were you saying?"

"Do you at least miss Beer Beakers...the way it was?"

None says, "That's not our life anymore..."

MJ rubs her wrists. "A shadow of death wiped away our happy memories."

None say, "I noticed." He eyes her wrists. Her goth steampunk seven-buckle gloves extend up to her elbows. They cover her scars, but they remain a symbol of pain. None points to his head. "My scars are in here."

MJ says, "You're president. You can help Beer Beakers."

None eyes the eagle on the carpet to avoid her eyes. "What happened at Beer Beakers wasn't our fault. It's survivor guilt talking."

"Our favorite bar become a crypt. I can't leave it like that." MJ steps inside his comfort zone. She blocks his gaze and clasps his hands. "I need this."

None meets her eyes. The warrior in her died in that place. "You used to be my rock. I've seen you struggle. I'll do what I can to help."

MJ grips his hands tighter and smiles.

None asks, "You know what I miss most?"

She beams at him.

"Coding," None says.

Her face drops. "You miss...coding." MJ slips her hands away.

None looks up, not at her. He smiles, nostalgic. "I miss all of it. Movie nights at your place." His eyes descend to hers. "Building a company and a life with you. That much doesn't need to be said."

"Yes, it does. I need to hear it. Your brain doesn't come with printouts."

None spreads his hands over the Resolute Desk. "I really miss a computer at my desk. Meetings are my life now. Look at my empty desk. It doesn't look like I work here." None pounds on the desk. "I'm not using the study anymore. I want the American people to see their government doing something."

None rushes down the hallway into the study, MJ on his heels. He hands a stack of paperwork to MJ and grabs an LCD monitor with its cords. They lug the items back to the Oval Office and deposit them on the Resolute Desk. None organizes stacks of paper and plugs in his laptop. "Now that's a busy desk," None says.

When the camera crew is ready, None steps up to the podium in front of his desk. "My fellow Americans. The country mobilized quickly to implement the vision I shared with all of you. We focused on three core areas: more jobs, improved education, and the economy. Unemployment is down. A tech boom now grows beyond Silicon Valley and the tech hubs.

You don't have to come to California to start a tech company. The Invest in America Program launched thousands of new businesses and showed a profit in less than six months. Amazing. Startups we sponsored failed, but some large successes made up for them. Our education platform will be ready next school year, and graduates from national colleges will transform government a few years after that.

"Months ago, we asked Americans to prioritize a national backlog. I showed your list to some Congress members and none of them will help me work on your goals. Their priorities come from donors, not the people. Will they talk about lowering the deficit? No. Will they talk about health care? No. Voters, remind them who their boss is." None points at the camera and nods. "Yeah, you."

"Should Congress read the bills they write? Yes. Should Congress write bills that are short and clear? Yes. Today I proposed a new law to Congress, the TL;DR act. It's Internet slang for too long; didn't read. For you pre-Internet types, brevity is the soul of wit. That's from Hamlet. I wonder how many people just Googled Hamlet.

"If a bill is longer than 100 pages, the entire bill must be read aloud in a congressional session before a vote to become law. That's the whole law.

"I want to reverse the trend of longer, more complex laws. In 1933, an important law, like the Glass-Steagall Act, was only 37 pages. A partial replacement, the Dodd-Frank bill of 2010, was 848 pages, which spawned over 22,000 pages of regulations. Big laws hide loopholes and create much bigger regulations. If the Constitution were 1,000 pages, there would be no America. States would revolt under the burden of all that complexity.

"Congress can still write 1,200-page bills if they're willing to self-filibuster. I'll sleep better knowing someone reads every new law. I would love to explain how we can improve the process of creating laws with version control and consistent electronic formats, but in honor of TL;DR I will keep this speech short. Goodnight."

49. Intervention

His short speech leaves time for more nightlife. None daydreams about his evening plans. Ten paces outside the Oval Office, Shen, and MJ stride toward him, flanked by agents.

MJ carries a stack of magazines and newspapers. "Bro, we need to talk." MJ opens the door to the adjoining Cabinet Room.

It's never good news when you hear those words, especially on the way to a hot date. "But I'm going out with Zelda."

Shen says, "And four hours after Zelda Remington, you have a date with Sarah North. I'm familiar with your schedule. It'll take five minutes."

"You have three." None hisses air and follows Shen into the room. He stands by the long table where the entire cabinet meets.

MJ closes the door behind her. "Voters elect married people, so they don't have to see their president on TMZ." MJ lays down periodicals at each seat, reading headlines. "Player President. Dater-in-Chief. American harem. To date a president. The race to first lady."

None grabs a magazine from the table. "I'm on more covers than Astrid." He slams down two fingers with a sizzle sound and flips through a magazine.

"What we're trying to say is…" Shen chokes up. He turns an uncomfortable red trying to admonish his brother.

None talks into the magazine. "This is a great picture of Zelda. She's going to love it."

MJ says, "If you're going to be this promiscuous, be discrete. Keep it in your pants, or you'll compromise everything."

None folds down his magazine. "Discrete? You mean like how you put my dating profile up on the Internet?"

"You fell apart after Astrid, I had to do something," MJ says.

"I wouldn't be mass dating right now if Corella and Astrid hadn't made my love life a public disaster," None says.

MJ says, "Your man slut routine raises red flags with the Christian right. They judge on morals first, platform second."

"This round I'm down to only 15 women."

"Only," MJ says.

The conversation mortifies Shen. He sits and turns the back of the chair towards them.

"We're all consenting adults. I'm a completionist. I have to explore every scenario."

"That works for video games, but not people," MJ says.

"What worries me is, what if it's all for nothing?" None dead stares at MJ. "What if the one I'm supposed to be with isn't one of them?"

MJ avoids his eyes and pretends to read a magazine.

Did MJ assume None meant her? Maybe he did. No, he must have meant Astrid. The Astrid wound sears deep. None self-medicates with other women. It dulls the pain but disconnects his feelings. "It'll be over soon."

MJ puts down the magazine and looks him in the eye. "How do you know?"

"Either I fall for someone, or I run out of choices." No choices. None shudders. "Not to change the subject."

"Please do." Shen turns to look at them.

None says, "Now that I'm president, can government lawyers help me get back Slam Dunk Jobs?"

Shen gasps. "That's a massive conflict of interests."

"Can they investigate Slam Dunk Jobs or Renquist?" None asks.

Shen says, "Unless we act in the official capacity of our government roles, it's a misuse of position and government resources. We'd break the law."

"There must be some government reason to justify an investigation." None mulls over ideas for several minutes, but nothing fits.

Shen can only wait so long. "Any ideas?"

"Not about Slam Dunk Jobs, but I've got other ideas," None says.

"About?" MJ asks.

None turns to MJ. "I know how to help Beer Beakers. You're going to sleep better, MJ. Pack your bags. We're going on a guilt trip."

The Secret Service readies Air Force One for a trip to L.A.

50. Guilt Trip

The guilt trip lands at LAX and travels North to Santa Monica via motorcade. Police block freeway on-ramps along the route, a get out of traffic free card. None and MJ exit from the bulletproof, bomb-resistant presidential limo, The Beast. A canopy covers their approach inside.

None waves for Shen to follow. "Are you afraid of ghosts?"

"No. I'm afraid you'll ask me to pay another bill," Shen says.

None laughs at the memory. "Pay the man, Shen."

His brother rushes to close the armored limo door.

None and MJ leave Shen behind. It's not his guilt trip anyway.

The Blacksmith waits at the entrance with crossed arms. "The agents you sent to chase off more customers are doing a bang-up job." The Blacksmith golf claps. "Bravo."

Major bar sections lie fallow. Few steampunks remain. A score of dark tourists crowd around a daylight seance at one of the memorial tables. The dead aren't particularly talkative at the moment.

None says, "We're sad about what happened. There has to be a way to turn it around."

The Blacksmith chuckles derisively. "I have an offer on the bar. The land is worth more than the business, thanks to you. Once I sign, this will be hipster condos."

None says, "You can't sell. We came to help."

"And make amends," MJ says.

"Fine, I get it. You want to ease your conscience. I forgive you. You can leave now."

"Join the L.A. tech incubator," None says.

"Does this look like a tech company to you? This is a bar. We serve alcohol."

None asks, "What about your brewery?"

"I had to stop production. Look around. Do you see happy customers?"

None says, "I see a brood of brooding customers drinking your American brewed home blend. Happy, or not, they drink your beer."

The Blacksmith scowls at None's wordplay. He strokes eight black armbands on his left arm. "This place is dead and so is the vibe. You destroyed my bar. I don't need your advice."

MJ speaks with conviction. "That burn station was real. If that were decorative, my scars would be worse. Your imagination and attention to detail created a special atmosphere here. Maybe all you need is a fresh start."

"Here's an idea for the incubator." None's face lights up with the wonder of a child. "With some automation, you could do customized brewing. Customers could create their own recipes and drinkers can enjoy their experiments."

The idea sparks a grin on The Blacksmith's face. After 15 minutes of brainstorming, he overflows with excitement. Mission accomplished. The Blacksmith struts towards his office to research his new venture.

None and MJ pile back into the limo. Upbeat, they stop for a photo op at the Los Angeles incubator. The motorcade picks up authentic Chinese food in San Gabriel Valley and stops at Mrs. Wong's house. The surprise visit provides the president ammunition for next time his mother guilt trips him for not visiting.

The agents in the advance team secured the house before arrival using None's house key. They have ten minutes before Mrs. Wong arrives from work. The shock of open spaces greets None as he walks in. The dining room table. Gone. The matching chairs. Gone. The TV. Gone. Was his mother robbed? No. Her new roommate took over None's room and expanded into the living room. His bamboo mats. His Buddhist shrine. His coffee table. It can't replace the dining room table he got decades ago to feel more American, more like MJ. His things, his taste, dominate the room. Only the plastic covered couch and tattered Bible remain of the living room he remembers. His mother lets a man discard his childhood memories.

None surveys the rest of the house. His mother's room feels barren. Her missing bed left carpet indentations. This controlling man, this boyfriend, couldn't let her keep her own bed. His mother's words replay in his mind, "No one will replace your father, but I still get lonely." Why does her dating again disturb him so much? None imagines what kind of man it would take to subdue a strong woman like his mother. Is he a brute, or master manipulator?

MJ says, "A minimalist aesthetic, very nice. I need to redesign my house when we get back from Washington."

Her words wake None from his thoughts. None swivels to Shen. "Aiya! Can you believe she's living with her boyfriend?"

Shen's eyes widen. "That doesn't sound like Ma. She said roommate."

"Friend doesn't always mean friend." None points to the missing bed repeatedly. "Two people, one bed. His bed."

It's clear None made his point from the discomfort on Shen's face. "It's been a long time, so I see...We should head back to Air Force One for an urgent..."

None says, "It's hard to think of Ma having needs, after all these years."

"That thing...very important." Shen garbles words as he flees the conversation and the house.

"Your mother has a right to happiness, even if that works differently than you want." MJ speaks with a consoling tone but keeps her distance.

None spent most of his life without a dad. Why would he need one now? He kneels down to examine the wilted money tree in the Southeast corner. Her new relationship distracts her too much. She neglects her plant. The money tree might die, a casualty of love.

A new man took over his mother's life. None doesn't want to discuss that topic with MJ. He doesn't even want to think about it. None used to talk to MJ about everything. He noticed a subtle change, since Christmas. Dating became an uncomfortable topic. None waits out the silence, a few more minutes until his mother arrives.

Mrs. Wong opens the door and immediately scolds None. "I saw your brother on the way in. What did you say to make Shen acts so strangely? He came all this way and can't spare time to talk to his mother."

"We can't stay long," None says.

"You came so far. Let me cook for you."

"That sounds wonderful." MJ grins at Mrs. Wong.

Mrs. Wong glances inside the bathroom and slams the door shut. She peers back at MJ. "If you need a bathroom, try the restaurant down the street."

MJ curses Astrid under her breath. "I'll go get Shen." MJ leaves mother and son alone.

None fixates on the spot where the TV used to be. Can he get his mother to talk about her boyfriend without bringing it up? "Did you see my speech?"

"Yes. Why was your speech so short? Did you forget the rest of it?"

"I was making a point about short laws." She saw the speech. He caught her. "How did you see it without a TV?" He asks with a smug grin.

Mrs. Wong holds up her phone. "Did you know you can see video on phones now? I don't need a TV."

None crosses his arms. "I did know that." Of course, he knew that, he's an expert. Anyone else who questioned such basic knowledge would immediately join his enemies list. None follows his mom into the kitchen.

Mrs. Wong places dumplings and cabbage in a steamer. "The magazines at the nail salon say you have a big, big harem now. A harem might bring me grandchildren, but it won't bring you love."

"I don't have a harem. You have to sample many dishes to know your favorite meal. Do you understand, Ma?"

"Sample too much and you'll get a stomachache."

He doesn't have time for this. His mother won't open up unless she wants to. Be direct. "Enough about my dating. What's going on in the living room? Your roommate took over like it's their house now."

Her mood chills the room. "When two people share a home, they must each have a voice. Would you have me silence theirs?" Mrs. Wong speaks with finality. The question is not meant to be answered.

MJ returns with Shen in tow. They join None on the floor to eat.

Mrs. Wong finishes the meal in silence and places it on the coffee table.

MJ and the Wongs dip their dumplings in soy sauce. Steam rises off the cabbage. A cold pall falls over the conversation that a hot meal can't thaw. They find comfort in the food, but not each other.

51. **Anomalies**

American economic data continues to impress for a few more months until anomalies appear. Durable goods orders and manufacturing numbers slip, forecasting an economic slump. Since companies fire workers and freeze wages when business slows, the next dominoes to fall will be jobs and eventually wage growth. Workers don't accept pay cuts easily. It's normally easier to fire employees than lower their salary.

None charts the latest data onto a 55" 4K Monitor at the Resolute Desk. The numbers don't add up. None looks perplexed. "Unemployment and wages both dropped. More competition for fewer workers should raise wages, but it's not happening. It doesn't make sense."

Shen says, "Mr. President, you can delegate. We have staff to analyze the economy."

"I can't wait for reports. My algorithms work faster."

Shen says, "Maybe the wage drop is what Senator Garfield meant when he said 'the American oasis is just another patch of desert.' He's your 2:30 appointment."

None reads the time from his computer. "We have five minutes. I want answers before the senator arrives." None puts his feet on the Resolute Desk and leans his chair back. He ponders scenarios. "National colleges shrank the workforce. We know incubators created new companies and new jobs. Why aren't wages going up?"

"Maybe the new companies create mostly low paying jobs," MJ says.

None says. "Nope. Tech jobs pay well above average, even startups."

MJ gasps and covers her mouth. "Robots?"

None drops his feet to the floor and jerks up in his seat. Fear spreads across his face. His eyes dodge back and forth. He calms and swipes his hand dismissively. "No. Not yet. We'd see a spike in unemployment. It's something else."

No more ideas present themselves before Senator Garfield enters the Oval Office. His strut and smile ooze with bravado. "Mr. President, I wanted to talk to you about the H-1B reform bill."

None says, "It's a great program. It lets us bring the smartest people from the rest of the world to America on H-1B visas. American companies should hire these workers directly, not through outsourcing firms. Does the new bill crack down on that loophole?"

Senator Garfield pulls a piece of paper from his back pocket with "TL;DR" stamped on it. He unfolds it. "There wasn't enough room. You wanted a short law. You got a short law." He winks at None. "It's one sentence long." He flings the paper on the Resolute Desk.

None grabs the paper and reads it. "All you did was triple the number of H-1B visas. How is that a reform?"

Senator Garfield mocks outrage. "You are downright ungrateful, Mr. President. This is the second time you've been delivered what you asked for. You lit a fire under the entire tech industry, with software wages in the stratosphere. Startups can offer stock options, but older companies can't compete. Didn't you expect a backlash?"

None says, "We're working on the supply side with the national colleges. You'll have more engineers."

"Tech giants can't wait, or they'll be slaughtered in the marketplace. There aren't enough American workers with the required skills." Senator Garfield stares None down from the other side of the desk.

MJ grabs the paper. "You can't sign this. Do you remember what I changed my name to?"

"More Jobs," None says.

"Right. My name is More Jobs. You'll sign away American jobs with that bill. Jobs is our platform."

Senator Garfield gives MJ a wide berth. "If you don't sign, you'll show a lack of faith in tech, the centerpiece of your recovery plan. You'll publicly say all those engineers and scientists aren't necessary, after all. The last bill I gave you was the rope to hang yourself with. This one is the noose. I suppose it's time for your hanging."

At the mention of hanging, Agents step from the background towards Senator Garfield to make their presence known.

"We're done here," Shen says.

Senator Garfield wags his finger. "That's a metaphor, not a threat. Whatever you decide."

None says, "You run the committee. I know you could add protections for American workers."

An unseemly grin spreads across Senator Garfield's face. "I'm a sporting man. Beat me in an immigration debate, and I'll put something together. We can do it at ground zero for your shenanigans, the L.A. incubator."

None extends his hand to shake. "Sure, we'll do it at high noon."

"Now you're getting it." Senator Garfield shakes hands. He winks at None on his way out. Silence prevails until the door shuts behind him.

"Promise me you won't sign it," MJ says.

"He can spin this against us either way," None says.

"See, it's a good thing you're not signing it," MJ says.

"I have to sign it. I'm the child of immigrants. Do you want to keep the next *me* from being American?"

52. **Judged**

None and MJ prep for the next episode of *The Internet President* backstage at the D.C. studio, surrounded by their entourage. MJ wears her curve-hugging power suit. Her scandalous black bustier attracts attention from agents and Corella's staff.

Agent Vincent seems like he wants to say something but remains silent.

MJ looks over her shoulder. Nothing there. "I feel like I'm being watched."

"You're always watched when you wear that," None says.

MJ says, "Corella fought against a version control show for weeks. Do you know what changed her mind?"

"I guess I wore her out." None gives a thumbs-up.

"No. Fixing the law is a snoozefest. She said my power suit would help the ratings enough to do it. I even filmed promos in it." MJ curtsies. "You're welcome."

"Thanks, I guess." The crowd stomps in anticipation, louder than ever before. Excitement greets her outfit, not his words. She can have their eyes if he gets their ears. What would it take to get the average American engaged in the political process? Bikini girls from a boxing match?

What changed at Christmas? None yearns for the days where no topic was off limits with MJ. He pushes his luck. "I'm down to five girls now. I could see a future with any of them."

"Great." MJ scrunches her nose and grows distant.

"If I can't tell you about my hookups, you're not much of a bro," None says.

MJ straightens up and braces for every gory detail. She hides behind a facade of smiles. "Give me the rundown. Who's left?"

"Sarah is my anchor. She's a no drama girl."

MJ says, "Sarah, the yoga instructor? I like her. You need someone to calm you down if you want to survive the next eight years."

"Lashay. Every time—"

MJ's eyes bulge out. "The Lashay? Another one name supermodel?"

"Yeah." None hangs his head. He feared she might freak.

MJ drops the happy facade. "No way. Don't marry a party girl. Did you learn nothing from Astrid?"

"It's chemistry."

"That kind of chemistry explodes in your face." MJ pretends to ring a bell.

"You aren't ringing any more bells." None grits his teeth and resumes the list. "Betty."

MJ blank stares.

"Betty, the waitress."

"There wasn't any Betty that stood out," MJ says.

"She has a particular set of skills." None beams.

MJ grimaces with disgust.

None teases her. "I have a mystery lady I'm not telling you about."

"Now you have to tell me." MJ smiles, intrigued.

"Maybe I just made it up. Last is Zelda. She's pretty, smart and funny. She likes steampunk. I bet she knows how to kick butt."

Corella sneaks nearby and eavesdrops.

"In other words, she's the knockoff version of me." MJ turns sullen. "I could see you together."

"Maybe she's a better version of you, one that knows how to say yes."

MJ purses her lips. "I'm surprised she isn't first choice."

None says, "I finally understand what you mean by 'trying too hard,' like I'm Zelda's mission in life. Is that how I came across my whole life? I've never had someone obsessed with me before."

None clues into Corella's presence.

Corella invades None's personal space. "Change is the only thing you can count on. I get it now. Everything is a version to you, every little change. The world is just a bunch of numbers in your version control system. That's why tech people talk about web 2.0 and government 3.0. Don't tell me what version I am. I'm not a number."

"I told you I was being watched," MJ says.

"It wasn't me. I just got here." Corella disappears onstage through the curtains.

Corella announces the episode, but the intense audience reaction blots out her words.

None lets MJ go first. Once she captures their eyes, he can capture their ears.

Agent Vincent says, "No limits, Mr. President?"

"As we discussed, the more shocking, the better," None says.

Agent Vincent nods. "Mr. President, I'm quite confident you'll beat that ham sandwich today." The agent tightens his lips.

None passes through the event horizon of the curtains.

Faces. So many faces.

His brain parses the images like a blurred slow-motion home movie.

His chest tightens. Awareness.

Three rows of folding chairs replace the guest couch.

Three rows of lovers and ex-girlfriends, a jury of 15, sits in judgment.

Cold sweat. Cramped muscles, frozen in place.

A metallic taste of panic curls his lips.

Stiff legged, shock and panic immobilize him.

Supercharged with adrenaline, the spectators jump to their feet. His reaction feeds the same primal fascination of rubbernecking motorists at a tragic freeway crash.

Corella swings her arm towards the screen behind her filled with even more female faces, some blurred. "Someone said booty call. The president changed his name to 'All of the Above.'"

Catcalls and heckling blanket the studio.

Contests have losers. It's easy to rationalize when they don't hunt you in packs. What will his mother say? Or Shen? He struggles to move. Shen and MJ warned him. The intervention felt safe, without the bitter taste of fear. Dread engulfs None.

"You're a feral dog that humps everything in sight," Corella says.

The women onstage glance at each other, clap and stand united. They judge the cheater, the man-slut, trotted out for shaming. Thankfully, they left their rocks at home.

Externalized. His face drops into a flat affect. Not here. Gone. He floats above his body, his empty shell, judging it with the others.

"You claim to be an honest man," Corella says.

A word summons him back from ethereal form. Honest. "I am." The mere suggestion otherwise enrages him.

Corella points to Zelda. "Did you have sex with this woman?"

None says, "Yes, next question."

Corella points to Sarah. "Did you have sex with this woman?"

"Yes."

Corella points to Lashay.

None swings his arm in a wide arc across the rows of chairs. "Yes!"

Silence spreads like a disease. He admitted everything. Is it over? The moment of calm disappoints the audience. They came for a spectacle.

An athletic brunette named Yolanda steps from the lover's jury to pass judgment. She implores Corella, "Don't let him get away with it."

Corella closes towards None. "This is the biggest sex scandal going back to the Clinton White House. You're turning the White House into a whore house."

None says, "In the 90s, whitehouse.com was a porn site. So, someone turned the White House into a whore house, but it wasn't me."

Corella says, "Are you talking about Bill Clinton?"

"No. The whitehouse.com porn site was one of the reasons for cybersquatting laws. Whitehouse.gov is the correct site. I remember Bill Clinton. I remember thinking 'If the best the president can do is Monica Lewinski, what chance do I have with women?'" None points at his lovers. "You know, I did all right."

"You're a dog." Yolanda thrusts her fist in the air as she chants, defiant and bitter. "Woof. Woof. Woof."

The audience chants along with her dog sounds.

None puts out his hands to quiet the audience, his back to Yolanda. "It's OK for a president to kill a hundred thousand foreigners in Iraq, but get a blowjob and it's Armageddon? The gun is good. The penis is evil." None shakes his head. "Zardoz must have been American. Why are we unfazed by violence, but preoccupied with sex?"

Yolanda steps closer to None. "When you slept with me, I thought it meant something."

Agents assess her threat potential and sneak closer.

None turns to face her with a puppy dog eyes. "How do you know it didn't?"

"You didn't pick me."

None wishes he could hold Yolanda's hands and comfort her. He never meant to hurt anyone, but her pain stares back at him. Any act of kindness could set off a cascade of jealousy and insecurity. He could lose them all. His pulse quickens, stuck in an emotional minefield. Any misstep spells disaster.

Corella betrayed him. Again.

Did MJ know? When he signed the immigration bill, MJ called it a betrayal. Is MJ mad enough to get even? He can't look at MJ to read her face. Someone might get jealous. MJ is his best friend. She would never hurt him. He must be paranoid from spending too much time with Astrid.

Yolanda grows impatient with None's silence. "Why didn't you pick me? Answer me."

None says, "I want to tell my child, I could have had anyone, but I chose their mom. I'm sorry. Competitions have losers."

Yolanda pulls back her arm to slap None.

Agents move in and restrain her before she can connect.

Yolanda lurches forward and spits on None. "I'm not a loser."

The moist wet spray splatters his face. "That's not what I meant. Just because you didn't win, doesn't make you a loser." None wipes her spit off with his sleeve.

As they carry her away, Yolanda shouts, "Woof! Woof! Woof!"

The audience chants along but doesn't really support Yolanda. They feed on her humiliation like a chocolaty dessert.

Sarah breaks from the group and saunters towards None. She puts her palms together to greet him. "Namaste."

None's breath deepens involuntarily, as Sarah nears. His neck hairs bristle. His skin tingles with unseen energy. None returns her greeting.

"I knew there were others, but I thought I was special. I acknowledge this new truth."

"Sarah, you're in the last five. You are special."

"Five isn't special." Sarah points to None's other lovers, still calm. "Intimacy and trust are harder to share with more people. If you were undecided, you should have waited to become physical."

"I had my reasons," None says.

Sarah channels her mastery of body and emotion as a source of strength. A subtle pitch change in her voice reveals how unsettled she is. "I'm going to meditate and find clarity on my feelings." Sarah widens her arms for a hug.

The pain-thirsty crowd waits eagerly for the next relationship smashup. The calm makes them antsy. They didn't come for hugs.

The presidency seems less like a power-up, more like a cursed magical item. Power alone isn't enough to prevail in his growing struggles. He needs Sarah's calm in his life. Even hurt, she stands willing to comfort him. It's one of the things he loves about Sarah. Her embrace is an invitation. None could choose her right now. He could whisk her off the stage and tell her he loves her above all others.

How will the others react if he hugs her? Will it detonate a storm of explosive anger? The engineer in him wants to optimize to exactly the best choice. Zelda looks good on paper. She's the perfect woman, but he has weaker feelings for her. Sarah's ability to calm him will add years to his lifespan, years that stress would steal. But what if he eventually finds serenity and peace boring? Now is overrated. Just a few more dates and he could really know.

Sarah remains present in the moment, a fleeting moment. Her arms droop.

Could a hesitation, a few lost moments, lose her? None rushes to reconsider. Choose. Choose now. He hugs Sarah.

The hug feels like a broken promise, a wilted embrace. MJ fully extended her arms twice as long to restore mere friendship. Where was Sarah's patience? Was the hug a goodbye?

His first landmine. An explosion of butterflies, of love and fear, rips out his stomach, exposing his entrails to the audience.

The Jury of Lovers quiets. What happened is nothing to be jealous of.

Sarah pulls away from None abruptly and steps back towards the Jury of Lovers.

None watches her go. He should have listened to MJ. Sarah was her favorite.

Glitter Girls in the first-row cheer, as None barely resists tears.

Sarah walks past the Jury of Lovers.

MJ chases after Sarah. They exchange words and a quick hug beyond earshot. MJ and Sarah pull out phones.

Why does MJ need Sarah's phone number? Is she meddling in his love life again?

Corella strides to the Jury of Lovers and addresses them. "Who wants to dump this player president?"

That bitch! None fixates on Corella, with fury in his eyes.

Four lovers strut past None in a reverse perp walk. They blow sarcastic kisses. He eliminated all four women, but they pretend to follow in Astrid's footsteps. They do a burn sizzle and say, "You've been Astrified." None of them have the originality to use their own names. They leave the stage.

None heckles the exiting lovers. "I know the real Astrid. You're not fit to imitate her shadow."

He golf-claps and approaches the remaining nine women to cross-examine them. "Aiya! I did nothing wrong. I wasn't cheating, or dishonest. Lashay, did I ever say the word exclusive?"

Lashay stands tall, a six-foot Amazon and towering beauty. She grimaces at being put on the spot. She reflects, then smiles back. "Yes, you said exclusive."

The crowd claps when Lashay catches the so-called honest president lying.

None looks horrified. "Can you repeat back the exact words I said to you?"

Lashay says, "You sound like a lawyer. I thought you were a hacker."

"Just answer the question."

"Yes. You did say the word exclusive...that we weren't exclusive," Lashay says.

Murmurs filter through the audience as her response vindicates the president.

"You call me a liar? You've got the wrong guy. We're done." None points for her to leave. He snaps his fingers.

"I'm Lashay. The Lashay. I could have any man I want."

Astrid's words sound like a farce in her mouth. At least this non-Astrid uses her own name. "I'm not any man. I'm None of the Above."

Lashay wanders into the audience and picks out a harem of men. She leads them away like the Pied Piper and sneers back at None.

None clears his throat and defends his honor with the eight remaining women. "Did any of you think I wasn't dating other people?"

Zelda contrasts from the others in her saloon girl outfit. "I didn't know there were so many."

"You stood in lines to date me," None says. "You saw your competition."

"The lines weren't that long," Zelda says.

"We had a schedule. Zelda, you weren't even in the main line. You were an alternate."

They yearned to feel special, chosen. His lovers could ignore evidence and lie to themselves until a word shattered their illusion. Alternate. Their faces register how unremarkable they feel, among the unseen, overflowing throng of potential mates. None never mislead them. The Jury of Lovers renders a silent verdict, not guilty. He never tried to hurt and embarrass them. They turn their eyes to the one who did, Corella.

None shifts focus to the screen behind Corella. He counts rows, 7 x 4 = 28. He points at the screen. "You're missing a few. Where's your picture, Corella?"

Corella recoils, caught by surprise. At the Inauguration Ball Corella danced with None, but so did MJ and his mother. People assumed all their appearances together were work-related.

The astonished mob revolts against its ringleader, with yells and boos.

None addresses the lovers, opposite of Corella. "She forgot to mention she was dating me too, didn't she?"

The lover's faces contort with shock and bitter resentment. Corella played them all.

"You fuck with my bro. You fuck with me." MJ charges at Corella from offstage.

Corella's bodyguards and MJ's security detail scramble onstage.

Corella knows MJ. She knows to run. Corella climbs up on her desk.

MJ barrels through a row of chairs, shoving aside obstacles. She knocks over Zelda and two chairs.

Zelda stays down, nursing her knee.

MJ jumps onto a chair. With another leap, she lands on Corella's desk.

Corella jumps on the host couch.

MJ pounces on Corella. They go down hard on a soft couch.

Corella's face smooshes into the brown leather under MJ's weight. She struggles to free herself.

Spectator chant. "Fight! Fight! Fight!"

MJ elbows Corella in the ribs several times.

Two bodyguards yank MJ from the couch.

Corella climbs over the couch to escape.

MJ kicks back and to the left. A hard kick to the groin takes the left guard down.

She stomps on the right guard's toe and breaks free.

MJ vaults over the couch to chase Corella.

Bodyguards protect Corella in a wedge.

Corella yells. "You didn't want him! Why can't I have him?"

MJ has no response, other than fists. She charges Corella's defenders. Three guards grapple to immobilize her.

Agent Reynolds leads MJ's security detail into the melee. He makes quick work of the three guards and pulls MJ back.

Agents form a wall around her.

Violence between Corella and MJ plays out through their proxies. Their security teams shove and push each other.

It's daytime TV at its worst. The only thing missing is a paternity test. Who's your daddy? No one.

Outmatched, Corella retreats to the back of the stage. Her bodyguards deescalate and regroup there.

MJ switches her ire to None. "This is your mystery girl? I warned you not to trust her. Why don't you listen to me?"

"I could barely make out what you were saying with that toothbrush down your throat," None says.

MJ extends her tongue in disgust at the memory.

The eight remaining women in the lover's jury mill around Zelda. They help her to stand.

Zelda says, "This is the last time I date a guy with female friends. Come on ladies. We better go, before our self-respect leaves without us." She limps offstage with None's lovers in tow.

None watches the last of his hopes for happiness slip beyond the curtain. His eyes linger at their departure. This isn't like when MJ beat up his bullies in elementary. "Aiya! MJ, next time you want to help, don't."

A security detail surrounds None. They keep a watchful eye on MJ. She acts more like a crazy ex-girlfriend than a vice-president.

None marches towards Corella, his agents in tight formation. "Was this all a ruse to scare the other women away?"

Corella stares back, insulated in her own security bubble. "When Astrid left you for dead on that couch, it was my kiss that revived you. Then you went out with every skank on the Internet. My reporting saved your life. I should be your first lady."

"I let you know you were in the final five." Hearing his own words aloud, he knows the only number Corella could accept is one.

Corella says, "I'm not a number. For the record, version control is not pillow talk. Stop trying to fix America when you're in the bedroom. You should choose me and cook me breakfast."

"Cook breakfast? I have people for that. This is the last episode of *The Internet President*."

Her pulse races in panic. Corella appeals his verdict. "You're six episodes short. We have a contract."

None says, "I'm so done with contracts. I'm done with TV. I'm done with you, Corella."

"If today never happened, would you choose me?"

"You pushed the dating angle, for ratings. You helped Astrid dump me, for ratings."

"I'm not sad to see her gone, but I had nothing to do with Astrid," Corella says.

None scoffs at her lies. "Then who did? You expect me to believe it was just some staffer at your show? You're going to miss me."

Corella pretends she won't. Her face tells a different story. "I won't miss your long ratings-killing speeches. If people want to learn, they can switch to the History Channel." Corella tries one final excuse to make him stay. "You can't leave now." She whispers to him, "Imagine the ratings."

Ratings. Corella has different priorities. It's clear to him now. "You're a fame whore. I bet that's why you became a newscaster in the first place. Goodbye, Corella." None chokes up. She betrayed him twice, but only to keep him for herself. Things would be different if Corella understood trust.

"Sarah got a goodbye hug. What about me?"

None considers for a moment. He waves for his agents to let him pass.

The two security teams part to let None and Corella embrace.

None whispers in Corella's ear. "How far would you go for ratings?"

She whispers back. "I'd marry a president." Corella cries on his shoulder.

She didn't say *his* name. Corella wasn't dating him. She was dating the presidency.

53. **American Brood**

None leaves *The Internet President* D.C. studio to go directly to Air Force One. None, MJ, and Shen board his airplane for Los Angeles to face Senator Garfield in the promised immigration debate at high noon tomorrow.

None practices talking points with Shen for the immigration debate. It helps keep his mind off the ambush he just suffered on TV. They sneak in a meal and several hours of preparation before Air Force One begins the gradual descent towards Los Angeles.

MJ enters the president's office on the plane abruptly. "Garfield built his gallows. I know what he's going to hang you with." MJ turns on the wall TV.

LDR News reporter Tim Houser with the Word on the Street crew interviews picketers in front of Zentripity Energy headquarters. An outsourcing firm has replaced the entire IT staff, hundreds of jobs. A system administrator in his 50s tearfully recounts the story of how he trained a young H-1B visa worker to replace him.

MJ snickers. "I guess Tim Houser isn't first on the scene anymore."

The news switches to a Senator Garfield sound bite. "These immigrants don't just steal well-paying American tech jobs. They destroy American families. The president has allowed—"

None clicks off the TV.

"I told you not to sign," MJ says.

None says, "Zentripity Energy powers sections of Los Angeles. Aiya! Garfield wanted the debate here."

"This will dominate the news cycle," Shen says.

MJ says, "L.A. is a big city. There must be a counterexample."

None snaps his fingers. "We'll look up the most successful companies from the L.A. incubator. Search the list of founders for foreign-sounding names and research their backstory."

None, Shen, and MJ split up the list and scour through company profiles.

"American Brood." None does a double take. "There's something familiar about this company. The CEO is Jaya Kumar."

MJ peers over his shoulder. "That address...it's Beer Beakers. Bro, the owner was white, but that name sounds Indian. What's going on?"

The pilot announces Air Force One is about to land.

"Let's have a beer and find out," None says. "Strap in."

The plane lands in Los Angeles.

Last minute changes to the itinerary irk the Secret Service. None won't allow agents to clear the bar and perform security checks beforehand. He demands a minimum footprint security detail. None can't disrupt the new venue when he might need a favor. The agents bristle at his instructions but maintain professionalism. They remind him of the assassination attempt at this very location, but None remains undeterred.

The motorcade reaches its destination. The new American Brood sign features four animated figures. None muses on the symbolism of the sign. The first figure hints at the original Bear Beakers: a steampunk girl with exaggerated proportions wearing a corset and goggles. A redheaded goth figure commemorates the dark changes since. She wears black leather and lace leggings, a contrast to her pale skin and red dreadlocks. The last two figures, a hunky jock in a football jersey and a party boy hipster, round out the sign.

None and MJ leave the safety of the presidential limo, The Beast, surrounded by a minimal security detail. Inside the revamped bar, a scoreboard hovers over a well-lit center table. In four corners around the table, life-sized statues of the four characters act as mascots for quadrants of the room. A diverse collection of customers packs the bar: goths, steampunks, college kids, and sports fans. Between the noise and focus on the center table, few bar patrons notice the president.

Eight drinkers surround the center table. Five wear jerseys that match the jock mascot. A computerized voice comes over the loudspeaker. "Crazy Eight Ball." The crowd cheers and laughs, except the eight. A waist height robot on track wheels carries a tray. It places eight shot glasses and eight full glasses of beer from the tray onto the center table.

A brawny 20-something downs a shot. He winces. "Nasty."

In quick succession, the rest of the team gulp down shots and trash talk it.

MJ whispers to None. "If it's so terrible, why drink it?"

The brawny man hits a red table buzzer. "Double down for the win."

The audience responds with gasps and a murmur.

An overweight man with a beard pulls off his jersey and tosses it on the table. "I'm out. This batch is worse than Jalapeno Crotch. It's puke in a cup."

The burly man grabs the jersey and hands it to the overweight man. "Harry, it's for the win." The rest of his group implore him not to quit.

The jock section chants, "Harry, Harry." It's like the whole room is daring him to drink the foul beer. Harry relents. He slips the jersey back on. A few torturous minutes and a dozen retches later, all eight finish a full glass. The room overflows with laughs and excitement. The eyes on the jock statue light up. The eight drinkers run back to their original table, arms raised in victory.

Without the distraction of the center table, awareness of None and MJ spreads. The Blacksmith strides towards them. He wipes a hand on his leather apron and extends his hand to None. "I'm glad you stopped by. You got here too late for Beer League Night though."

None shakes his hand. "Let me guess. The first rule of Beer League is you can't talk about Beer League."

The Blacksmith belly laughs. "No. It's all we talk about."

Agents give None impatient looks.

"Is there somewhere we can talk?" None asks.

None and MJ wave to their constituents as they follow The Blacksmith to his office.

The Blacksmith drones on for 20 minutes about the rules of Beer League. Brewing experiments fail. The league was designed to dispose of repulsive beer from failed beer recipes. The combination of fantasy football, drinking game, and beer tasting did more than achieve that goal. It took on a life of its own.

The Blacksmith wasn't kidding. His religious zeal about Beer League turns the conversation one-sided. None looks for opportunities to interrupt.

A well-dressed Indian woman peers into the office. The Blacksmith rushes to introduce her. "This is one of my two partners, Jaya Kumar. She runs the business side of things so that I can design the American Brood experience."

None immediately likes her comforting smile and high energy. She must have caffeine for blood. Jaya is The Blacksmith's MJ. None reminiscences about MJ running the business, so he could focus on writing software for Slam Dunk Jobs and his educational software prototype, Adaptonium. None tunes out The Blacksmith as he prattles on.

MJ notes None shift focus to Jaya and rolls her eyes.

None interrupts The Blacksmith. "Jaya, how did you two become partners?"

Jaya bounces around, happy to tell her story. "I founded a social network for beer."

None snorts involuntarily. "A social network, just for beer?" Every idea exists on the Internet. None can't keep up, especially during the biggest tech boom yet.

Jaya says, "The Blacksmith told me his idea to experiment with small batches of beer to find the world's tastiest recipe. I told him he'd need a chain of bars to test enough recipes. He came up with the franchise program. By the end of lunch, we merged our startups. The missing piece was automation." Jaya and The Blacksmith exchange looks of mutual admiration.

None remembers those looks with MJ in another life. None suppresses jealousy and nostalgia. How can you miss someone who stands a few feet away? None laughs at himself with a hard-fought smile.

Jaya says, "I'm friends with Ning Bo, creator of Bobot Enterprises. He couldn't sell ice cubes in a desert. Great tech, but no investors. Ning completed the puzzle as our final partner with his automation."

"I'm thankful I listened to your advice, Mr. President," The Blacksmith says. "The incubator changed my life. We've expanded to 70 franchises. I could never repay you." The Blacksmith chokes up with a blissful smile.

"Actually, you can. I need a favor," None blurts out.

The Blacksmith sours on his praise. He fiddles with eight black armbands on his left arm. A moment of silence follows with a bowed head.

Jaya eyes his change of mood with concern.

The Blacksmith flips his head up. "Bravo. You had me fooled." He raises his finger in air quotes. "Amends. You're just another quid pro quo politician looking for favors. I'll do whatever you want...if you endorse our bars."

If None were a politician, he wouldn't have been blunt. He chooses not to correct The Blacksmith. Surely, shilling for a company as president must be against the law. None grins. "Deal, as long as it's not illegal." Only a politician would enter into a deal they planned to escape. None shudders at the thought.

The two men shake. Their female partners exchange glances in approval.

54. **Gallows**

Warning: This chapter covers controversial topics: immigration and race relations. It recounts historical events and terminology that may be upsetting to some readers.

Under the smokescreen of removing super PACs, Congress weakened campaign finance laws near the point of legalized bribery. None's only victory against special interest money was increased transparency. Any TV ad or appearance by public officials would require they disclose major donors and company ownership.

Today's debate implements the new laws for the first time.

Shen updates None on campaign finance paperwork on the way to the debate. "American Brood donated enough beer to show as a major donor. Good news, you can make that commercial after all."

None doesn't take it as good news. "That can't be legal. What about conflicts of interest?"

Shen hands None paperwork. "They passed new rules before you took office. Just fill out a form that says you won't be influenced."

"I can shill for anyone, with a pinkie swear?" None shakes his head and fills out the form. Under the form is the script for his commercial.

The Beast arrives near the L.A. incubator, with None, MJ, and Shen inside.

The Blacksmith waits nearby with a camera crew and some beer.

None gets out of the presidential limo and heads to the Blacksmith. He takes a beer, smiles, and says his lines. "When I have a rough day, I need a drink." None holds up the beer. "American Brood is American brewed." None promised a commercial, but he didn't promise multiple takes. He sets the beer down and continues with his security detail to the L.A. incubator.

A cluster of revamped warehouses East of downtown provides a home for the incubator. Two podiums stand on a wooden stage in a courtyard between warehouses. Sound and lighting equipment string along beams overhead.

None remembers Senator Garfield's warning. They'd hang None with his legislation. A piece of rope is the only ingredient missing, to turn the stage into gallows. Is the H-1B law that piece of rope? None imagines hanging from the rafters, suffocating from an extension cord around his neck. It's anxiety talking. A creative imagination has its downsides. None collects himself and moves forward.

Senator Garfield waits behind his podium.

MJ and Shen join the audience.

In elections, there is no prize for second place. If it weren't for None, Senator Garfield would be president. Instead, he settled for Senate Majority Leader: third in the line of succession, and two rungs below MJ. A grudge is all that remains of Senator Garfield's run for the presidency.

None walks up the steps to the wooden stage. He's ready. It's a mano-a-mano fight, with no moderator, no teleprompters, no earpieces.

He steps behind his podium and surveys the audience. None sees Ma and Lan in the front row. After the ambush on TV with the Jury of Lovers, None dreads his mother scolding him over his harem. He chose not to visit her while he was in L.A. None dreads hearing about that too. He'll delegate to Shen to make excuses later.

Senator Garfield talks first. The TV networks cover half the screen with corporate logos and pictures of billionaires. He's got more sponsors than a NASCAR race.

None wonders if displaying the vast array of special interests funding Senator Garfield receives will remind voters that he is an establishment Senator, in spite of his membership in an insurgent political party. Only a complete reboot of Congress can wash away the corruption in Washington.

Senator Garfield addresses None. "Government never solves problems. Their interventions cause new ones. Mr. President, your focus on the tech sector lit a fire under STEM wages. You've made the shortage of skilled engineers even worse. Companies can't afford these high STEM wages, so American workers are being outsourced. Your policies played out here at Zentripity Energy, where the entire IT staff was outsourced."

When None speaks, logos for Slam Dunk Jobs and American Brood take up a little corner on TV. It looks lonesome compared to Senator Garfield's overlay. A company Renquist controls, and a bar None almost destroyed isn't much of a roster. At least voters can see he isn't bought and paid for.

None chuckles. "The outsourcing deal was signed last year, well before my policies. There's an important aspect of STEM I didn't cover in my inauguration. STEM is always boom *and* bust, as skills and technologies fall in and out of favor. In sales, the motto is: 'always be closing.' In tech, it's: 'always be learning'. Without learning, workers become disposable and get discarded like an old phone. Tech sector layoffs are normal, even during tech booms. Keep your skills updated and tech remains the golden ticket."

The shift between so many logos to a nearly clear screen with each change in speaker jars the audience.

Senator Garfield says, "Hand out too many golden tickets, and they become worthless. You're a tech sector shill. You pushed everyone into technical fields, which will lower wages. By the time these new engineers graduate, it'll be a bust with no jobs. What a disaster. You let down the American people. The American miracle is an American mirage."

None says, "For a demonstration of the American miracle, I'd like to introduce a rising star at the L.A. incubator, Jaya Kumar."

Rage turns Senator Garfield's face red. "This debate is between us. You can't invite guests."

None asks, "Why debate here unless you want to showcase the incubator?"

Senator Garfield won't admit the real reason. He puckers his lips and flashes a bitter smile. The senator goes silent, as he plots his next attack.

Jaya walks onstage and recounts the founding of American Brood.

Beer bots roll through the crowd on track wheels with serving trays of beer. The crowd delights in the novelty. The bots serve top beer recipes, not the garbage from Beer League Night. The drinks impress the audience, earning new customers.

Jaya brings Ning Bo and The Blacksmith up to continue her story and explain Beer League.

"American Brood, make us your designated driver." The Blacksmith explains how their newest program brings fun to the ride home.

After their presentation, the beer bots and trio of American Brood founders depart towards the back with applause.

Senator Garfield takes a parting shot. "What I see is, two foreigners come in, and now the American is third string."

The Blacksmith scowls back at him. "My partners are Americans." The founders and their phalanx of bots continue their exit.

Senator Garfield shifts his focus to None. "The President sold out the American worker. He tripled the number of foreigners that can take jobs from hardworking Americans."

None knew that attack was coming. He begins his prepared response. "The tech industry doesn't just see the future. We build it. That requires the best minds from around the world. One-third of individuals contributing to major advances are immigrants, but they only account for 13% of the population. That's a triple power-up. Forty percent of the Fortune 500 companies were founded by immigrants or their children. Would you keep out Elon Musk, one of the most impressive innovators on the planet? No Tesla? No SpaceX? No Hyperloop? He's from South Africa. He wasn't born here."

Senator Garfield pounds the podium. "What about Americans losing jobs? You claimed jobs as your core issue, but I don't hear solutions." Senator Garfield pauses to let the attack hit home.

None says, "You're stuck in old patterns, Senator. Don't fight over crumbs. Make a bigger pie. Create new companies, to create new jobs. This incubator proves that works. That's my fix, what's yours?"

"Ask anyone on a border state, we're being overrun," Senator Garfield says. "We've got 11 million foreigners in this country illegally. Something has to be done. I'll ship them back where they came from and strengthen our borders with a wall that puts the Great Wall of China to shame." Senator Garfield seems proud of his idea. "You know from your heritage that a wall works."

None glares back and shifts to a sarcastic tone. "As long as people don't develop ways to tunnel underneath or invent airplanes."

"You mock me, but violent people, murderers, and rapists, come over that border," Senator Garfield says.

"Don't blame drug cartel violence on immigrants," None says. "We screwed up Central and South America with our drug war."

"You blame Mexican violence on America?" Garfield shakes his head.

None says, "We saw this movie before. Prohibition, the war on alcohol, gave rise to the mob and organized crime. Once the crime infestation began, it took the 21st amendment legalizing alcohol and decades of law enforcement to weaken the grip of the mob."

"You admit a government intervention made things worse?" Senator Garfield's mouth opens with a huge grin.

"Yes. The war on drugs and the war on alcohol, before that, were failed interventions." At least None found something to agree on.

None says, "You want to secure our borders, but what about borders on the way out? All these corporate-backed trade deals make it easier to move jobs out of the country."

"That's free trade," Senator Garfield says. "Government shouldn't withhold freedoms from job creators."

"It's even worse to ship away jobs and call it freedom," None says. "I checked your record, Senator Garfield. You haven't met a trade deal you didn't like."

Senator Garfield swings back to the H-1B law. He rants about immigrants stealing jobs.

None tunes out. Instead, he imagines being among the first wave of Chinese immigrants. They were brought to America as a source of cheap labor in the Old West. He'd shave the front of his head, with hair in braids down his back to his waist. The queue hairstyle signaled submission to the Qing dynasty. The Chinese penalty for not wearing that hairstyle was beheading for treason. Hundreds of thousands died in China fighting against the hairstyle and what it represented. In America, it was another thing that made the Chinese alien.

As Senator Garfield recounts stories about families going through hard times after losing jobs to immigrants, the sad tales strike an emotional chord with the audience.

None shifts his attention back to the debate. He turns to Senator Garfield with a glare, then nods to the cameras. "I *have* learned lessons from my heritage as a Chinese American, and it has nothing to do with walls. Chinese immigrants came with Gold Rush dreams to America. Many stayed in California, where unjust laws took away their ability to own land and gain U.S. citizenship. Some went to work in the mines of Wyoming.

"What happened when, as a matter of policy, Union Pacific Coal these paid Chinese laborers less?" None asks. "Did miners protest the coal companies for discrimination? No. They vilified the immigrants for accepting lower wages. In Rock Springs Wyoming, white miners robbed them and burned down their homes. At least 28 Chinese miners were shot, went missing, or were burned alive."

Senator Garfield pounds on his podium. "This isn't Wyoming, last I checked. Jobs were lost here, in Los Angeles, under your watch. You let immigrants take jobs from decent, hardworking Americans."

None turns from the TV cameras to face Senator Garfield directly. "Do you know what happened in Los Angeles?"

The tone of his question unnerves the Senator.

None points Northeast. "Not two miles from here, in a place called 'N*gger Alley,' was one of the biggest lynchings in American history. In 1870, California passed a law against immigration of unmarried Asian women. The next year, a fight over one of the few Chinese women escalated into a gunfight. A white man got killed in the crossfire. A race riot ensued. A 500-person mob cut a swath of destruction through China Town. They looted, burned, and killed anything and anyone they could find."

"The N-word and lynching have nothing to do with Asians." Senator Garfield scowls in disbelief. "You got your history mixed up."

"They beat and dragged their first victim several blocks." None looks up at the overhead beams. He points up, visualizing the events. "The vigilantes strung him up, cheered on and assisted by women and children. They ignored his cries for mercy."

Senator Garfield's eyes bulge out. His shock turns into skepticism.

None flings his hand down. "The rope broke. With hardened hearts, they choked the life from him with a second rope. Their bloodlust would not be sated. Even Dr. Tong, a well-liked doctor, who offered a ransom and pleaded for his life was shown no quarter. They stole everything he had, cut off his finger to get his ring and hung him."

"Fake news, fake news." Senator Garfield waves his arms.

Undeterred, None continues, "In the end, at least 17 Chinese men and boys died. Corpses rotted on the end of their ropes in the moonlight before they were buried in shallow graves."

Senator Garfield pounds the podium several times. "I've never heard of such a thing. First, you interrupt the debate. Then you make up stories."

None shakes with rage. There are Asian Americans born in L.A. that don't know this history. "Made up? Try cover up.

"Robert Widney, the hero of the massacre, who later helped found the University of Southern California, judged the murder trials," None says. "By law, the Chinese couldn't testify against whites and Latinos. Despite such injustices, there were convictions. All of which were overturned on a technicality." None doesn't mention the convictions were overturned by the California Supreme Court, the same one that made him president.

"The massacre was not to be spoken of," None says. "Los Angeles had to protect its reputation, or risk losing its place on the Transcontinental Railroad. Local newspapers omitted the massacre in their year-end recap. Even a book on L.A. history, co-authored by Robert Widney's brother, failed to mention the massacre. They renamed 'N*gger Alley' from Calle de los N*gros to part of Los Angeles Street, and whitewashed it away."

Senator Garfield pounds his pulpit. "A Wild West conspiracy tale, now I've heard everything. I can't continue the debate without a fact check." He waves over one of his staff.

None says, "That's not the only time Los Angeles renamed places after race riots. South Central became associated with the Watts Riots and the Rodney King Riots that took place there, so they renamed it to South Los Angeles."

An assistant hands Senator Garfield his notes.

Senator Garfield reads. His eyes bulge out again. "It happened? In a place called…I'm not saying that."

None says, "You like to talk about families, Senator Garfield. What kind of families can you build without women?"

Senator Garfield shakes his head. "Families without women? This debate isn't about non-traditional families. It's about protecting American jobs."

"America has built walls before," None says. "The Chinese Exclusion Act was created to keep out people like me. Before that, the Page Act of 1875 was so effective at walling out Asian women that, in the last year before the Exclusion Act, only 136 out of 40,000 Chinese immigrants were women."

None glares at Senator Garfield. "What kind of families can you make without women? None. How much do you have to hate someone, to prevent them from being born?"

Senator Garfield says, "It's not like they couldn't marry American women."

"Until 1931, women who married Asian immigrants would no longer be American," None says. "It's a pretty high ask to lose your citizenship." None imagines coming to America back then. Who would become a foreigner in their homeland for him? MJ? Corella? No. The Jury of Lovers would be rows of empty chairs.

Senator Garfield tries a conciliatory tone. "The massacre was a tragedy, but that's over 100 years ago. Let's stick to modern problems."

"Immigration isn't a modern problem," None says. "Across the Old West, the Chinese were rounded up and driven from their homes with threats, boycotts, and violence. Chinatowns were burned to the ground. For their crime of accepting lower wages, they were vilified as a threat. Then, it was the Chinese. Now, it's Mexicans. Big Money people have pitted us against one another for centuries. You're attacking the wrong villains."

The true villains, Garfield's owners, stare back at None in little boxes on TV. The cameras capture Garfield's dismissive reaction.

None says, "The Slaves. The Chinese. The Mexicans. It's not about race, but power and the exploitation of cheap labor to feed the American machine. The next cheap labor won't be human. What happens when the powerful elites have no further use for the poor?" The thought worries None.

Garfield tries to interrupt.

None waves him off. "You want to hang me for defending immigrants? Get the fucking rope. I won't go down without a fight. My people survived every injustice and flourished. I'll do the same. I've heard it said, 'only in America does one immigrant call another immigrant a foreigner.'"

55. **Second Chances**

None, MJ, and Shen strategize their next move in the president's office on Air Force One. They rush back to D.C. from the debate in L.A.

None looks to MJ. "You were right about Astrid. I'm sorry I pushed you away, but my administration takes a credibility hit with every one of your cat fights. I need you to work on self-control."

MJ scrunches her nose. "You're going to lecture me about self-control, Mr. Harem? Fine. Then I want my own dojo, at the White House."

Shen looks horrified. He rubs his ankle.

None blinks in disbelief. "Aiya! You fighting more is my worst nightmare."

"Martial arts aren't just fighting. Training includes the meditation and self-reflection, which I need for self-control. I put my training on hold to help you build Adaptive Unlimited and this administration."

None looks indignant. "Are you blaming me for your lack of control?"

"I remember enough to be dangerous, but I haven't set foot in a dojo in years," MJ says.

"I never asked you to stop," None says.

MJ pauses with a weary sigh. "It was implied. Startups require sacrifices, as do presidential campaigns. I chose you, both times. I could have said no. If you want me to control my impulses, I need a dojo."

None look at MJ with a thoughtful expression. "I can tell you've thought about this more than five minutes. What brought this on?"

MJ says, "You go down easy, but you bounce right back. I'm not like you. I go down hard and stay down. It's time to get back up. I saw the limits to my training in Beer Beakers. The bar reinvented itself. So, I will, too. I decided to become invincible again."

Shen scoffs. "You can't just decide to become invincible."

"If None can decide to be president, I can decide to be invincible," MJ says.

None and MJ exchange laughs.

None says, "Martial arts are your power-up. I can't say no. Shen has bureaucrats on staff that should know proper procedures to create your dojo."

MJ asks Shen who to talk to. She leaves to follow her new mission.

Left alone with Shen, None moves on to the next topic. "Find a way to get out of *The Internet President* contract. There's no way I'm trusting Corella with my sanity again."

Shen says, "I scheduled a meeting with the owner of the network when we get back to D.C."

"You want to know why I slept with all those women?"

Shen shifts in his seat. "Wouldn't you rather cover this with MJ, when she returns?"

None says, "Not particularly. I rationalized so many excuses, but I know the truth now. Shen, I can't survive another MJ."

Shen acts like he is allergic to the conversation. "Shouldn't you tell MJ how you feel?"

"I can't. MJ is the reason I slept with all those women. I had to be sure I wasn't in their friend zone, too. I can't be hung up again on someone who can't be physical with me."

Shen accepts he can't escape the conversation topic. "Are you still hung up on MJ?"

None takes a reflective pause. "Not anymore."

Shen says, "You always force me to talk about *your* relationships. You don't see me complaining that Lan doesn't want to come to D.C., do you?"

None asks, "Did Lan say she wasn't coming?"

"No, but she took on lots of new clients in L.A.," Shen says. "That's not how you wind down a business."

"I'm sorry, you should have told me earlier," None says.

Shen tries to distract None from his own relationship with dating advice. "Pickup dating where you left off and apologize. If apologies don't work, online dating is still an option. Your software isn't the only thing powered by failure. I've seen you fall and get back up, stronger every time. Why should this be different?"

"I'm lucky to have such a wise brother." None smiles at Shen. How many times did Shen give him second chances? Maybe someone else will too.

Shen withdraws to avoid further relationship discussions.

None was down to five women before *The Internet President* episode that destroyed his love life. During the show, he eliminated the supermodel Lashay, the Astrid wannabe that called him a liar.

None owed Corella for what she did on election day and providing him a pulpit to millions of Americans. Yet, how could he forgive her for the ambush and shaming he was forced to endure? None said he was done with Corella and he meant it.

Of the three that remain, Sarah is first pick. None brings up her number but doesn't enter. Less than a week ago he was so cocky. Today he panics, worried what she might say. None remembers the way his neck hairs bristled around Sarah, the way his skin tingled. He takes a deep breath and ignores the butterflies in his stomach. He presses enter. The dial tone quickens his pulse.

Sarah picks up. "Hello?"

Fear grips None so tight. He's unsure how to introduce himself. "It's None…of the Above…The President."

"I know who you are." Her words sound cold and artificial, almost like a different person. The uncomfortable pitch change in her voice is even more pronounced than the last time he saw her.

"I'm sorry about what happened. I want to see you." None trails off.

A few moments of silence pass. Dread surges in None with each second.

"That is a path I choose not to follow. I wish you well on your journeys without me. Namaste."

The call ends.

None redials in a panic.

Sarah's voicemail comes on. The recording talks about a vacation to reexamine her life. The unsettled voice on the machine sounds like a ghost of the Sarah he knew. None wonders how deeply he hurt her. He knows what betrayal and humiliation in front of the nation feel like. Could that be enough to crack her?

None has probably lost her forever.

If None hadn't hesitated to embrace Sarah onstage, would she be his? How many seconds cost him Sarah's love? One? Two?

None makes his next call to Betty, the waitress. Her number's been disconnected. She chose to hide from him. Can he blame her?

One name remains: Zelda. If he crafted the perfect girlfriend, it would be her. She loves science fiction, video games, and steampunk. Zelda even knows her way around a computer.

She's perfect on paper. Why doesn't he love her? None never believed there was such a thing as trying too hard until Zelda did it. Having someone devoted to him isn't a bad thing. Why does her clingy nature turn him off? Maybe *he* needs to try too hard, to love her back.

Is picking Zelda a rationalization? Does it matter? Zelda is the only one for him, the only one left. If she gives him a chance, None will return the favor. He calls Zelda.

Zelda picks up. "I was hoping you'd call. I wasn't sure where we left off, after that disaster on TV." She laughs to ease the tension.

It comforts None that Zelda still wants his call. He recounts the whole embarrassing episode with Zelda. They bond over the shared humiliation and laugh together. They talk for hours until the plane lands.

None tried an Olympic-sized love life, but he's no athlete. It's a relief to be dating one woman at a time.

56. **Bistro Valmont**

None and Zelda arrive at Bistro Valmont to enjoy fancy French cuisine. None only brought Astrid here, but Zelda insisted on coming.

The bistro is Astrid's territory. She proclaimed Bistro Valmont her favorite French restaurant, not just in D.C., but the entire East Coast. Given that Astrid was French and known for her discerning palate, her endorsement made Bistro Valmont an instant sensation. A poster-sized autographed picture of Astrid hangs on the wall, overshadowing pictures of the bistro owners.

"Your table is ready Mr. President." The waiter directs them to a table right next to Astrid's photo.

Zelda tips the waiter ten dollars and sits down.

None looks up. Astrid looms over them with a haughty and judgmental scowl. Only a super snooty restaurant would have that expression on their wall. It feels like restaurant patrons must prove to Astrid that they are worthy of the food.

None doesn't feel worthy of the food. "Zelda, are you sure this is where you want to sit? Do you know the history of Bistro Valmont?"

Zelda takes a deep breath. "I have to know you're over her. She has some strange hold on men. I watched a two-hour documentary on Astrid's men. I've watched documentaries on heroin addicts that had happier endings. I'll help you forget about her."

None looks up at Astrid's picture. "You're not doing a good job so far."

Zelda says, "What I meant was, she's like heroin, and I'm here to help you come down more gracefully, without the withdrawal."

Her description of heroin addicts isn't far off. Some addicts use replacement drugs for heroin to help them come down. "You're offering to be my methadone?"

"I'm methadone plus." Zelda stretches her hand across the table. "I don't want to see you end up just another notch on her lipstick case."

None clasps her hand in his.

They stare into each other's eyes, forgetting the world for a few long moments.

None skips all the pretentious French cuisine and orders fondue. Their famous eight cheese recipe doesn't disappoint. None's palate isn't sophisticated enough to detect the various cheeses. He remembers Astrid could rattle off every ingredient when she came.

Sharing a pot of cheese gives the evening a more intimate feel. None and Zelda dip broccoli florets, cubes of oven-hot French bread, and a selection of foods designed to trigger every taste sensation. They take one luxurious bite after another.

None and Zelda hold hands through an hour of innuendo and small talk until MJ's name comes up.

"I'm glad you like fondue," None says. "When I go out, MJ never lets me order communal food."

Zelda retracts her hand when she hears MJ's name. "The one time I met her, she knocked me down into a row of folding chairs."

None feared she might remember that. Zelda led the other women offstage after she fell. Zelda warned them not to date men with female friends. The MJ problem unsettled every relationship his whole life. How do you solve a problem like MJ? You can't just dump your best friend or your vice president. None invested his life into that relationship. Ma told him his wife should be his best friend. Life would have been simpler if MJ gave him that option.

None doesn't just need an Astrid replacement. He needs an MJ replacement. Zelda seems like MJ without the baggage, his methadone plus plus.

"Look I've thought a lot about MJ since our last date," Zelda says.

Aiya! When a woman thinks about MJ, it usually signals the two-minute warning a relationship is over. None cringes and steels himself for bad news.

Zelda says, "She's a lifelong friend. If she's stayed a part of your life this long, I don't expect that to change."

None's love life is in overtime. Not again, MJ. None prays silently to let this time be different. Beads of sweat form on his forehead. Winter is coming, probably seconds away.

Zelda says, "MJ tripped over me. I forgive her. I have to leave room in your life for her. I'm not here to replace her. Hell, I have a lot in common with her. In time, I hope I become as close a friend to her as you are."

None feels like bouncing off walls, ready to sing a song of joy and relief. Tears well up. None feels butterflies. He sees a real and tangible path to happiness. None extends both arms across the table. "I'd like that."

Zelda holds both hands. Her smile melts None's heart.

None says, "I want to commemorate this epic night. You loved those magazine spreads. I could invite the paparazzi outside for an impromptu photo shoot."

"If you think it's a good idea." Zelda tilts her head towards Astrid's picture.

"You're right, here is not the best place," None says. "We'll do it when we leave."

The dessert arrives, chocolate mousse.

Zelda gives a disappointed shrug and ignores the dessert.

None picks up on the disappointment. Non-verbal communication isn't his strength. MJ usually filled him in on those things. He has to figure out the source of disappointment before he leaves the table. It's too bad he can't run algorithms on human emotion.

None plays detective without MJ to help.

Zelda tipped the waiter to sit next to Astrid's gigantic head.

Cross-reference Astrid.

Rewind conversation stack trace.

Addiction. Heroin. Documentaries. Zelda has to know he is over Astrid. It's a test.

None smirks. "You know what would make a great picture? Let's get the paparazzi in here and kiss in front of mega-Astrid."

Zelda perks up. "What a great idea."

None releases her hands and gets the agents to bring in the paparazzi.

They immediately snap pictures.

None and Zelda French kiss in front of Astrid's disapproving scowl. None should send a message that he's over MJ too. He'll share food with Zelda. What if Zelda subconsciously translates chocolate mousse into feeding each other wedding cake? It's a dangerous game to conjure up such symbolism with someone already obsessed with you.

None dips a spoon into the chocolate mousse.

Zelda takes the cue and fills a spoon with mousse.

They feed either other with intertwined arms as flashes light the wall.

None and Zelda savor the moment. They revel in the fact that the Astrid behind them is powerless to stop them.

One of the photographers plays reporter. "Is this your official reaction to Astrid's new boyfriend?"

None says, "It's not my business who Astrid's dating."

The photographer asks, "It doesn't bother you she's dating your business partner?"

"What business partner?"

The answer dawns on None moments before the photographer answers.

"Alvin Renquist."

The news hits None like several stab wounds to the gut, but he downplays his shock and hides behind smiles. If he shows any other expression, those will be the photos that will be seen in public. The narrative will be about how Astrid hurt him, not how he's gotten over her.

Business partner. Only someone with access to Adaptive Unlimited financials would refer to Renquist that way. The very thought makes the news cut deeper. Renquist bought 5% of Adaptive Unlimited before be betrayed None and stole his company. Technically, Renquist will remain his business partner, until he lets the Adaptive Unlimited sale go through. None dances a duet with Renquist he can't seem to escape.

None mutters to the paparazzi, "Thank you for coming." He gestures to the agents and sighs relief when the paparazzi leave.

He shifts his attention back to Zelda. "You already knew, didn't you?"

Zelda gives None a sympathetic look. "I'm addicted to celebrity gossip. I figured you knew. *The Daily Reveal* is on the same cable network as your show."

57. Temp Nation

None feels everything shifting against him. Garfield lured him to L.A. with a sham debate and ambushed him. He did his best to counterattack, but Congress resists None even more. Worst of all, None still hasn't figured out what's killing the economy.

None, MJ, and Shen sit in the Oval Office staring tensely at one another.

"I expected the insurgent parties would help me take back America." None hangs his head and sighs. "Instead, the Tea Party and Occupy Party are failed revolutions, controlled by career politicians like Senator Garfield."

Shen throws up his hands. "Congressional leaders aren't even taking my calls. I don't see how we'll pass any more legislation."

"I hope it wasn't my TV show that was keeping them in line." None gulps. He had no choice after Corella ambushed him. None had to quit the show.

MJ asks, "During the debate, did you see how many special interests sponsored Senator Garfield?"

None says, "Hard to miss. Let's hope voters do something about it, or the taint of corrupt politicians will spread through the new parties until you can't tell any of them apart."

"It's already too late." Shen says. "What now? Start more political parties?"

None says, "We'll cut out the middleman and go straight to voters. Instead of lobbyists, we'll appeal to the American public to get important projects done. Thousands of volunteers on the IWantMyAmerica Portal already turned into grassroots activists. We just need to replicate our success on a higher scale."

MJ says, "Our TV show used to push voters to the IWantMyAmerica Portal. Volunteer numbers already declined and we've only been on hiatus a week."

Shen hands the president a stack of folders. "There's one more unintended consequence of ending the show. Without a daily overload of information, the press will have questions. You'll need to select one of your volunteers to be White House Press Secretary."

"I'll call for a new revolution," None says.

MJ says, "No. Revolutions aren't reliable. Fix the economy, and you'll get all the backing you need. We need more jobs."

None says, "More jobs aren't the answer to everything. My policies already created more jobs. Something is wrong with the numbers. Unemployment is down. But, so are wages. We didn't stop the crisis. We just pressed the snooze button. Whatever brings down wages gets stronger by the day. It overwhelms every improvement I make like I'm pissing into the ocean."

"Find the root cause and fix that too," MJ says. "If anyone can figure it out, it's you."

None has government data and corporate filings. He expands data sources for his algorithms, adding news and stock market data. None expands his search with every source of information he can think of. None suspects a complex web of reasons sinks the real economy to new lows, while the stock market hits new highs.

"Senator Garfield blames more H-1B workers for the economic slump," Shen says.

"Of course, he does." None shoos Shen and MJ away. "I need to work without interruptions."

They take the hint and leave None alone in the Oval Office.

None assumes whatever drives wages lower isn't luck.

He tries to find the root cause by finding companies profiting from lower labor costs. None writes code to parse and tokenize all corporate filings into a searchable index. He uses a faceted search to bring up the most common keywords in lower-wage companies.

None finds the usual suspects, higher profit, higher profit margins, lower cost. But, a couple of keywords jump out at him.

Dunk.

Slam.

None feels shivers down his spine. Dunk and slam aren't common words to find for quarterly results. Maybe they had a slam dunk quarter. None goes back to the original document to check them out in context.

Why would Yager Food Palaces thank Slam Dunk Jobs for saving their quarter? None does a double take and reads it again. Slater Dynamics refers to them as a competitive advantage.

None indexes the documents by date and finds a pattern. As wages went down, Slam Dunk Jobs got mentioned more often. Correlation is not causation. There could be other explanations. How could an app bring down the economy? Nonsense.

What did Renquist do to his company? Slam Dunk Jobs was the first brand to launch from Adaptive Unlimited. None wrote it so people would get training and once certified, they could pick jobs they wanted. With one assessment of their skills, they wouldn't waste time on interviewing. They'd set their pay rate and instant hire at a company. Without experience, they could take a lower rate to get some.

Adaptonium was meant to be the adaptive of Adaptive Unlimited, but his vision wasn't completed before Renquist tricked him out of his company. Adaptonium wasn't just supposed to fix the education system. It was designed to improve skill training. Better skills with Adaptonium would lead to better jobs with Slam Dunk Jobs.

Renquist murdered None's vision, replacing it with some abomination. Renquist didn't believe in investing in employees. He preferred externalities: corp speak for "let someone else pay for it." None knew training would be the first piece cut. Slam Dunk is the only part of Adaptive Unlimited Renquist cared about.

It's lucky Renquist never saw the importance of Adaptonium. Help someone to learn and remember better, and you change everything. The mind is the center of everything, not the dollar. Someday Renquist might understand that.

None researches all the changes Renquist made to Slam Dunk Jobs online. None remembers how excited he was when he saw the first commercial for Slam Dunk Jobs inside Beer Beakers. The commercial mentions that you bid on jobs and keep the job if you like it. What they don't mention is you have to outbid people every week to keep your job. In one design choice, Renquist turned tens of millions into permanent temps with no job security. Renquist created a Temp Nation. It's like the app transformed average workers into desperate day laborers looking for work. It's only slightly more dignified when the app hides the desperation.

Only two industries haven't suffered wages drops: tech and medical. There still aren't enough skilled tech workers to keep up with the tech boom, and the medical field has the ultimate leverage: fear of death. People will pay anything to save their own lives. It's lucky his mother's a nurse. None even got her a two dollar raise. Most Americans aren't so lucky.

Slam Dunk Jobs focuses on keeping workers at a bargaining disadvantage so that employers can lower salaries. If Renquist were a piece of code, that would be it, software to make low paying jobs pay even less.

With Americans bidding on their own salaries, the race to the bottom puts on a jetpack. Salaries might go up temporarily, but the trend is always down. A big part of the secret sauce was lowering wages just a little each week, so workers were less likely to leave. If they raise the heat too quickly, all of the frogs will fight their way out of the boiling pot of water. This advance drives wages to minimum wage for professions that used to be many times that.

The way things are going, soon minimum wage might be the only wage.

Slam Dunk Jobs took over HR in more and more corporations, taking on a virtual monopoly in dozens of industries. Companies sometimes tell people they aren't qualified to do their own job as a negotiating ploy.

None remembers his mother talking about her friend Marjorie from work. When Slam Dunk Jobs fired her and took over HR functions, every employee in the company became their client. Their salaries were replaced with bidding wars.

None realizes, by taking all the existing job holders into their system, Slam Dunk Jobs guarantees there are always more applicants than jobs. Those job postings aren't new positions. They're advertising jobs that are already filled. Advertising existing jobs as job openings, can you get any more despicable?

Wait. This all feels familiar. His memories feel jilted and out of sync. They're just out of reach, like a forgotten dream.

None greets memories of a day he had blocked out with guilt and disgust: the day Renquist asked him to add vile, ruthless features to his software. That day, Renquist went from partner to mortal enemy. That day, None wanted to sell Adaptive Unlimited and hide Adaptonium, but he couldn't remember why. Memories of *that* day burden him like a medieval torture session. Even now, None wants to distance himself from those thoughts. He foresaw the potential damage the changes might wreak on Slam Dunk Jobs users and their families.

It's more urgent than ever to take back his company and undo the damage Renquist inflicted. None gets on the intercom. "The law of unintended consequences just bitch slapped me. Get back in here." Unintended consequences, yes, but not unanticipated. He anticipated it all, but he can't tell MJ or Shen that. None would unknow his role in all this if he could.

After MJ and Shen return to the Oval Office, None proceeds with the unhappy chore of explaining how an app he created brought down the economy.

Shen says, "Forget what I said earlier. We should have enough for an official investigation into Slam Dunk Jobs. I'll schedule a meeting with the Attorney General."

None fears what other secrets he could have buried in the vast expanses of his mind.

58. **The Daily Reveal**

Astrid exploded in and out of None's life when she dated him. His new live-in girlfriend, Zelda, latches on and moves into the White House. She officially lives in the Queen's Bedroom, but everyone knows where Zelda really sleeps.

Zelda tries to kindle a friendship with MJ whenever they cross paths. Since the Queen's Bedroom is right across the hall from the Lincoln Bedroom, where MJ stays, that happens a lot. She invites her to every event she can think of, but MJ never accepts.

MJ encounters Zelda in the hallway again. Zelda's persistence wears her down. MJ promises to spend time with her when she's not busy. MJ makes her way to the Oval Office, where None and Shen prepare to meet the owner of the Reveal Channel.

They go over the background of the owner, Olivia Ford. She got a small role on a disastrous sitcom that only survived four episodes. Olivia was a terrible actress, but a captivating personality. Her unintentionally funny catchphrases, "that's a reveal" and "I knew that" made her a household name. She parlayed popularity into a reality TV show, then a hit talk show culminating in her own fledgling media company, the Reveal Channel.

Olivia's a walking tabloid, obsessed with celebrities. She became known as the celebrity hookup whisperer. On her talk show, Olivia covers the day's gossip, then muses over who would make a good couple. Her fame grew years ago when several famous people took her suggestions seriously and ended up happily married. On Olivia's flagship TV show, *The Daily Reveal*, she recounts past hookup suggestions each show and rates them with "that's a reveal" for ones that surprised her, or "I knew that" for ones she called correctly.

None, MJ, and Shen take the presidential motorcade to the Reveal Channel D.C. studio. Once they arrive, Olivia escorts them to his old set where *The Internet President* was filmed.

None misses being a daily part of life for millions. To be effective, he needs to rekindle that relationship with the American people. Is he just rationalizing to soften the very real threat he'll have to continue the show?

A new show, *Relationship Nightmares*, replaces his show but reuses the same furniture. Corella sits at her desk in the middle of the set. When Corella sees MJ, she calls security guards to the stage.

None scoffs. "Relationship Nightmares sounds like the last episode of our show."

Olivia remains attentive, but silent.

Corella says, "I'd like to start things back up."

None does a double take. "Are you talking about us, or the show?"

"Both."

MJ says, "Corella, the answer's no...to both."

"MJ, I'm not talking to you. You assaulted me. I should press charges. It was good TV though." Corella laughs. Not everything has healed yet, and the laugh makes Corella's chest sore.

"My new friend Zelda won," MJ says. "You won't get your grimy ratings-loving hands on him. Corella, I only trust you half my throwing distance."

"I wish life came with an editing room, so I never had to see you again," Corella says.

MJ whispers to Shen, "Make sure any new contracts have a clause against editing me out."

Shen nods.

Olivia says, "Ladies, please. We're here to resolve our differences, not fight. Mr. President, what's your main concern with resuming *The Internet President?*"

None says, "I've come to the conclusion it doesn't behoove a sitting president to star in a reality TV show."

Corella says, "I can't embarrass you too much. No one wants to watch the loser president."

Olivia motions for Corella to be quiet. "We have a contract."

None bows his head slightly. "I can't get things done without voters on my side. My approval rating tanked when you exposed my personal life."

"The minute you became famous, your love life became as important as your accomplishments," Olivia says. "I've built my career on that. Celebrity gossip is a hunger you'll never satisfy."

"Accomplishments are all that should matter," None says.

Olivia says, "Dream on, Mr. President. If magazines ever disappear completely, the last ones to go will be supermarket tabloids covering celebrity scandals. I'm a generous woman. You can take the summer off. The remaining episodes will be in September."

None knows Olivia isn't generous. Most TV shows do summer reruns and fall premieres.

Corella cracks her knuckles. "I'm not generous. You created Slam Dunk Jobs. I know what it's doing to the economy. Your promises to stop Wall Street from strip mining the economy got you elected. If I link you with Renquist and an app that's gobbling wages, you'll be an outcast in your own movement. We're talking revolution over. Six episodes aren't enough. If you want me to bury the story, I need another season of *The Internet President*."

"If you really wanted me back you wouldn't blackmail me," None says.

"I was willing to give you another chance, but you let MJ speak for you," Corella says.

None fumes silently. He needs more time. "I wished you worked for the DOJ. I'm still waiting for my report. How did you find this out?"

Corella smirks. "Did you forget I was a reporter?"

None shakes his head. "Bury the story. I'm in with one condition: no surprises."

Olivia says, "I can agree to no on-air surprises. We'll update the contracts when you come back."

None shakes hands with Olivia. "See you in September."

59. **Dojo**

MJ and several White House staffers walk from the White House to the South lawn to check out her new dojo.

MJ memorized a speech to make her case for the dojo. She had a mixed martial artist president as precedence. President Teddy Roosevelt lined the White House basement with training mats to spar in wrestling, boxing, and judo over a century ago. MJ didn't have to argue. None just said yes. Today she tries out the dojo for the first time.

A chain link fence with a green tarp surrounds the White House Basketball Court. Obama repurposed it from a tennis court. MJ repurposes it again as the White House Dojo. The Washington Monument juts up above the trees outside the fence, a reminder of where she is. 2500 square feet of green mats cover half the basketball court.

MJ asks the staffers for a volunteer to practice Kendo.

A confident male staffer volunteers.

MJ and the volunteer don protective gear modeled after samurai armor. They each grab a shinai, a bamboo sword made from four bamboo pieces fastened together.

She teaches the volunteer a few basics and begins a Kendo match. MJ focuses her breathing and channels her ki into her attack with a battle yell. Her scream startles the volunteer. MJ quickly lands her strike on the top of his helmet for a point.

After her first couple yells, the volunteer yells back with his attacks.

MJ easily deflects the volunteer's attacks and lands her own. He provides little challenge, but she appreciates the practice. Kendo is about more than winning. It's also about focus and training the mind. MJ feels a peace she hasn't in a long time.

Her volunteer gets tired after 20 minutes of sparring.

MJ looks for a replacement. Fernando Rivera, the Secretary of Defense she helped pick, watches with the other staffers. He came! Fernando was the first one she invited to her makeshift dojo.

Fernando asks, "Why build a dojo at the White House?"

MJ needed a place to train and reflect. The off chance she might lure a world class fighter like Fernando to train with didn't hurt. "Build it, and they will come."

"Did it work?"

"You're here aren't you." The words come out flirtier than she intends.

Fernando says, "When you told me about your dojo, you said your focuses were mental discipline and being combat ready. I know some training you're missing."

MJ asks. "What did you have in mind?"

"I brought you a dojo warming gift." Fernando hands MJ four arm length rattan sticks tied together with a bow. "Kendo is great for discipline and mental focus, but rules go out the door in combat. You could add some Filipino Martial Arts, like Kali, to your training."

"Is it stick fighting?"

Fernando says, "The sticks stand in for bladed weapons like machetes, just like your shinai represents a katana. Each strike in Kali is designed to be used three ways, open-handed, bladed weapons, or blunt weapons."

MJ unties the rattan sticks and hands two back to Fernando.

"I didn't bring my protective gear," Fernando says.

MJ smiles at the volunteer. "You can use his."

The volunteer was too tired to continue anyway. He takes off his gear and gives it to Fernando. The exhausted volunteer and other staffers leave for the White House.

Fernando goes over Kali basics and demonstrates stick attacks.

MJ devours Fernando's training like a ravenous wolf. Her lifetime of martial arts accelerates her learning. She harmonizes with her body, each action more fluid. The new moves and sticks in her hand feel natural.

Fernando adapts, going faster as MJ picks up a move, a block, another angle. He thrills at how quickly she learns. "You sure you haven't studied Kali before?"

MJ smiles back.

They drill with combinations of attacks and counters, reacting off each other. They breathe in sync. The rhythms of banging sticks combine into the music of combat. Their energies intertwine with intensity and sweat, a connection that is almost spiritual, even though it's rooted strongly in the physical. They grow closer with each attack and counter. Mutual respect strengthens the bond.

MJ doesn't think. It's intuition and instinct. She flows. Her moves become meditation, a seamless connection between mind, body, and spirit. The clashing sticks become a wooden chant, resonating deep within her. The sticks and their music extend as part of her.

Even without a kiai, a battle yell, MJ senses the strength of Fernando's ki. His energy draws MJ closer. His muscular frame and skill represent the strength that pulls MJ towards Fernando and reflects the strength within herself she holds dear. An unfamiliar heat within her surges. It disturbs her.

MJ pulls back. The bond unravels.

Fernando lets the connection dissipate. He transitions into a new lesson. "Now, these are blades." Fernando explains how all the moves she learned translate with edged weapons. Normally, Fernando would demonstrate with 15-inch bolo knives, but it was hard enough to get rattan sticks approved with the Secret Service. MJ will have to use her imagination and remember which side of the stick represents the blade.

MJ visualizes her water serpent katana as the stick. She has to think. Thinking is too slow. She can't get her groove back. She misses the flow she had moments earlier. Her frustration grows. Nothing syncs up. MJ makes mistakes with moves that were natural before.

Frustration and disappointment spread to Fernando. He whacks the inside of her right wrist. "Keep your wrists on the inside. Otherwise, your opponent can cut an artery. Remember where the cutting edge is."

MJ visualizes a knife slashing the exposed artery on her wrist. In her mind's eye, blood spurts out, exposing her tendons. She cringes in disgust.

Fernando restarts the drills with sticks as placeholders for bolo knives. He slows down the exercises to give MJ a chance to adjust.

MJ gets worse. She focuses her thoughts on remembering the imaginary blade. A vivid image of a bloody wrist and severed tendons won't leave her mind. MJ can almost feel the pain. She feels searing burns up her wrists and arm.

Every move feels wrong. MJ lashes out. Her attacks don't connect.

Flashes. A smoldering gun barrel. Stomping boots. Stuck on her back.

She flails around with desperate swings.

Fernando backs away, perplexed by the decline in her skills. "Are you alright?"

Glass cuts her feet. The flashes coalesce into a detailed memory of her near-death experience in the bar. "He had me. It was over."

"We can stop."

"No. I have to push through this, or that day will own my future." She struggles to layer the pain from the cuts on her feet onto her wrists. MJ can't imagine cut wrists, only scorched wrists. The memory is too strong. It crowds out her imagination.

Fernando shows MJ patience, slowing down even more. He telegraphs every move.

Scars of that day aren't just in her flesh. Her imagination can't beat the memory. Does she have to feel a cut to imagine the blades? All she needs is the pain. MJ quits fighting the burn memory. She uses it and visualizes pain when the imaginary blades cut. MJ looks at her sticks and sees real blades. Blades that burn.

MJ quickens her attacks. The fake blades in her hands slash and parry. Everything feels natural again.

Fernando grins at her newly found skills and confidence. "That's enough for today."

He undresses into street clothes. "Something clicked for you. I felt it shift."

MJ takes off her Kendo armor slowly and reflects. "I had to quit fighting my pain and use it." She unties one of her seven-buckle gloves to show off her scars.

Fernando motions for her to stop. "Don't start with that. It'd take 20 minutes to inventory all my scars." He softens and winks. "I know a better way to spend 20 minutes."

MJ parries the innuendo. "Thanks for reminding me. I haven't done my daily security briefing yet."

Fernando looks deep into her eyes with confidence and smiles. "Let me walk you back to the White House."

MJ agrees. They stroll through trees onto the path towards the White House.

"How to adapt and improvise is part of Kali training. You can use everyday weapons. Look for improvised weapons in your surroundings." Fernando stops to points at a tree branch. "We'll count weapons on the way to your office. That's one."

Fernando resumes the walk. "Be prepared for combat. You don't have to be James Bond or Jason Bourne to use situational awareness. In the military, we use the OODA loop. Observe, orient, decide, and act.

"Observe," Fernando says. "Be attuned to your surroundings. Map out exits, cover, and lines of sight. Be careful about blind spots and attack vectors where you can be surprised, like from behind."

MJ self-assesses her performance during real combat. She mapped out the window exit in Beer Beakers, even if she didn't use it after Corella upstaged her. MJ used cover and lines of sight to approach the gunman's blind spot. She even exploited the element of surprise from behind. MJ looks smug. "When it mattered most, I did all of that."

"Good. Make it a habit everywhere you go," Fernando says.

MJ nods.

Fernando says, "Orient. Put your observations into context. Create a baseline for what's normal. If you don't know what normal is, it's much harder to detect subtle changes. Look for people, or things, that are out of place. Potential threats often show up as anomalies."

MJ snickers. "That's a bad mood word. None complains about anomalies every time he's mad at his computer. I hear that word a lot. What's an example?"

"Like someone who isn't shocked by gunfire, or an explosion because they knew it was coming."

MJ remembers None's story about jail. "Or someone wearing a trench coat on a hot day?"

Fernando nods. "Exactly. Let's say, a criminal cases a jewelry story, pretending to be a customer. It takes extra effort to pretend to be something you aren't. Look for people who aren't acting naturally. Anomalies aren't always threats. But, pay attention to them."

"None and I feel out of place all the time, but am I a threat?"

Fernando tilts his head towards MJ and smiles.

MJ laughs. "Don't answer that."

Fernando says, "Decide. Map out the potential plans ahead of time. Visualize it, to make your plan actionable. If you brain isn't prepared, you're more likely to freeze up. Sometimes boldness carries the day, in spite of a poorly executed plan, but don't rely on it."

When she decided to attack the gunman, boldness got MJ through that day. It almost got her killed too. She lets his point stand without comment.

"Act. Put your plan into action and use any takeaways in your next OODA loop." Fernando shows MJ a diagram of the OODA loop on his phone.

MJ lights up. Her understanding deepens. "I saw a diagram like this when None first tried to explain Adaptonium. His example was how many decision loops professional gamers make in complex games like *Starcraft 2*. I don't relate to keyboard ninjas doing ten actions a second. Your combat scenarios I understand."

Fernando says, "I remember Adaptonium. I tuned out the software bits, but the phrase 'powered by failure' stuck with me."

MJ smirks, "You remember that phrase because he repeated it like three, or four times."

Fernando lights up too. "No. It's more than that. He's using the OODA loop in Adaptonium. That's why it's familiar. Observe wrong answers. Orient failure against your baseline and find patterns. Decide a teaching strategy. Act, teach with the new approach."

MJ finishes his thought. "Then loop through again and see if you got it right. Learn and improve with each loop."

They smile at each other like they just unlocked ancient knowledge. They enjoy their shared epiphanies and resume their walk.

"Observe. How many weapons have you seen so far?"

MJ says, "Including stick worthy branches I could break off, within high kick range, I'm up to at least 24 weapons."

They enter the West Wing. Fernando rattles off improvised weapon possibilities as he moves through each room. He stops in the hallway by the Cabinet Room to admire some paintings. A red velvet couch sits below four western landscapes. Well-polished end tables with a lacquer finish and skinny wooden legs adjoin both sides of the couch. Lamps stand two feet tall on the tables. Fernando asks, "How about here?"

MJ asks, "The lamps?"

Fernando picks one up and tilts it. "It's awkward, not well balanced. You could throw it. You'd probably get one swing, but there are better weapons."

MJ looks closely at the end tables. "I could snap off the legs on the tables as fighting sticks. That's eight weapons right there."

"And?"

"I could break the glass over the paintings and use the glass shards?"

"Right, most things can be weapons. There are diminishing returns though."

Their tour reaches MJ's office in the West Wing.

MJ adds up numbers from the notes on her phone. "You're going to ask me how many weapons, but I need a minute."

Fernando smirks. "Close. I know you love Kendo, but how many katana did you see?"

"None. I'd like to show you the katana in my bedroom."

Fernando lights up at the word bedroom. "Yes, Ma'am." It doesn't take situational awareness to notice Fernando likes her.

MJ sighs. "But I can't. It's been stolen."

Fernando deflates. "I'm sorry to hear that."

MJ flashes a coy smile. "Someday I'll show you the two matching swords, but not today. Same time tomorrow?"

Fernando smiles back.

MJ escapes into her office, where her security adviser delivers her briefing.

60. Worse Than No

In the beginning, mostly ruthless companies use Slam Dunk Jobs to lower wages. When those businesses succeed, more companies jump on the Slam Dunk bandwagon. Business leaders rationalize their reasons for using Slam Dunk Jobs. It's capitalism, the markets will decide. If they don't use the app, their competitors will. The markets speak, Slam Dunk Jobs wins.

With the network effect, Slam Dunk Jobs becomes more effective with every company and job seeker that joins, until it has an overwhelming advantage. It reaches critical mass in one industry and profession after another.

Copycat job auction sites try to adopt the same business model. Slam Dunk Jobs ties them up in patent litigation and outcompetes with them. Renquist crushes his opponents like ants.

Slam Dunk Jobs moves from competitive advantage to business necessity. It acts as the performance-enhancing drug of the stock market. It wrings every inefficiency from the job market. It becomes the job market, a market that trades jobs like stocks. Wages drift lower, corporate profits soar. The same process spreads like a virus over the next couple months. Slam Dunk Jobs moves from dirty little secret to open secret.

Professions and industries dominated by unions resist the wage drops. Their employers lose market share, competing against lower labor cost companies. The stock price of Slam Dunk holdouts and union shops plummet. Still, unions might be the best shot at fighting back.

A handful of hot tech specialties with labor shortages remain unscathed, the last holdout of None's technology revolution where both company and worker continue to do well. The incubators of the Invest In America Program enjoy much higher profits, but overall tech wages decline.

Stocks and corporate profits hit record highs, while America and the world sink into quicksand.

Like climate change deniers, those profiting from Slam Dunk Jobs live in denial that it harms the economy. Renquist sends his media teams and enforcers to attack anyone so bold as to suggest the possibility. Despite the campaign of fear, uncertainty, and doubt (FUD), it won't be secret for long.

None considers options in the Oval Office. Until he finds a viable solution, he decides to keep quiet.

Shen and MJ enter the Oval Office with three large binders.

"Good news." Shen smiles at his brother. "We've got copies of the final report for the Slam Dunk Jobs DOJ investigation."

None takes a binder and skips to the end. "This doesn't read like good news."

"The investigation remains sealed, and Renquist's FUD campaign worked to our advantage to keep it secret," Shen says.

"Which laws are we going after them for?" None asks.

Shen says, "That's just it. Every law Slam Dunk Jobs would violate was weakened, repealed, or had an exclusion added to give Slam Dunk Jobs a competitive advantage. Anti-trust laws, labor laws, employer regulations all just washed away. It's like Renquist colluded with Congress and state legislatures to get all this done. Honestly, it's impressive. Even the RICO statutes were updated. So, if you had some crazy idea of arresting Congress as an ongoing criminal enterprise, they covered that too."

MJ says, "The good news is the Attorney General made a timeline of all the legal changes so that you can skim that section."

None looks at the beginning of the law change timeline. "Renquist put all this into motion the same day I agreed to sell him Adaptive Unlimited. This is all according to plan, and we're playing catch up." None slams the binder down on the Resolute Desk. "I want better options. Let's go over every inch of this report until we find them."

Shen goes over sample contracts in the report. "The contracts give Slam Dunk Jobs ongoing bonuses based on how well they lower wages."

None scoffs. "They've made it their business model to bring down salaries across the board."

MJ puzzles at another section of the report. "Renquist bought surplus cargo planes from his defense contractor, Renquist Aerospace, to build an airline called Slam Dunk Express. None, it sounds like the sort of crazy ideas you have."

None says, "It makes perfect sense. Renquist can dropship workers anywhere in the country, shipping them like commodity goods. If the workers have no place to stay, he can rent them corporate housing. Renquist always has a profit angle. With cloud computing, you provision new computer resources as needed. Slam Dunk Express is the people version of that. Any company can quickly expand or downsize their workforce."

"People aren't interchangeable though," MJ says. "You're the poster child for that."

None says, "I don't think companies who are focused on bottom line thinking care. Slam Dunk Job clients are poster children for that."

MJ displays a video on her phone of work tourists. The adventuresome of the Temp Nation condense their belongings into a Slam Dunk Express container and move from city to city. Rootless young people embrace the impermanence as a lifestyle, a new generation of drifter.

Shen gasps and holds his chest. He rereads the same section of the report. "None did you say you got Ma a raise?"

None puffs his chest out proudly. "Yeah, a big one, a full two dollars."

Shen runs his fingers through his hair with a scared expression.

None focuses on his brother. "What's wrong?"

Shen struggles to choke out a few words. "Do you know what you did?" Shen points to a section of the report.

None reads intently. It's a snippet of internal marketing material that says, "Never pay a raise again. Our app prominently features jobs where workers asked for more money. The job gets more bidders, and the worker gets a pay cut. They'll never ask for a raise again."

None blankly stares as he digests the implications. "It probably wasn't a big pay cut or Ma would have said something."

"She's a private person. She wouldn't say anything. I need to check something." Shen turns to MJ. "Can we get privacy? We have family business."

MJ picks up her binder. "I'll leave you to it."

Shen follows MJ out and comes back in with a laptop. He works on it for several minutes.

Shen's eyes go wild. "Since I know real estate, Ma asked me advice about a friend going through a foreclosure. I could tell she was upset about her friend, but it was more than that. Her house has a Notice of Default filed. That means she's at least three months behind on her mortgage."

None says, "That has to be a mistake. She's owned that house over 30 years. Why isn't it paid off?"

Shen says, "All her friends made huge profits flipping houses in the last real estate boom. She knew I'd tell her not to, so she kept me in the dark. She mortgaged her home to invest in some rental properties in Phoenix at the height of the market. She lost everything, with only the mortgage to remember her experience by."

"She's been in foreclosure before?" None looks horrified. "Why didn't I hear about it?"

Shen says, "You were too busy with MJ and your company to ask. Last time she ignored the foreclosure and let them take her rental. She handed me a box with paperwork afterward and shushed me when I tried to say anything."

"Maybe we're jumping to conclusions. Call Ma and find out what happened." None holds out the phone receiver to Shen.

"When it comes to family, you're not my boss. You call her, Mr. President. This is your mess." The words spray from Shen's lips with frustration.

Is it None's mess because he got her to ask for a raise, or because he created Slam Dunk Jobs? Either way, he's guilty. None gathers his thoughts and dials his mother. He goes through pleasantries with her. They both pretend everything is alright.

Mrs. Wong can tell he knows. A mother knows. She bursts into tears when None mentions the Notice of Default. "I can't keep my house on half my salary."

"Half?" None gulps hard. When None got his mother a raise, he thought the worst they could do was say no. He was wrong. "Why didn't you tell me earlier?"

Mrs. Wong stifles her tears and summons her strength. "I brought you up to be self-reliant."

"You also taught me family comes first. Given how often I hear that from Shen, I know you taught him that too." None grins at his brother.

Shen resists the urge to respond.

None says, "You have a *roommate* now. Don't they help with the mortgage?

"My roommate is out of work. There are others worse off."

"I'm president. I'm sure I can pull some strings."

Mrs. Wong says, "What will they say of my son if he helps his mother and no one else?"

"We'll keep it in the family then," None says. "Shen and I can help."

Shen presses mute. "Lan didn't liquidate the business yet. Everything I have is tied up in real estate that I can't sell."

None unmutes. "Have you tried to sell the house?"

Her tears return. "I've lived in that house your whole life. Don't let them take my home."

"Ma, Shen, and I will come up with a plan and call you back. It's going to be OK." None ends the call and shifts his attention to his brother. "Out with it, Shen."

"Lan is spooked. This real estate death spiral is worse than the last crash. Lan fired her assistant Cathy today. We still owe her back pay."

"How bad?"

Shen says, "I thought Lan wanted to stay in L.A., but the real reason she took on a bunch of clients was to offload properties quicker. For every new client, two more disappear."

None gets irate. "You didn't think this was important to mention?"

"Lan told me this morning, and we were kind of busy." Shen holds up his investigation binder.

None asks, "Do you think it's Slam Dunk Jobs?"

Shen says, "What do you think? Would you buy a house, if you didn't know how big your paycheck was, or if that paycheck could be your last?"

"Welcome to Temp Nation." None's facial expressions turn gloomy. The connection is obvious, after the fact. The media will crucify None once they connect the pieces. "Why isn't it showing up in the economic data?"

Shen says, "A Notice of Default shows up months after mortgage problems start. It should show up in the next month's foreclosure data." Shen checks the date on his phone. "In three days."

None says, "Let's address the one foreclosure we can fix. I'm still paying back debts. Maybe we can get a loan modification."

Shen says, "We don't have enough to get her current on the mortgage, but we could swing a payment plan. I'll call the bank."

Shen leaves to handle family business and sends MJ back in.

MJ returns to the Oval Office. "Everything work out OK?"

None scowls back at MJ. "I don't want to talk about it."

It's quieter than a library, as None and MJ read the investigation report.

Shen barges in with a smile. "I worked out the mortgage payment plan with the bank. Everything should be fine."

None says, "I can't just save Ma, I have to save everyone. Shen is there anything in this report, any law, that can take down Renquist, or Slam Dunk Jobs?"

"No," Shen says.

"Government can't stop Renquist, but I will," None says.

"We're back to where we were before you were president," MJ says.

None says, "A hostile takeover of my own company doesn't sound crazy anymore. Heck, it might be one of the more normal things I've done this year."

MJ asks, "Unless you joined the billionaire super friends since the last time we talked about this, who's going to buy Slam Dunk?"

None says, "That was before Slam Dunk Jobs had a hundred million customers. A billion dollars is nothing for a company like that."

Shen types up some numbers on his laptop. "The financials on this company are amazing. The hard part is picking who to sell to."

None, MJ, and Shen vet potential billionaires late into the night. They have to be sure they don't replace Renquist with someone just as ruthless. Shen writes up a contract that gives None 10% and MJ 5% stakes in the new company, with None as CEO. The terms comprise his "finder's fee" to the investors. With a $1.4 billion offer, the plan is to find two equal partners with None and MJ as tiebreakers. The number of partners must be small to keep the agreement secret.

The next morning None calls Rick Slater of Slater Dynamics. Rick developed a good working relationship with the president when he helped build the IWantMyAmerica Portal. He's an ideal partner.

When None brings up a business partnership, Rick is dubious. Once Rick hears Slam Dunk Jobs can be purchased, he begs to sign the paperwork the same day.

None says, "Olivia Ford already knows the secret about Slam Dunk. What if we made her a partner."

MJ says, "Can we can trust her?"

None scowls. "I don't know."

Shen says, "It's September already, we're overdue to sign new contracts anyway. We can hit two birds with one stone."

The sleepy trio cram into the motorcade to meet Olivia and Corella back at the studio.

Shen stays in The Beast to work on the contract. The presidential limo is set up like an office.

Olivia and Corella greet None and MJ as they enter their former TV studio.

None says, "Olivia, we have a once in a lifetime offer that I want to talk with you about."

Corella asks, "Why don't we discuss your offer in my dressing room?"

None flags over the lead of his security detail. "Agent Vincent, can you please sweep Corella's dressing room for cameras and electronics? The less shocking, the better. I don't intend to beat any ham sandwiches today."

Agent Vincent grins back. He sends several agents to the dressing room.

Shen arrives with a briefcase. He hands documents to Olivia and Corella. "Sign an NDA to cover this meeting."

"Why?" Olivia asks.

Shen says, "You blackmailed us. That doesn't inspire trust. We need to feel safe divulging secrets."

Olivia smiles back. "I prefer to call it leverage in the contract negotiations. This NDA covers our secrets too?"

Shen nods.

An agent reports to Agent Vincent. "We disabled some hidden cameras. The room's clean."

None smirks at Corella. "Lead the way."

Corella scowls back and leads the group into her dressing room.

Shen makes sure everyone, including agents, sign the NDA forms. He and Olivia get a copy of each form.

Olivia says, "Everything is signed. What's your offer?"

None asks, "Can you and/or your business partners come up with half, which amounts to $700 million?"

The question shocks Olivia. "How did you know? Oh. You don't know anything. You're fishing."

None says, "In return for discretion, I'm offering you the business deal of a lifetime that will recoup your investment many times over. Can you do it, or not?"

Olivia looks to the president. "To answer that question, I need to know that no one outside this room can hear the conversation. Can your agents go dark and turn off their communications equipment?"

None gets confirmation from his agents.

The agents turn off their earpieces and put them in their pockets.

"I can get the money if the deal is amazing," Olivia says. "What are we buying?"

"Slam Dunk Jobs."

Olivia stands silent for a few moments, stunned. "I deal in people and relationships, not money. Slam Dunk Jobs typifies the greed I try to ignore on my shows." It takes a few more moments for curiosity to overcome her self-righteousness. "There's no way you can buy it for just $1.4 billion."

None points at Corella. "Ask her. Somehow she got a copy of our contract with Renquist."

Corella nods. "Exploiting a loophole Renquist left for himself, smart. I'm in."

"Sorry Corella, we don't trust you," Shen says.

Olivia says, "I can get money from my family, but I need absolute secrecy."

"Your family isn't rich." Corella looks at Olivia strangely.

Olivia says, "I confess. I was born rich. As a child, I was surrounded by famous people, rich people. I didn't know that wasn't normal life. I was born to be famous. I understand the unique pressures of fame. It's why I'm such a good celebrity matchmaker."

Corella gasps in shock. "Your rags to riches story was a lie? That's a reveal."

"I did live a rags to riches story," Olivia says.

"Why would you need to hide wealth?" None asks.

Olivia says, "When I was growing up, I loved the story *The Prince and the Pauper*. Most people identify with the pauper. I identified with the prince. I wanted to understand life from the other side of the red carpet. Anything I achieved in life would be attributed to my family and their money. My favorite actor, Nicholas Cage, showed me the way out. He was born as Nicolas Coppola. His uncle is famous director Francis Ford Coppola. He changed his name to Nicholas Cage to avoid the appearance of nepotism."

"You changed your name, just like MJ and I did," None says.

Olivia nods. "Yes. I won't tell you my birth name, but if everything checks out, I'll put you in touch with surrogates, a couple of steps removed, for my family's business people."

Corella goes into reporter mode. "Did you ever take any family money?"

"Just once, as seed money for the Reveal Channel." Olivia looks deep into Corella's eyes. "If you're thinking of using this as leverage against me, consider that you don't know what powerful family sits on the other end of my secret."

"To sweeten the deal, I want off the hook for season two of The Internet President," None says.

Olivia says, "Let's compromise. We'll go back to six episodes."

The compromise irritates Corella, but she remains silent.

Olivia says, "No more surprises on air. Corella will still control show content, but you'll know what's coming, as we agreed."

Corella holds out her hand. "We need something stronger than a contract. Give us your word. Make an oath to go with the contract."

None smiles at Corella. She understands his hatred of contract trickery. Corella knows he's a man of his word, an oath keeper. Maybe Corella understands trust after all.

Shen creates a simple new contract.

None, MJ, Corella, and Olivia sign the contracts and make an oath to uphold the provisions of the contracts.

Corella hands None and MJ envelopes. "Here's your tickets for the season two opening episode of *The Internet President*."

"Why would I need tickets to my own show?" None asks.

Corella moves behind her desk and backs away. "They're not tickets to your show. These are tickets to where we're filming the episode: the Astrid and Renquist wedding."

None's face contorts into a mangled disaster of shock, fear, and anger. Corella does understand trust, how to play his trust.

Corella scowls. "I wish I had that look on camera."

MJ drops her ticket. She prepares to lunge at Corella.

Corella watches MJ closely, preparing for her to pounce.

"Self-control." MJ smiles at Corella through clenched teeth. She stoops down to pick up the ticket.

None glares at Olivia. "You said no surprises."

Olivia smiles back. "No. I said no surprises on air."

None wonders whether Olivia will be a trustworthy partner. Did he have a choice?

MJ stares at the ticket. "This says $1,000. Who pays a grand each for wedding tickets?"

None sighs. Renquist forces wedding guests to buy expensive tickets. Leave it to Renquist to hold an expensive wedding at a profit and Astrid to use it as a marketing opportunity. Perhaps such soulless creatures deserve each other.

He ought to unsubscribe from Astrid news and take a date to the wedding, to prove he's over her. "I need a plus one for Zelda."

"Jealous girlfriend, that'll make good TV." Corella pulls out a handful of tickets and gives one to None.

"I'll take one too," MJ says.

Corella puts the other tickets in her purse. "Sorry. I ran out of tickets."

No is Astrid's word. She took it back from None when she dumped him. By making None give an oath, Corella stole the word "no" from him as well. Even if he has to attend the wedding of his nightmares, None won't break an oath.

61. **Grapple**

MJ has to tell someone. It can't be None, or she'll put him in danger. Who can she trust? Even in the comfort of her dojo, she no longer feels safe.

Barefoot, MJ stands in a white Brazilian Jiu-Jitsu gi with a black belt. She prefers it over a Karate gi for the double reinforced knee padding.

She schedules her security briefings at her dojo to make them bearable. Each new security briefing lasts longer than the last. She's thankful for the distraction from her worries.

MJ marks a start position on the mats with intersecting pieces of yellow tape. She practices karate katas in her dojo, while her adviser drones on. Katas are a sequence of choreographed moves, like the forms of Kung Fu.

The adviser covers the civil unrest and riots in Spain. The State Department releases an updated travel advisory, warning U.S. travelers to avoid the country.

MJ bows with clenched fists to begin her next kata over the yellow X. Slow graceful movements, swift kicks, and punches act as musical notes in the composition of action. MJ shouts with a forward fist strike.

The loudness of her kiai startles the adviser.

A few more moves, then she returns to the start position to end the kata.

The adviser resumes talking about austerity protests in Italy. The declining economies around the world grow the ranks of the desperate and needy.

Another kata, another country in despair.

The memorized movements help MJ endure the infinite loop of bad news. Almost 200 countries populate the world. How long will the briefings get if all of them fall? The thought chills MJ. After covering 23 countries in Europe, North Africa, and the Middle East, it already feels like forever when the briefing ends.

Normally her katas bring MJ calm and peace, but she remains on edge. MJ continues to practice katas until Fernando arrives.

Fernando approaches barefoot in a red gi with a black belt. He carries a bag with protective gear. "This dojo is just what I need to keep my head clear. I wasn't sure what your dress code is here. I often use street clothes when I practice Kali."

MJ says, "I want to practice some grappling, so a gi is perfect. Since you're training me, what should I call you? Guru Fernando? Sensei? Secretary of Defense Rivera?"

"Fernando is fine. We don't need fancy titles in the dojo Madame Vice President."

"Call me MJ." She returns his gaze with a playful smile.

Fernando says, "176-pound Brazilian Jiu-Jitsu black belt, Royce Gracie, won three of the first four UFC fights, proving grappling techniques can overcome size advantages."

There's nothing he can teach her about UFC history. She recorded every fight. MJ grins and imagines her rows of UFC recordings at home.

The grin doesn't last. MJ remembers the break-in when Renquist's henchmen tore through her media collection looking for her video surveillance recordings. MJ scrunches her nose. They stole the UFC fight videos along with her katana and everything else. She feels dirty and violated reminiscing about more stolen memories.

Fernando picks up on her mood shift. "Did I say something wrong?"

MJ forces a smile. "No. I just want to hurry up and see how long it takes you to get me down on the mat."

Fernando asks, "You sure grappling isn't an excuse to get close to me?"

MJ pretends to be insulted, then flirts. "Maybe." Her playfulness covers an underlying urgency.

They bow and circle each other in fighting stances.

Fernando sweeps her left leg and pins her on the mat in two seconds. He straddles on top of her. "Was that fast enough?"

MJ whispers a little too loudly, "Help me."

Her Secret Service detail closes around her.

Fernando releases her immediately and rolls off with hands raised.

MJ speaks to Fernando loudly to fake a misunderstanding. "Help me *learn*." She waves off the agents. "This might get rough, don't intervene."

Her security detail disappears into the background.

"Same positions," MJ says.

Fernando straddles over MJ and shoves his right arm underneath her neck.

Her back on the mat, MJ signals she's ready. MJ tries to push Fernando's chest away with her left arm.

Fernando shoves her left arm aside. He locks his own head next to hers. They face opposite directions only inches away. "I'm going to choke you using my arm and your shoulder."

MJ whispers, "Act natural."

It's pretty clear to Fernando that whatever is going on with MJ, it's definitely not natural. He completes the choke hold with one hand on his bicep and the other on his head.

MJ whispers again. "Stay whisper close. Don't let go."

Fernando can't see her facial expressions facing down. He loosens his grip to read the situation.

She whispers urgently. "Tighter. Sell it."

He tightens all the way.

The pain is intense. MJ resists the urge to tap out. She whispers. "Can't breathe. Too tight."

Fernando loosens up enough to relieve most of the pressure while looking convincing. Without any context, Fernando is confused. He whispers back. "What are you talking about? I like you, but not enough to play mind games."

MJ says, "Can I trust my agents? I don't know who I can trust."

"Is that why we're whispering?" Fernando clues in.

MJ used grappling as an excuse, to get close enough for a secret conversation. She would enjoy his hulking presence pressed up against her if the armlock didn't hurt. "Do you know what a dead man's switch is?"

"Of course. I'm Mr. Espionage. I used to go on covert ops as a SEAL. I'm Secretary of Defense. The NSA reports to *me*." If Fernando whispers any louder, the agents will hear.

MJ needs to calm him down. She whispers calmly without emotion. "OK. OK. I had two dead man's switches. If anything happens to me, they release evidence against Renquist. That doesn't work when your dead man's switch dies first."

Fernando asks, "How did they die?"

"Accidental drowning. She's aquaphobic. She took sponge baths instead of showers. No way she went to a beach willingly."

"I'll investigate. What's her name?"

MJ chokes up. She remembers the day she handed Thelma the packages that marked her for death. "Her name was Thelma Livingston, an executive assistant at Adaptive Unlimited."

MJ can't hold back tears any longer.

Her crying attracts the agents to check on her again.

MJ waves off agents in irritation. "Pain is part of the lesson. Let me practice some damn grapples." MJ wipes tears away awkwardly with her free arm.

To help her maintain her cover, Fernando yells at her to submit.

The agents relent and give MJ space again.

"What's the other name?"

"He's a scientist. No. I can't tell you."

"Do you trust me?"

MJ thinks about it. "I don't know."

Fernando whispers, "Listen carefully. Make a go bag, to survive on the run. Put emergency supplies in your go bag, so you're ready to go on a moment's notice."

MJ whispers back, "Isn't that for doomsday preppers?"

"Anyone can have emergencies. If your scientist disappears, instant emergency."

"I know at least one cabinet member infiltrated this administration." MJ catches herself before she raises her voice to audible levels.

Fernando struggles to keep his voice down too, but the anger still bleeds through. "I hope you're not accusing me."

MJ says, "Anyone might be working for Renquist. I trust you enough to have this conversation."

"Ask to grapple if you need to talk. Watch your six." Fernando makes his last whisper.

MJ taps out. It's been long enough the agents might be suspicious.

Fernando teaches MJ some Brazilian Jiu-Jitsu headlocks and armlocks to make their conversation seem more natural.

62. **Strike**

Shen sends the counteroffer to Renquist's law firm.

None, MJ, and Shen celebrate in the Oval Office.

Filing the paperwork to take back Slam Dunk Jobs feels like the biggest victory yet for None. He gobbled a major power-up. The anticipation surges through None's veins like an overdose level of drugs.

MJ acts so crazed with happiness that she hugs None without him asking.

Shen joins a group hug.

They have little time to celebrate before the worsening economic climate captures their focus.

The foreclosure data comes in much worse than expected, triple last year's numbers. On a normal day, skyrocketing foreclosures might be the lead story, but much bigger crises loom.

Ukraine defaults on its debt, taking down four large European banks as collateral damage. Ukraine remains the fault line of growing tensions between Europe and the Russian sphere of influence. They use the specter of closer ties with Russia to scare a coalition of American and European interests to bail them out. Ukraine is too great a prize to give up at any price.

Governments bailed out failed banks in the previous economic crisis. Those bank debts became government debts. When combined with otherwise normal levels of government overspending, many sovereign nations found themselves on the edge of bankruptcy and collapse. When the European Central Bank bailed out EU member nations, they enacted harsh repayment terms, called austerity.

Cyprus and Portugal become the latest two countries to vote to leave the European Union over the harsh terms. Their economies struggle more than countries not forced into austerity.

Because of the interconnected nature of global finance, the sovereign debt crises become contagious. All the talking heads on TV discuss is which nation will be the next domino to go bankrupt.

Lower interest rates, negative interest rates, qualitative easing, stimulus: none of it works anymore. Currency wars move the pain around, achieving nothing. The economy is a junkie that's had one fix too many.

Things are so terrible that an American foreclosures crisis barely merits a mention. None finally caught a break. The international problems might distract the media and buy him more time to take back Slam Dunk Jobs.

The American economy might outperform the global mess, but the cracks show. Just when Americans could use more dollars in their pocket, they have less because of Slam Dunk Jobs. Unions prepare to fight back.

The main union contract for Zentripity Energy in Los Angeles ends today at midnight. All the utility workers vote to strike. They vow to hold the line on wage cuts. The battle lines are drawn.

After the immigration debates featured Zentripity Energy, the union strike holds personal relevance to None. "If Renquist defeats our counteroffer, a union strike is the best attack against Slam Dunk Jobs."

MJ asks, "How big a crisis is this? Do we need to get on a plane?"

None says, "I'll give a speech at Zentripity. We can close off the streets with the Secret Service. It'll block the entrance better than a picket line."

"We can monitor the strike on Air Force One," Shen says.

None, Shen, and MJ board Air Force One and leave for L.A. They watch media footage on the plane.

The D.C. news anchor says, "We're going live to our Los Angeles affiliate LDR News."

LDR News reporter Tim Houser with the Word on the Street crew interviews picketers in front of Zentripity Energy headquarters. "I'm here again with striking workers from the Utilities Workers Federation local 306. Will your union offer concessions to Zentripity Energy?"

An electrician, wearing a hardhat, answers. "No. All my non-union friends get small weekly pay cuts. Only strong unions can fight back. I'm confident they'll give into our demands. They need us. Otherwise, entire portions of L.A. will be without power."

Tim asks, "Zentripity energy outsourced the entire IT staff. How can you be so sure they'll work with your union and not hire replacements?"

"It took them months of planning and coordination to replace less than 300 IT jobs. There are close to 3,300 union members working at Zentripity. There's no way they could replace us."

Tim says, "Zentripity Energy just released a prepared statement. There will be no contract negotiations. All union workers will be permanently fired at midnight when the contract ends."

The electrician throws his hardhat to the ground. "How?"

Yells, screams, and crying work their way down the picket line.

Tim taps his earpiece. "I'm told 25 Slam Dunk Express airplanes are in route to Los Angeles airspace."

The electrician screams, "They're flying in scabs. Shut down the airport."

None gasps. "Of course. Slam Dunk Express is a union buster. When you have the entire country to pick from, it's easy to find desperate Americans willing to cross picket lines."

Shen says, "Still, it'll be hard to organize that many people."

None shakes his head. "A subsidiary of Renquist Aerospace, Renquist Logistics, subcontracts to U.S. Armed Forces. Renquist can coordinate armies."

"It sounds like he just did," MJ says.

None says, "If we beat them to L.A., Secret Service will clear airspace for us. That should slow them down."

None, MJ, and Shen eat dinner on Air Force One while they watch the strike play out.

Workers from various unions join in solidarity to shut down the airport. Teachers, nurses, public employees, they all know what's at stake. Unions organize picket lines across Century and the 96th street bridge to block access to the LAX airport.

Union members pile into cars by the thousands and overload traffic around LAX. They park in the terminal, until the police force them to leave, then repeat the process, circling the airport.

Airline passengers leave shuttles behind and hoof it into the airport on foot. Few will make their flights. The airport fills with stuck passengers unable to easily leave the airport.

The first Slam Dunk Express plane lands and taxis from the runway towards a gate in terminal two to unload its passengers.

Unionized ground service workers surround and block the aircraft with baggage carts and pushback tractors. They completely trap the aircraft.

Tim interviews happy workers in the picket line. They celebrate the stuck plane.

The TV shifts to the LDR anchors Julie Reed and Bruce Cannon in the studio with the caption 'Breaking News.'

Bruce says, "We just received an update on the 25 airplanes entering Los Angeles airspace. Only two of them were destined for the main LAX airport. They're scheduled to arrive at every airport within 100 miles. We have Burbank, Santa Monica, Palmdale, Van Nuys, Ontario, even small airfields with one runway, like Lancaster. They've rented private airstrips. This is like a military operation."

Shen asks, "We're still two hours out. Will we get there in time?"

"Strikes can last for months, we'll be fine," None says.

On TV, Julie says, "I don't recall ever seeing shock and awe tactics used against a union strike before."

Bruce says, "Let's check in with the LDR News Copter."

The helicopter-mounted camera films a convoy of buses escorted by black vans speeding southbound down the 405 freeway. The co-pilot says, "Slam Dunk Express offloaded its passengers at the Van Nuys airport. Replacement workers are headed to Zentripity Energy."

The news copter follows the convoy like a police chase without the police. News crews shift between dueling updates on Slam Dunk Express replacement workers and union members trying to stop them.

Union members abandon their blockade of the LAX airport. They head for the Zentripity picket line. Unions double down for a final confrontation.

Convoys of buses converge in a parking lot a mile from Zentripity headquarters. Slam Dunk Express didn't just deliver replacement workers, but security as well. The news copter watches while hundreds of security guards don riot gear uniforms with helmets and plastic shields.

The buses of security guards empty a block from Zentripity headquarters. They close off the street and march in phalanx formation. They wield pepper spray and interlocking shields like intimidating Roman soldiers. The security guards earn their danger pay. Rocks and bottles bounce off shields as they approach the picket lines.

Buses with the new Zentripity employees fill the entire street behind hundreds of guards. After a brief skirmish, the guards create a wedge large enough for the buses to enter headquarters. Bus after bus goes inside.

It's over. The unions lost. With union membership at a 100-year low, they aren't strong enough to fight back, at least for now. Demoralized and shocked, workers drop their picket signs.

"Aiya! We were only 30 minutes away." None puts his head in his hands.

MJ asks, "Why did Renquist have to fly them in? There are millions of people in L.A. They could have hired most of them there."

None perks his head back up. "This was a demonstration of power. Union, or otherwise, there's no industry he can't touch."

Shen says, "The battle is over. Now what?"

"We don't want a photo op with an abandoned picket line," MJ says.

If they land in L.A., they'll have to answer uncomfortable questions about the strike. None will be ready next time. "Agent Vincent, turn the plane around."

Lasting repercussions follow the Zentripity strike. After the victory of Slam Dunk Express, small unions give concessions, including wage cuts, in their new contracts. Large unions still think they can win.

A few days later, Renquist uses a company he controls as the next test case. He offers the 85,000 workers in the Galaxor Telecom union a 60% pay cut to renew their contract. They decide to strike, confident in their numbers. Besides a much larger headquarters, Galaxor Telecom has 4,023 kiosks, field offices, and phone stores spread across the country. Defeating the strike requires the ability to simultaneously deploy new employees to thousands of locations and take down the large central headquarters.

Companies choose where to list job openings. That simple fact defeats boycotts and picket lines. In spite of hating the app, most workers have no other choice than to go where the jobs are, on Slam Dunk Jobs.

Renquist runs tearful ads of the long-term unemployed recounting how Slam Dunk Jobs ended years without work. The unemployed who had given up, install the app, willing to accept any job, even sales jobs where they cross picket lines.

The Slam Dunk app replaces workers at the small stores nationwide without a problem. The challenge is the 8,000 workers at Galaxor Towers, a 39 story skyscraper headquarters in Baltimore.

Before the strike, Renquist built brick walls around Galaxor Towers. It keeps the protesters at bay but provides the protesters with a choke hold to close off the building, at the front entrance.

Rumors warn Renquist will break the strike tonight.

When Renquist busted the smaller union, he deployed his own security guards. Against the larger union, there's much more at stake. How far will Renquist go?

None must stand with the workers, but this confrontation could get violent. As a precaution, None convinces MJ to stay behind. It's too dangerous for them both to go.

None and Shen travel 40 miles to Baltimore in a motorcade at sunset. Renquist shouldn't have picked a fight so close to D.C.

Renquist utilizes all his resources to break the strike. It pushes Renquist Logistics to its limits. Slam Dunk Express deploys all planes in its fleet to deliver new employees to the main headquarters.

Unions stand ready to lock down every airport in the entire Baltimore area against the Slam Dunk Express fleet.

Renquist lands his planes at surrounding cities instead, Washington D.C., Philadelphia, Wilmington, and Richmond.

Secret Service agents and police block off the roads in front of Galaxor Towers for None to address the crowd. Only emergency lights glow in the darkened skyscraper, providing a foreboding backdrop.

None delivers a speech about income inequality.

Spectators and strikers stretch for a few blocks in front of Galaxor Towers.

None talks about the dangers of corporate control of the government and media as roadblocks to fixing income inequality.

Murmurs interrupt his speech. A phalanx of guards in riot gear approaches with buses of people behind them. They stop 50 feet from the crowd and deploy plastic riot shields.

The crowd strips every pebble from nearby landscaping. An eerie silence spreads as both sides prepare for conflict. The crowd readies bottles and rocks to throw.

None sets down the mic and picks up a picket sign. Shen follows his lead. Surrounded by a security detail, the brothers work their way through the multitude towards the stalled rows of security guards.

The two sides stare at each other in a standoff.

None wonders what the guards are waiting for. Maybe there are too many strikers, or they worry about being arrested.

Loud booms reverberate behind Galaxor Towers.

None and Shen ignore the noise and press on. They swim through a sea of people, towards Renquist's security guards.

None steps out a few feet into the gap between both sides. He gambles that the security guards aren't paid to risk their lives against Secret Service snipers. He waves his picket sign. "I'm the line you cannot cross."

The guards don't advance, even though his Secret Service agents on the ground look ominously outnumbered to Renquist's security guards.

None strides out by himself in the middle, alone with his sign.

The guards inch backward.

Cheers ring out behind None.

Replacement workers stay in their buses and drive away.

It looks like they won this time.

None feels elated until he turns around and notices the building.

Lights shine in the first five floors of Galaxor Towers. Then the sixth floor.

Agent Vincent informs None that the brick wall behind Galaxor Towers was demolished with C4.

Replacement workers march in the back by the thousands. That must be an exciting day at work.

The guards were a decoy. Renquist blew up the back wall. How did he sneak C4 inside the security perimeter? Renquist will go to any lengths to win. How far will None have to go to defeat him?

The success of Slam Dunk Express terrifies unions. Other unions make large concessions to their employers to quietly renew their contracts.

Slam Dunk Express crushing unions brings an unwanted spotlight to the economic repercussions of Slam Dunk Jobs. Mainstream reporters follow the connections from Slam Dunk Jobs to wage compression to deflation to the mortgage crisis to America joining the global economic collapse.

None and MJ lose major support from their political movement for their association with Slam Dunk Jobs and the collapsing economy. It will be hard to earn that support back. Taking a stand with striking workers helped None's image, but Renquist spreads conspiracy theories that None helped

Renquist distract striking workers so their replacements could sneak in the back.

The weekly wage auctions on Slam Dunk Jobs act like death by a thousand cuts. The wages of every profession with the slightest surplus compresses towards minimum wage. There remains only one minimum wage, with a college degree, or not. The value of a college degree in liberal arts drops to ten cents an hour above minimum wage.

Only practical degrees like engineering, or medicine keep some of their value. If Renquist ever takes Slam Dunk Express global and weakens immigration law, those fields won't fare any better.

Resentment builds among workers with so little to separate them. The industrious and skilled resent the lazy and incompetent for making the same. Their hard work and skill seem unappreciated. The experienced resent the newly trained for making the same. Their experience becomes almost worthless. Electricians make barely more than unskilled receptionists. The difference in social status between professions feels artificial when so little separates their salaries.

Wage compression moves up the food chain. CEOs force managers to use Slam Dunk Executive, the high-end version of Slam Dunk Jobs, which includes lots of perks, at least initially. Slam Dunk Executive jobs are deemed too valuable to be easily replaced. Their boss can adjust their salary, or press the red button labeled "Open Market." Managers live in fear of the red button. They see what the red button does to their subordinates every day. Despair travels through corporate headquarters with every quarterly result or sales projection that misses the mark. The red buttons are coming.

Only the CEO and board of directors remain untouched in corporate America. They reward themselves with bonuses and heave a sigh of relief that stockholders don't get a red button.

Large scale drops in wages begin a cascade of deflationary forces. Americans have less purchasing power, so prices drop to compensate. A more valuable dollar sounds great unless you have debts in those more valuable dollars. Mortgages, car payments, and student loan debt become lead weights around the throats of consumers. Any debt becomes toxic. Cash becomes king.

Central banks propped up home and stock prices with easy money in an asset bubble. Slam Dunk Jobs pops the bubble and keeps going. Real estate prices drop in a vicious cycle creating a mortgage crisis like never before. Even at reduced prices, buyers are too afraid to purchase a home, when they don't know how much it'll be worth next week, or how hard the debt will be to pay off.

Finally, stocks themselves capitulate, in spite of record corporate profits. Stocks trade on future value, and the future looks very gloomy. American stock indexes finally follow the direction of world markets, down. It's the last straw for the global economy.

The minimum wage puts a cap on Renquist's profits because he can't get bigger weekly bonuses for pushing wages lower. His lobbyists pressure Congress to repeal the minimum wage under the claim this regulation is preventing further "innovation" from his company. Renquist threatens to run attack ads if they don't vote for the repeal, but then again, attack ads will come after anyone voting for a minimum wage repeal.

Without a bottom, how low could wages drop? None uses his vanishing political clout to push back against lowering the minimum wage, much less a repeal. Congress hates None's strategy of using voters as leverage. They hate the heavy-handed threats from Renquist even more. None becomes the lesser evil, which earns him the backing of unlikely allies, including Senator Garfield. Members of Congress balk at removing the one thing keeping wages from dropping further and easily defeat it.

Renquist is one of the most powerful people in the world, but he has overreached. Defeated in his repeal attempt, Renquist takes satisfaction in one simple truth: most wage growth used to go to the top one percent. Now, it goes to the top one man: him.

The government can't beat Renquist. Unions can't beat Renquist. It's up to None and his partners to beat him.

63. **Burned**

To get married, Astrid ends all her casual relationships. Astrid throws away the discardables, the men in her life, to burn them in effigy across the globe. She turns her wedding into a breakup concert series in the world's fashion capitals. Spouses renew their vows to celebrate years of marriage. Renquist and Astrid will renew their vows dozens of times as newlyweds in their concert series of weddings. The first four scheduled stops are New York City, Paris, London, and Rome.

Demand to hear original Astrid songs sells out every concert date, in spite of exorbitant prices. Leave it to Renquist to have a wedding at a profit.

Renquist relies on Slam Dunk Express airplanes to move the wedding materials and staff across the world. He delays the dates of his weddings by a week because he used every plane to defeat the Galaxor Telecom strike.

The public has a love/hate relationship with Renquist. They admire his success, while they loathe his path to it. Astrid and Renquist act like American royalty, an unstoppable power duo. If Americans respect anything, it's winners.

The New York City wedding takes over Renquist Stadium, an 80,000-seat football stadium that he owns. Renquist originally named every building and company after himself, until he realized the advantage of not being associated with his companies. Scandals used to hurt all his brands, and it was hard to hide self-dealing between his companies.

None, MJ, and, Zelda enter Renquist's turf backstage at the stadium, accompanied by Secret Service agents, Corella and *The Internet President* film crew. They wait in a locker room, while wedding preparations continue.

None drowns in anxiety at the wedding of his nemesis and the woman he loved. He agreed to film a new episode of *The Internet President* at the wedding, but he'll control himself to deny Corella any juicy footage. None, MJ, and Zelda wear live microphones, but they know how to turn them off to avoid anything too embarrassing.

MJ notices Renquist in the same locker room. His bodyguards wear heavy body armor. They outclass the Secret Service details None and MJ have.

MJ moves towards Renquist.

His bodyguards close ranks and block her path. They scan MJ for electronics and force her to turn off her mic.

Renquist traces his finger along the scar her stiletto heel left on his cheek. "You won't get me alone again. If anyone leaves with scars, it won't be me." Renquist motions for them to let MJ through.

She leans in close to whisper in Renquist's ear. "Don't hurt anyone else and I'll give you all the packages."

Renquist smiles back and whispers. "I have Thelma's package. You'll be buried with the last copy." He smirks and glances at the ground.

An aid with a clipboard approaches Renquist. "They're ready for you, sir."

Renquist and his bodyguards leave the locker room.

Corella steps up to MJ and shoves a microphone in her face. "Your mic is off. What was that about?"

MJ flashes a false smile. "I was complimenting him on his scar."

Corella points at MJ's mic. "It's still not on."

MJ shoves the mic in Corella's face and turns it on. "Can you hear me now?" She nudges past Corella to follow Renquist.

None and his entourage join MJ in her pursuit.

They follow Renquist to an interview room with 150 folding chairs and plate glass walls. Financial analysts and major shareholders of Renquist Aerospace fill all the seats. They give Renquist a standing ovation.

None and MJ stand in the back. It's surreal for them both to see the man they despise so highly regarded. Many in this room owe Renquist for their wealth.

Renquist covers Q2 results for Renquist Aerospace. He made his first billion dollars with that company. Its most profitable division remains private military contractors fighting America's wars. His mercenaries cost far more than soldiers, but their names don't show up on casualty reports. That makes them a bargain to politicians, who lose votes for every dead soldier.

Renquist squeezes out extra profits by selling military upgrade to his contractors. They have to pay for anything beyond basic flak jackets and helmets. Unlucky soldiers return home in coffins, still in hock for their battle armor. That callous greed fits on a line item called internal sales. Renquist talks openly about a bill he's pushing through Congress to outfit soldiers with his battle armor. It's easier to get money out of Uncle Sam than dead employees.

One thing bothers None about military privatization. Where do their ultimate loyalties lie? To the company that hires them, or the government which ultimately pays their salary?

Renquist says, "Behold, the future of warfare. Introducing, Andraste Battle Armor, named after the Celtic goddess of war and victory. Her name means the invincible one. Our troops will be invincible wearing the latest protection technology."

At his cue, a mercenary marches in, wearing black futuristic armor made of interweaving layers of exotic materials with miniaturized hydraulics and a reinforced exoskeleton. Only the mercenary's face is visible through the transparent faceplate. With three-inch protective layers, the suit towers a full seven feet.

Investors jump to their feet and clap at the sight of the Andraste Battle Armor.

"When fully powered, the augmented strength has a two-ton lifting capacity," Renquist says.

The mercenary grabs two of Renquist's bodyguards in full armor and lifts them overhead with each arm.

Against an Andraste, his existing armors may as well be homemade. Two hapless guards smack at the vice-like grip of the new battle armor. Their feet dangle.

The mercenary flings both guards eight feet through the plate glass walls.

The old armor protects the guards against the shattered glass, but not the concussion. They fly another 15 feet, before coming to a stop. They don't get up.

A roar shudders through the awestruck crowd. Shareholders see dollar signs.

A TV behind Renquist displays a recorded demo of an Andraste under attack.

Renquist says, "The armor is resistant to small arms fire, flame, and bladed weapons. If a soldier steps on a landmine, there are extra protection layers below the foot. Explosions can cause long-term brain degeneration from as far away as 300 feet, but the Andraste's advanced materials can absorb shocks from concussion grenades, or IEDs. With a roadside bomb survivor, it can pay for itself long term with reduced medical bills."

Enemy soldiers lob grenades in the video. They narrowly miss the Andraste, which continues unfazed.

Renquist smirks. "Heavy armaments sold separately."

Employees pass info packets down the aisle.

MJ grabs the promo literature enthusiastically. She would definitely feel invincible in armor like that. MJ scrunches her nose in sticker shock. "There are houses you can buy for this."

MJ's comment elicits laughter.

Renquist says, "That's why we have a much cheaper, thinner, unpowered version for civilian use." For the rest of the earnings report, Renquist rambles on about expansion in the civilian security industry. They consult and subcontract to police departments nationwide.

For high profile civilian events, Renquist has formed Invincible Juggernaut Security (IJS). Staffed by former members of the Secret Service and special forces, IJS premieres today as primary security for the wedding.

Renquist ends his presentation. He takes questions.

An analyst with rimmed glasses asks, "You're getting married today. Are you taking time off?"

"You don't get this rich by taking a day off." Renquist points to another analyst.

"What would you say to Congress if they don't buy your newest armor?"

Renquist puts his hand on his chest. "My heart goes out to the soldiers who died because Congress was too cheap to buy them Andraste Battle Armor."

None has to say something. "Hypocrite. You won't buy your own soldiers battle armor, but you expect taxpayers to?"

The shareholders glare at None.

Renquist says, "This meeting is for shareholders. How many shares do you have, Mr. President?"

"I am None, and I have none."

Calling out Renquist in front of his investors quiets None's nerves. He takes Zelda's arm and leads MJ and their security details outside onto 50-yard line to find their floor seats. Once outside, the immensity of the stadium sinks in. None sweats from the afternoon sun.

A big stage covers one of the end zones with a video wall backdrop to compliment several other gigantic LED screens throughout the stadium. From the main stage, a double-wide fashion runway extends to the 50-yard line where it ends in a 30-foot diameter circle.

A 250-foot tall spire rises from the middle of the stadium at the center of that circle. Near the top of the spire, three metal arms extend five feet out. The arms each connect into separate series of pivoting zip lines that spread outward in a web across the stadium at varying heights. A pair of white angel wings hangs from the center arm that faces the opposite end zone from the stage. Dozens of high tension cables keep the spire firmly in place.

Three tiers of seating surround the football field.

The Secret Service and IJS coordinate heightened protection for the many VIP guests backed up by police and private security contractors. Facial recognition software scans the audience for potential threats. Sniper nests in the nosebleed seats behind the stage and helicopter patrols stand ready to stop any threat.

IJS agents stop None and his group. "We're upgrading your seat, Mr. President. I'm glad you could join us. You can bring one guest. The rest of your party will need to stand off to the side."

None fist pumps. "Being president has its privileges."

Agent Vincent eyes his IJS counterpart with suspicion but goes along.

None avoids MJ's gaze. He picks Zelda as his guest. It was a big ask getting Zelda to come. None can sense her insecurity. It was one thing for Zelda stand up to a two-dimensional Astrid in the Bistro. He brought her to a place where thousands practically worship his ex-girlfriend. Does she care that he picked her, or will getting a closer seat worsen her insecurities?

An IJS security detail shadows the president and his agents. They lead None and Zelda to a cordoned off area a safe distance from the center stage. Two men and a woman sit in the special section.

Zelda stops cold. She directs None's gaze to the man in a second-hand suit. Zelda whispers in None's ear. "That's Craig Steel. He was an A-list action star before he fell apart over Astrid. I heard he was homeless. I wonder how he earned the grand for his ticket."

None cringes. He knows firsthand how easily Astrid shreds a man's heart. There probably isn't much Steel wouldn't do for one last glimpse of her.

The other man has blue, purple, and green feathered hair. He wears the kind of impractical outfit only worn in fashion shows. The shirt and trousers puff out to give him a larger presence. A Glitter Girl sits next to him.

Zelda recognizes the other man. "That's MaxPlume." Zelda spews out years of celebrity gossip in excited whispers. MaxPlume is the "male Astrid" famous for Peacock Style, with multicolored hair and oversized indulgent clothes.

MaxPlume has his own following, the Peacock Posse. The colorful hair and bloated clothes represent a peacock's plumage, to attract women and attention. It's mostly male models and men trying to lure the hottest girls from fashion culture.

None looks around and sees hundreds of men dressed in a similar aesthetic.

Before Astrid went mainstream, MaxPlume's fans thought he would be the one to tame her. The day she dumped him, fashionistas wore black in mourning. He was the first ex-boyfriend to burn in effigy.

Zelda eyes MaxPlume's guest. She whispers. "Have you heard about Glitter Girls? I hear they're very friendly to anyone who met Astrid."

None chuckles uncomfortably. It's better not to mention the crazy from speed dating. He whispers back. "Glitter Girls? They don't ring any bells with me." It's not a lie. MJ was the one that rang the bell. None pretends to listen to Zelda's Glitter Girl explanation.

Whispering back and forth for several minutes in the aisle earns odd looks from other concertgoers. Corella and the film crew approach to investigate.

"Laugh, so they won't know what we're talking about" Zelda whispers.

None chooses a different explanation for their whispers. He kisses Zelda on the neck.

She plays along and kisses back.

Over Zelda's shoulder, None glimpses MJ watching them with a scowl. He focuses back on his girlfriend.

Corella's film crew captures the kiss in a closeup.

Corella smiles at Zelda. "A decent kiss, but I know from experience he's had better. Try again. Take two."

"I do my best work in private." Zelda French kisses None.

He's caught off guard. She considers kissing him work? Way to ruin the mood. None ushers Zelda to their upgraded seats, leaving Corella behind. They introduce themselves to Steel, MaxPlume, and his Glitter Girl.

Zelda looks at MaxPlume, then Steel. She grabs None's arm. "You're getting burned today."

It rings true the moment she says it. The seat "upgrades" aren't because None is president. Astrid gave him a good seat to watch his effigy burn. None holds back tears. He's not giving Astrid, or Corella's camera crew the satisfaction.

Steel puts his head on his lap. He looks like he's on a plane about to crash.

None slumps into his upgraded seat. If None doesn't fix the economy, Astrid won't be the only one burning him in effigy. Watching Astrid achieve closure at his expense might be painful, but it's temporary. The breakup song that goes with it is permanent. The song will bind None to the dead relationship forever, lost in Limbo.

MaxPlume was one of the early ones, the lucky ones not linked to a breakup song. None looks at how easily MaxPlume copes and how fragile Steel is. Curse whoever taught Astrid to sing.

None listens to Zelda prattle on about the emotional devastation left in Astrid's wake. If he wanted to know all that, he'd watch the damn Astrid documentaries. None wishes Zelda would quit talking. She's traumatizing Steel.

The Plumage subculture had roots in fashion culture. It grew bigger than disco and became 100 times more ridiculous. MaxPlume popularized extreme hair and clothes, but Astrid was the poster child that transformed Plumage into a dominant mainstream style. Even cop dramas add Plumage fashionista CSI technicians these days.

All the Glitterati, the famous people in Astrid's universe, arrive for the concert. Attendees represent all the factions of the Plumage subculture.

Glitter Girls and the Plume Posse are out in force. So are hair purists: a sect of those who believe extravagant outfits distract from the hair itself. They wear only simple single-color clothes that contrast with their outrageous hair.

None gets up to people watch. He sees Afros with Christmas ornaments, six-color dreadlocks, artificial snake Medusa hair, and weaves with glowing neon wires. Every hairstyle and experiment imaginable is here, including programmable hair. The wedding is a place to be seen.

The only hairstyle that tempts None is the programmable 3-D printed hair, with actuators that move in preset patterns. None ponders what he would program his hair to do. None imagines hair dance moves. It's a welcome distraction from the Zelda commentary track.

Zelda finally quiets with the opening act. A fashion show. It looks like the unholy union of a science fiction convention and an avant-garde hair expo.

Next up, macho crooner, Dane Hadrian, sings about his many female conquests. Debauchery personified, he's a one-track mind, set to music.

Zelda turns into a chatterbox again, about Hadrian. A master player, Astrid broke up with him when he wrote a conquest song about her. She vowed never to date another musician.

None instantly dislikes Hadrian. He's the jerk that taught Astrid to sing and made her bitter enough to weaponize music. None shushes Zelda. "I'm a prisoner of my word. I'm not here by choice. If I hear one more word about Astrid or her men, MJ will take your chair."

Zelda turns ice cold. "Don't worry. I'm done talking to you."

None enjoys her silence. He'll deal with the fallout later.

The crowd stands at their seats when Hadrian leaves. They chant Astrid's name.

Stagehands push a barn facade and fence onstage. Cows and panoramas of the French countryside show on the video walls.

A family in farm clothes rushes onstage.

Astrid wears a plain polka dot dress and dirty shoes. A silk headscarf hides her signature hair. Astrid sings about leaving the family dairy farm behind to seek adventure.

None laughs to himself. A dairy farm. That's how Astrid knows so much about cheeses. She rarely talked about France.

An American tourist walks by. He asks for directions back to town.

Clips of American movies with French subtitles play on the video walls.

Astrid latches onto the tourist and peppers him with questions about America. She hopes to meet a celebrity someday.

He tells Astrid that she's so beautiful, she should be a model.

Her family laughs at the idea. Her place is milking cows.

Astrid announces she's going to be a model in New York City and convinces the American to take her with him.

The backdrop switches to video of New York City.

Astrid pulls a suitcase with an awkward walk. She sings with wonder about her new home and gawks at every little thing.

The video wall switches to fashion shows.

Stagehands push out racks of outlandish clothes.

Models dance around Astrid. They roll her suitcase away and steal her things.

Gullible and trusting, Astrid sings thanks for putting her suitcase away.

The models ridicule Astrid's uneven walk and rural background.

Makeup artists take off Astrid's glasses, uncover her hair, and give her a makeover.

None imagines an innocent Astrid. He muses over the different trajectories their lives would have taken if he met her then. President wasn't good enough for Astrid. With smaller dreams, she might still be his.

A stagehand holds up a plastic sheath around Astrid. They shake it for six seconds and drop it.

Astrid steps out transformed into runway fashion with her hair down. Her delight at wearing cutting edge fashion broadcasts across the room. Wide-eyed with a rapturous smile, Astrid emanates happiness. Her lopsided walk feels endearing and authentic. With a hint of a skip, Astrid rushes down the double-wide runway to the spire at the 50-yard line. The floor lights up with white LEDs as she goes.

None stares at Astrid, transfixed. It's hard not to miss her.

Her distinctive walk and delight on the runway make her a star. The audience yearns to feel the same untarnished joy that she does. Astrid sings an ode to clothes racks and big city magic.

The models sing how they wish they were Astrid, with her walk and wide-eyed bliss.

Shoppers rush to the clothing racks and leave with copies of Astrid's dress.

Dancing photographers with telephoto lens surround Astrid, flashing pictures.

A dancer meant to be Steel exchanges glances with Astrid. Dancer Steel parts his way in. They tango with seductive moves.

The middle LEDs of the runway turn red.

Astrid skips hand in hand down the virtual red carpet with Dancer Steel all the way to the spire. They wave to the audience as they circle the spire. They leave the spire and step a few feet onto the LED runway.

Cables from offstage tug at the runway. The LED floor panels become a moving sidewalk.

Dancer Steel kisses Astrid with closed eyes. They creep past the audience in a frozen moment as the LED tiles move towards the stage.

The audience hollers at the everlasting kiss.

For the first time since None met him, Steel chokes up at the memories.

A minute later, the kiss ends when the floor panel they're on slides back to the main stage. They walk off the panels onto the stage. The LED floor panels disappear offstage.

The video wall shows mansions, fancy cars, Rodeo Drive in Beverly Hills, and screaming throngs of fans.

MaxPlume shakes his head. "I made Astrid a fashion goddess. How could they skip me?"

His Glitter Girl hugs him. "She told your story years ago, without song. You had your burn day. You're remembered. Look at all your fans."

MaxPlume acknowledges how many people dressed in Peacock Style. He smiles.

A camera rolls across the stage with action scene dancers.

Dancer Steel mock fights with a set of ninjas who drop and play dead. A boulder comes his way. Dancer Steel does a rolling dive off-screen.

Astrid sings about Hollywood dreams, bigger dreams.

When Astrid said bigger, she wasn't kidding. A motorized crane with a car-sized base drives onstage controlling a 20-foot tall marionette that resembles Steel. Dozens of steel cables and pulleys control the steel body and papier-mâché head of Giant Steel. With only a driver visible, the marionette must be controlled remotely.

None imagines Renquist controlling an army of marionettes against him. When will Renquist step out of the shadows and attack him directly? He's president. Isn't he worthy of more than henchmen?

Astrid strolls next to Giant Steel, as they cross the stage. She sings to upbeat music about her Hollywood fairy tale, in love with an industry giant.

Giant Steel follows Astrid down the runway, with large steps.

Astrid skips ahead of Giant Steel. She stops below the angel wings, suspended from the center arm at the top of the spire.

The center arm lowers a cable with the wings down to Astrid.

Stagehands strap the white angel wing harness onto her.

The cable lifts Astrid up, five feet from the spire.

Astrid sings of her wings of love, soaring high. A love as strong as steel, it will surely last forever. Given all the men she's dated, Astrid plays up the irony.

Giant Steel waves at the crowd as it circles back down the runway towards the stage.

From top of the spire, Astrid zooms down the zip line attached to the center arm. She heads for the end zone away from the main stage.

The audience gasps in amazement at the spectacle of Astrid swinging overhead. She definitely wore underwear today.

"Woah!" None didn't realize Astrid was so brave.

Astrid slows to a stop at the very back of the top section of seats, near the lighting fixtures.

Her audience goes insane with cries and screams. Fans in the worst seats glimpse their fashion goddess. Astrid risked her life to dazzle them, and they adore her for it.

Astrid pauses her song to wave at her admirers.

Giant Steel goes offstage.

She pushes off the wall. Astrid pivots and zigzags down the series of ziplines. She stops at the front of the top seating tier, then the front of the middle tier. At both junctions, Astrid dangles herself over the crowd in the tier below and waves, before flying off.

The final zip line stops on the main stage.

Click.

Her harness disconnects from the cable.

Giant Steel returns onstage with a human-sized cage. He drops the cage over her.

She ducks as the cage crashes down around her with the thud.

With her angel wings, Astrid is too big to squeeze through the bars. She laments in song that Steel showed her the world then locked her from it.

She swears she will never love again. Her wings of love fall off.

Astrid squeezes through the cage bars to freedom.

Giant Steel backs off and falls apart. Steel panels drop with a loud clank to reveal a papier-mâché body underneath.

Astrid confronts the marionette. "You claimed you were steel, but you're weaker than paper."

The papier-mâché version of Steel slumps to the ground.

Steel clutches his chest.

The video wall replays Steel clutching his chest in slow motion.

None worries Steel might have a heart attack. He stands in front of Steel to shield him from the video wall cameras. None won't allow them to continue broadcasting Steel's misery.

Corella's crew films from a side angle, where they can still capture Steel's suffering.

Steel pushes None to the side. "I have to see this for myself."

None returns to his seat. Steel doesn't want his help.

Astrid raises one arm.

A torchbearer runs on stage and puts a torch in Astrid's raised hand.

She lights up the effigy without hesitation. Flames consume it with a wall of heat and plumes of smoke. Ashes float into the air.

Astrid sings about the death of love, which ends in ash and sorrow. She catches ashes from the air. It blackens her hand. Astrid slumps to the ground in tears.

MaxPlume's Glitter Girl tries to console Steel.

He shoves her away. "Glitter doesn't make you her." Steel pulls his rumpled suit coat over his head. He sobs uncontrollably.

Hadrian, one of the opening acts, comes back onstage. He strides towards Astrid and sings how music will get her through. The sun isn't broken. Light will come again.

Astrid perks up and smiles at Hadrian.

He helps her up and serenades her with stories of epic passion.

The video wall flashes snippets of music concerts, dance parties, and music videos.

Astrid and Hadrian sing a happy duet.

He rushes offstage. In his place, a motorized crane controls a 25-foot wooden marionette that resembles Hadrian. The steel cables move in time with the music. Giant Hadrian dances.

Astrid sings about finding love again. She caresses the bars of her gilded cage with her angel wings still inside.

Giant Hadrian lifts the cage.

With outstretched arms, Astrid announces she found love again. Anything is possible.

Stagehands strap the white angel wing harness back on her.

Astrid runs to the center of the spire. She promises herself she won't be a mere star. Astrid will be a supernova.

A cable lowers from the arm on the left side of the circle. Stagehands attach Astrid's wing harness to the cable.

She goes through another love song as she ascends the left side of the spire. Astrid flies down a separate set of zip lines through the stadium as the love song continues.

Giant Hadrian lip syncs to Hadrian's music, while a set of dancers swirl around the marionette, blowing kisses.

Stagehands set up a burn platform on the left end of the stage.

Fire reduces Giant Steel to embers and ashes nearby.

Hadrian comes onstage with his own dancers. "I'm bigger than big." He points to his marionette likeness. "That ain't me. I'm bigger than that." Hadrian performs his biggest hit, the Astrid conquest song.

She sneers at the first note of the song. Astrid flies through the stadium with the battle rage of a Valkyrie. She raises a clenched fist and chants, "Burn him."

Sections of the stadium go insane as Astrid zooms by them, chanting her battle cry.

None gets drawn in. "Burn him." He chants with at least 70,000 others. None feels the energy of being part of something larger. He imagines what it might feel like if the entire nation fit into one stadium. This feeling of oneness with the crowd feels out of place in a world of dissent and discord.

Astrid zooms down to the main stage. She shoos the flirting dancers away from Giant Hadrian. "Astrid doesn't share."

Hadrian finishes his song. He leads his dancers towards Astrid.

The motorized crane positions Giant Hadrian over the burn platform.

Stagehands douse Giant Hadrian with gasoline.

Astrid raises her hand.

The torchbearer puts the torch in her hand.

Hadrian stands between Astrid and his marionette likeness. He's not down with being burnt.

Astrid forces Hadrian and his dancers out of her way with the torch. She rushes to torch Giant Hadrian.

The marionette transforms into a bonfire.

Astrid releases her angel wings and heaves them into the bonfire.

Hadrian fights back the only way he knows how. He begins a rap battle by talking about how he made her a star when he trained her to sing. She's nothing special. He has ten more waiting at home.

Astrid takes the bait. She waves two fingers with a haughty sway in time with the music. Astrid retorts she was going places without him. His biggest hit is only popular because it's about her. She slams down two fingers with a sizzle sound.

Hadrian ridicules Astrid's hair.

None cringes. This could get ugly.

Astrid motions like she's cutting her throat. She explains he needs all those sexual conquests since he can't keep a woman. He's so obsessed with sex. You'd think he'd be good at it. Astrid does another burn sizzle at Hadrian along with her insult.

He tries to reply, but his mic has no power. Astrid cheated to guarantee she got the last word.

IJS security teams converge from all directions to remove Hadrian and his dancers.

Astrid watches Hadrian's removal with glee.

Stagehands carry a red disk and place it above Astrid's head. They move their hands towards the outside. Material drops around Astrid in the form of a cylinder.

Ten seconds later, they lift the edges back up into a disk.

Astrid in her Queen of flame costume holds up the insides of the wardrobe changing tube. She steps out from beneath it.

Stagehands roll a foam rock next to the spire. They place a set of blackened wings near Hadrian's flaming effigy.

An Asian dancer comes onstage.

Is that guy meant to be None? This isn't happening. Astrid sings some damn song. None hears words, a jumble he can't make out. Is this really happening? It sounds familiar. Stare at Zelda. Stare at MJ. Stare anywhere, but the stage. She's really going to burn him, isn't she? If only he had a suit jacket that he could hide under, like Steel.

None feels the watchful eyes of the video wall and Corella's film crew capturing his every reaction. His face droops. Gone. Not here. None feels like he's leaving his body, skyward. The mind can do anything. He wills himself away. This never happened. None wasn't even here. He focuses his mind on blocking out everything. None builds ice castles in his mind, invisible walls to hide behind.

Astrid zooms overhead.

He can almost feel the wind buffeting her as she glides by.

Is this what happened on that day, the day he agreed to sell his company? He blocked everything out, leaving a hole. He can't retrieve the missing memories. Those were important. What if something important happens today, while he's off building ice castles in his mind?

None returns to normal awareness. He remembers the dress from when she knighted him Sir Excalibur and professed her love for him.

Astrid holds a torch in her Queen of Flames outfit with blackened wings near a 30-foot tall wooden marionette that looks like None.

"A Queen needs more than a knight. She needs a King." Astrid walks the last few steps and stands there.

His effigy is on a burn platform ready to go. What is she waiting for? Five seconds. Ten seconds. She didn't hesitate at all for the others. None smiles. Maybe she still loves him.

Astrid mouths some words, then lights Giant None aflame.

None stares at his likeness burning with a dumb grin. He loses himself in the dancing fire and swirling smoke. None watches without seeing. She hesitated.

Corella seems disappointed None didn't have a meltdown.

Zelda tugs None's arm to get his attention. "I'm done being quiet. Do you still love her?"

Regardless of what happens with Zelda, he'll always love Astrid in some twisted way, like an echo haunting the back of his mind. She's imprinted in a section of his DNA. That's not an answer he can give Zelda.

He opens his mouth, but words don't come out.

Corella shoves her mic in his face. "Do you still love Astrid? How did it feel to be burned?"

None turns to Corella. "I'm not the story today, Corella. Why don't you go interview the happy couple instead?"

None ignores the rest of Corella's questions. He'd rather watch the concert.

Zelda looks like she wants to say something. She glares at him instead.

The motorized crane releases Giant None. It collapses into a burning heap. The motorized crane drives offstage.

Astrid trudges to the spire at the 50-yard line, with dragging feet. She sings a tearjerker ballet describing her ideal king, which she will never find, doomed to loneliness as queen. Astrid infuses the anguish of every former lover into her music.

The video wall displays a melancholy Astrid with gloomy clouds. In video, she strolls through dank castles, forlorn marshes, and wades into the darkness of the nighttime ocean.

Mist creeps across the center of the main stage. The floor sections in the middle lift up like steps, higher towards the back. Mist shrouds the top.

The motorized crane tugs two immense inflated stuntman airbags into place along both sides of the rising steps.

Astrid wishes for a king that could survive her flame. All others perished.

Flame jets point upward in a circle around the perimeter of the center section with Astrid and the spire.

At the end of the song, Astrid cries tears onto the foam rock. She bends over and collapses onto the rock with her head down.

The mist clears. Renquist sits on a throne at the top. He wears black Andraste battle armor with a gold crown painted on the breastplate, a sword in hand.

A dozen modern knights in black riot gear armor close the faceplate on their helmets. They climb the steps with plastic shields and drawn swords.

Renquist ascends from his throne. He closes his faceplate and sheathes his sword.

Astrid lies motionless on her rock, oblivious to the excitement.

Renquist stomps down the steps to meet his foes. He breaks swords and swats several of his attackers over the side.

The airbags puff up when fallen knights hit them. The knights quickly roll off.

Renquist raises his arms and intentionally lets them hit him.

The knights swing. Their swords fling back from the Andraste armor like bouncy toys.

Renquist lifts two foes and hurls them into the remaining men.

Of the dozen, none remain standing.

None shakes his head. The poorly choreographed fight achieved its purpose. The wedding will sell plenty of Andrastes for Renquist.

Renquist steps over the downed knights.

He marches towards the crown of flame that encircles Astrid and the spire.

Renquist stops at the fire ring. "I walk through fire for you."

Astrid springs to her feet.

He walks through flame to her. Renquist lifts up his faceplate and gives Astrid a dipping kiss.

The fire jets flare higher around the couple.

As the kiss ends, the spire blaze goes out. The only fire that remains is the three smoldering corpses of former loves.

The audience slurps up Astrid's melodrama like soup.

None feels closer to Astrid after hearing her life story but wishes she'd rewrite the last act. She could have anyone. Why Renquist?

"Put Sir Excalibur where he belongs," Astrid commands.

Renquist pulls out his sword. He holds it overhead and jabs it deep into the foam rock.

Astrid remembers Arthurian legends better than she let on. Did she know Excalibur was a sword the whole time?

She learned fashion from MaxPlume, stunts and fame from Steel, and music from Hadrian. She used his tech connections to make her cosmetic line cutting edge. None realizes he underestimated her. He re-examines his entire past with Astrid with new-found respect. He was too busy with his own genius to notice hers. This is exactly what MJ was talking about. Has he underestimated MJ, too?

She collects men and learns their skills and secrets. Astrid's a master manipulator. None wouldn't be surprised if a preeminent psychologist were in her relationship DNA.

"I will forge a kingdom worthy of you." Renquist marches back towards the main stage.

The cable on the right side of the spire lowers. Stagehands attach the cable to Astrid's black wing harness.

Astrid ascends the spire. She belts out an anthem about Renquist, her fire walker. Astrid zooms across the last zigzag series of zip lines.

The video wall shows knights going to war.

The knights cower near Renquist and pledge their fealty to him, their new king.

Astrid makes her final descent onto the main stage to reunite with Renquist, her new king.

The video wall shows elaborate churches from around the world.

A preacher approaches holding a Bible.

Five pairs of long-haired beauties and knights file in next to Astrid and Renquist.

A preacher begins a quick ceremony next to the dying embers of her three former loves. He skips the part where someone might object.

They're smart not to give that option. Part of None wants to storm onstage and whisk Astrid away from Renquist.

With an exchange of rings and vows, the golden-haired Queen of Flame marries the black knight who would be king. Until death do they part. It is done.

64. **Duet**

Astrid upgrades the famous, the beautiful, and the exotic to wedding reception guests. Steel whimpers under his coat, a broken man, unable to leave his seat.

Secret Service and IJS agents lead None, Zelda, MaxPlume, and his Glitter Girl backstage. They separate into male and female locker rooms, each lined with racks of French 1780s fashion. Astrid's themed wedding reception pays homage to the royal court of the last queen of France, Marie Antoinette.

The discovery of the hot air balloon in 1783 inspired clothes so exotic it took them a few hundred years to come back in style. Inflated clothes, globe hats, balloon inspired dresses, it's Plumage fashion heaven. Hairstylists stand ready to a construct a pouf hairstyle with wireframing for support. The style looks like a beehive with extra adornments of feathers, gauze, bows, or flowers. Astrid puts a three-foot height limit on the hair. Thou shalt not upstage the bride.

None feels the masquerade ball is tone deaf, given the collapsing economy. The extravagant hairstyles and fashion of the French royal court incited the early stages of the French Revolution. Is Astrid taunting the guillotines, or being ironic?

Corella confronts None when he leaves the locker room. "All you've done is watch. Do something, or you'll be left on the editing floor."

"That's not much of a threat, given I didn't want to be here." None strides past Corella to meet Zelda. He finds her waiting with MJ.

Both women enjoy three-foot tall poufs with feathers and flowers. They laugh at the novelty and silliness of their hair.

IJS agents lead None, MJ, and Zelda onto the main stage, to take their place with the other guests. They become part of Astrid's spectacle in their costumes.

Secret Service agents shadow the three of them on stage. The agents appear to be the only ones in the correct century.

None heard Renquist formed his own Praetorian Guard today, a detail of eight bodyguards in Andrastes. Renquist has powerful security because they're in heavy battle armor. However, his bodyguards won't take a bullet for him like None's would.

None sees a man in Andraste battle armor. Renquist must be nearby.

Astrid and Renquist come onstage to massive cheering. Astrid models her four-foot towering hair, a mash-up of Marie Antoinette hairstyles, complete with ostrich feathers, jewelry, and a model of a French frigate. Astrid outdoes Marie Antoinette on her flame-red wedding dress, a full 12 feet wide, but flat in front and behind. The red matches the one used in her Queen of Flame outfit.

Renquist walks next to her, but the dress is too wide to hold her hand.

They immediately begin their first dance to romantic music. Given the logistics of Astrid's dress, they perform synchronized dance moves at arms lengths. They rotate slowly, unable to spin with her absurd clothes. Renquist and Astrid dance down the runway to the spire.

At the spire, the music switches to upbeat electronic music.

Astrid presses a release button on the front of her red wedding dress.

She emerges from the wedding dress in a red evening gown.

The audience cheers her transforming dress.

Her detached wedding dress remains upright where she left it.

Renquist twirls Astrid in her evening gown. They dance back to the main stage and return to the spire.

Astrid docks back into her wedding dress and wears it back to meet the guests.

None sips champagne with Zelda. He sees Astrid talking with MJ. They're always insulting each other, but today they're cordial. Why are they so friendly?

Zelda excuses herself for the bathroom.

None makes a beeline towards Astrid. She got her closure, but he didn't. He turns off his mic again and sneaks close enough to eavesdrop on Astrid and MJ.

Astrid pulls out a fancy bottle from a compartment in her dress. "The President said you put on lipstick before a fight, to feel invincible. I created a battle worthy makeup line, with you in mind. Here's the first bottle of Invincible Lips, as thanks for your advice."

MJ's eyes light up. She grabs the bottle with excitement.

What did MJ do to earn such a big gesture from Astrid? None speaks up, revealing his position. "You should thank me. I found the tech talent that makes your cosmetics. Why thank MJ?"

"I've got a new line of douches coming out. I could name those after you." Astrid does burn sizzle with a smirk.

None suspects they're hiding something. Astrid insulted him to avoid his question. He interrogates MJ with a look.

MJ squirms under None's stare. She looks around for the nearest celebrity and points behind None. "MaxPlume, I haven't met him yet." She scurries past None to greet MaxPlume.

Astrid swivels her dress around, with her back to None. She won't face him alone.

His words sound desperate. "We weren't bad pie. Tell me you didn't mean it."

Astrid keeps her back to him. "Don't make this hard. If I admit we were good pie, will you go?"

"I have to know why you really left me. Why did you pause before you burned me? Astrid?"

She ignores him.

None grabs Astrid's arm from behind. "I understand all the parts of you now."

IJS agents in Astrid's security detail converge towards the president, followed by Secret Service agents and Corella's camera crew.

Astrid yanks her arm back and jettisons her wedding dress in None's path without a word. She retreats in her red evening gown.

The security details resume their distance. Nothing to see here.

Astrid abandoned None, just like she abandoned the dress that blocks his path. He hears footsteps. None doesn't have to turn around to know Corella is there. She and her crew lurked around him the whole day, waiting for some gotcha moment to pounce. He turns around.

Olivia accompanies Corella and her crew.

"Oh, great! It's the microphone police." None turns on his mic.

Corella gestures at None. "See what I've been dealing with?"

Olivia says, "Mr. President, I was expecting you to be cooperative. We need your audio, to get your unedited reaction to the wedding."

"Fine. *I* won't turn it off again." None rushes to the bathroom, to intercept Zelda before she comes out. Astrid hesitated. Why won't she talk to him?

The guests dance in their costumes across the whole stage.

The novelty of watching a masquerade ball wears thin for the audience. This is Astrid, so they expect something over the top to end the wedding. Like waiting for an extra scene after movie credits, the concertgoers remain in their seats. After all they paid, they don't want to miss out. They watch supermodels on giant LCDs to pass the time.

None takes Zelda's hand. He leads her closer to Astrid as they dance.

Astrid ignores None. She whispers in Renquist's ear. Astrid holds him tight as they sway to the music. Near the end of the song, None and Zelda end up close to Astrid and Renquist.

None lets go of Zelda. "Can you turn off my mic?"

She looks at None sideways. "Can't you do it?"

"No, I can't."

Zelda waits for an explanation, but none comes. She relents and turns None's mic off.

None says, "I'll be back. I need answers."

"How are you going to get answers?" Zelda asks.

"Dancing."

Zelda gasps as None ditches her.

He taps on Renquist's shoulder. "Can I cut in?"

"Sure." Renquist turns around and takes None's hands in his.

None's eyes bulge out. What is Renquist doing? None wanted to face off against Renquist directly, but this isn't what he meant.

The next song begins. Renquist whisks None away on the dance floor. "I haven't forgotten you."

None grins back. "I know. I keep finding your people."

Renquist laughs at None. "You have no idea."

Astrid shrugs and takes Zelda as her partner.

Zelda looks enraptured. She stares at Astrid with awe. "Do you watch *The Daily Reveal?*"

Astrid sneers. "I don't watch the news. I am the news."

Zelda loses her appetite to continue the conversation.

Renquist looks at None with thankfulness. "You made me the richest man in the world."

"No, your perversion of Slam Dunk Jobs did that."

"It was your brainchild," Renquist says. "I merely planted your seeds. I feared you'd take those ideas and create something better before I could corner the job market."

None fumes. He would never use his brainpower to wring every cent possible out of the nation's paychecks. Only a greedy shark of sharks like Renquist would do that.

"I'll go down in history with robber barons like Carnegie and Rockefeller." Renquist twirls None towards Zelda and Astrid. He seems to enjoy keeping None off balance.

None struggles against Renquist to lead the dance.

Renquist says, "Mr. President, stay away from my *wife*. I've taken your girl. I've taken your company. You're beaten. Accept it. End this futile attempt to take back Slam Dunk Jobs. After what you did for me, you get one last chance. Let it go. Look at how you became president. You impress me. You're the only one smart and creative enough to threaten me. Don't make me destroy you. Come work for me."

"I already work for someone, the American people." None stomps on Renquist's foot.

The dance turns into a struggle.

Renquist snarls in pain. "You've become too dangerous. I'm not just top of the food chain. I'm eating the whole damn thing. Out of my way, or I'll devour you next." He shoves None to the ground.

None lands on his butt and scrambles to his feet. "I'll defeat you with your own loophole."

Before None and Renquist can fight, their security teams separate them.

Groups of Secret Service and IJS agents scream at each with weapons drawn. "Stand down! Stand down!"

It's a tense Mexican standoff. One wrong move and their fight over a woman and a company could turn into a massacre.

Many of the Secret Service and IJS agents know each other. Firing their weapon would be like shooting a brother.

Agent Vincent says, "We're crashed. This site is compromised."

Secret Service agents provide a human shield to protect None.

Eight bodyguards, Renquist's Praetorian Guard, converge around Renquist in Andraste battle armor. Six of them form a defensive box formation. Two inside the box help Renquist into his own Andraste. With their better equipment, Renquist's Praetorian Guard outclasses the Secret Service.

Secret Service agents evacuate the president backstage.

A helicopter flies overhead. It drops a cable next to Renquist.

IJS agents attach the steel cable onto the airlift port on Renquist's Andraste.

The helicopter reels in Renquist until he clears the zip lines, then zooms off within seconds.

Renquist evacuates miles from the stadium before None reaches backstage.

The audience cheers. They think the dramatic exits are part of the show.

Agent Vincent rushes None into The Beast, where Shen waits for them. None feels safe behind the eight-inch armor plating on his limo. The motorcade heads towards the airport.

Shen says, "MJ and Zelda have a two-minute head start. They're on their way to Air Force Two."

"So much for me being left on Corella's editing floor," None says. "I'm glad the first girlfriend got a ride home. Imagine how pissed she'd be if she got left behind, especially after I ditched her to talk to Astrid."

Shen shakes his head with his eyes closed. "I wish keeping up with your love life wasn't part of my job."

None replays the wedding in his mind. Was the pause meant to lure him in? Astrid knows how to press his buttons. Was Astrid bait? Astrid conspired against None in whispers with her new husband. She's not just his wife. She's his henchwoman.

At the airport, Air Force One leaves for D.C. None and Shen watch news coverage on board.

On TV, Renquist upstages the president and the Secret Service. He flies into the sky like a superhero to make a big exit. Commentators talk about replacing the Secret Service with more impressive IJS agents in Andraste battle armor.

Renquist isn't just trying to sell Andrastes. He wants to control all government security, including the Secret Service. Renquist must see it as another market to conquer.

None checks online for more reactions. Pictures of Renquist and None dancing at the wedding feed all sorts of conspiracy theories on social media. Some say None works for Renquist. Others claim they're both members of the Illuminati, plotting world domination.

Lengthy editorials on mainstream news sites endorse privatizing the Secret Service. There's no way such long articles could have been written in the 20 minutes since the showdown against Renquist's Praetorian Guard. He seeded the articles to the news sites ahead of time. The standoff was intentional.

Renquist laid a trap and None fell for it.

65. **Dead Man's Switch**

MJ flees from Renquist Stadium to Air Force Two in such a rush that she doesn't have time to change from 1780s fashion. MJ sits in her flying office. Without overhead bins, the cabin ceiling looms almost ten feet tall. It's lucky she's not flying commercial, or MJ would have to lean her three-foot-tall beehive into the aisle.

Her expansive French ballgown takes up most of the three-person couch in her cabin. MJ decides to take off the dress to get more comfortable. She reaches around to the back, but the intricate drawstrings are difficult to disentangle. It took two people to weave her into the dress. Why did she think it'd be easy to get off?

The phone next to the couch rings. MJ picks it up.

On the phone, Agent Vincent says, "Zelda Remington says she has an appointment with you."

MJ promised to spend time to get to know Zelda. Stuck on an hour flight back to D.C., she's out of excuses. MJ sighs. "Send her in."

Zelda enters with a canvas bag. She looks around and searches the cabinet drawers in MJ's cabin. "What kind of alcohol do you have in here?"

Trapped with this rude woman, who rummages through her office without asking, MJ could use a drink. "Air Force Two doesn't serve alcohol."

Zelda plops down on the couch next to MJ. The two huge ballroom gowns overflow the couch. Zelda pulls a champagne bottle out of her canvas bag. "Good thing I grabbed this on the way out."

Zelda pops the cork. Frothy champagne oozes over the side of the bottle. Zelda tilts her head to slurp up the spilling alcohol.

No way. MJ is not down with sharing a bottle. She yanks the champagne from Zelda's hands. "Air Force Two does have cups."

A flight attendant answers MJ's summons. She brings two cups, fills them, and leaves.

MJ drinks hers so fast that she feels bubbles up her nose. Zelda imitates MJ and chugs her cup of bubbly.

The carbonation builds up discomfort. MJ drank too fast. Who cares? It's only Zelda. MJ lets out a huge burp. Zelda tries to outdo MJ with her own burp. It breaks the ice. They laugh together.

Zelda lays her hand on MJ's hand and peers into MJ's eyes. "I want you to know that I forgive you for knocking me over on TV."

What? MJ hides a gasp with her cup. MJ barely tapped her. Zelda acts like she crashed into her. Whatever. MJ pulls down the cup to reveal a forced smile. "I appreciate that."

"I know I come on too strong sometimes." Zelda taps MJ's hand as she talks.

"That's for sure, Cling Wrap."

"Is that my nickname? You're so funny." Again, with the hand. Zelda is one of those touchy girls, that has to put her hand somewhere, while she talks to you. Maybe they have less in common than MJ realized. MJ understands Zelda wants to be her friend, but she's getting way too comfortable.

Zelda says, "To be honest I'm ready for the L word, but I'm not going to say it until he's ready to hear it."

MJ hides her emotions behind her cup. None isn't hers, so why does it bother her? Zelda clings too tight. It's a matter of time before None dumps her.

He plays a major part in both their lives. MJ and Zelda bond, sharing stories about None for the rest of the trip home.

When MJ complains how hard it is to undo the dress, Zelda suggests they help each other undress when they return to the White House. How helpful.

Maybe MJ really could be friends with this girl. "We'll totally hang out again when I have more time."

The champagne bottle empties just before the plane lands. They return to the White House with their old clothes in bags.

Zelda wants to change clothes first. MJ and Zelda bring their old clothes to the Queen's Bedroom, where Zelda sleeps. MJ's protection detail waits outside.

With pink walls, a flower bouquet on the table, floral prints on the canopy bed, and matching furniture, it's a bit girly for MJ's taste. MJ notices ropes next to one of the chairs. She's not going to ask what None and Zelda use that for.

Zelda sits on the bedroom bench at the foot of the bed with her back to MJ.

MJ untangles the puzzle of a dress. She recaps the craziest things None did in the last 20 years, including particulars she would skip if she were sober.

Zelda isn't as forthcoming with details about her own life. She's more content to listen.

MJ unties all the drawstrings on Zelda's dress.

Zelda stands up to let the ballgown drop to the floor. She struts in her bra and panties. "You like what you see?"

She must be one of those insecure women that fish for compliments. MJ takes in every inch of Zelda. "You look fantastic." MJ decides not to mention Zelda's biceps are underwhelming. "What's your workout routine?"

Zelda stares into her eyes. "You really are gorgeous in that outfit."

Gorgeous. Not that word. MJ groans. Is Zelda flirting? Could MJ's booze-soaked brain be misreading things? Maybe she just needs to learn how to take a compliment.

MJ doesn't care anymore how the ballgown looks. She feels trapped, in a dress designed for people with servants. "This dress isn't going to take itself off. Cling Wrap, we had a deal." Her words slur a little.

MJ sits on the bedroom bench for her turn. She spaces out while Zelda undoes the endless drawstrings.

Zelda sits down next to MJ and strokes the back of her neck. "Everything is loosened up."

The touch focuses MJ's attention. She swivels her head towards Zelda and purses her lips in irritation.

Zelda sees her pursed lips. She closes her eyes, puckers up and leans in for a kiss.

Instinctively, MJ blocks the incoming kiss with her hand, as if it were an attack.

Tongue.

Zelda must be pretty drunk to tongue her hand. MJ imagines the tongue probing her mouth. It's enough to make her nauseous. MJ resists the urge to vomit.

She's tempted to shove Zelda's head back. MJ nudges Zelda back and withdraws her hand. "I'm not into that."

Zelda opens her eyes. "What, Girls?"

"No, sex. I draw the line at penetrating conversation." MJ grimaces with her tongue out. "Nothing is more disgusting than having your body *invaded*."

Zelda says, "Sex with a woman is different. Maybe you just hate men."

MJ trembles with rage. She can tolerate hugs, holding hands, and affection, but no one goes inside her. Man, or woman, the only difference is the *weapon* they'd use to invade her. Nobody crosses that line. Nobody.

"After what just happened, I'm not sure what kind of friends we can be." MJ holds her dress tight. Zelda will never see her naked. MJ heads for her bag of clothes.

MJ grabs her phone. She'll tell None about this. Zelda will be out on her ass.

Three missed calls from Fernando. They're not scheduled to meet today. MJ freezes when she reads his text: Want to grapple?

Zelda picks up on her sudden dread. "Is something wrong?"

What an understatement. Everything feels wrong. "I'm going to the dojo to clear my head."

MJ goes across the hall to her room. She dresses in a gi and rushes to her dojo. The buzz is wearing off. Today was a total buzzkill.

Fernando waits with a joyless face.

They jump straight into grappling, without foreplay. It must be serious. Fernando enjoys his innuendos. He pins her in seconds. Her free arm flails as she attempts to reposition.

Fernando whispers at the first opportunity. "Is Gerald Hopkins your guy?"

MJ sobs. She got someone else killed. Her tears draw her security detail. MJ waves agents away with her free hand. "I'm training to increase my pain tolerance. Expect crying, until I master it. Can you nannies please not jump at every little sniffle?"

Her detail takes the hint and gives MJ space.

MJ fights the sniffles to collect herself. She returns to her whispers. "I hired him as a consultant on human memory, for Adaptonium. How did he die?"

Fernando says, "Fatal weightlifting accident."

MJ struggles to keep her voice down. "He never worked out. I can't believe this."

Fernando whispers, "From the way it was staged, it was a professional hit. Based on your level of surprise, it's not just murder. It's a message. They want you to know it's no accident."

MJ taps out.

They separate.

She takes out her frustration on the messenger. MJ charges at Fernando with fierce eyes. She knocks him down.

Fernando squirms out of an armlock and pins MJ again. He resumes whispering. "Who else would know his identity?"

MJ whispers, "Just me, and now you."

"Is there anything else besides Adaptonium that ties you to Gerald?"

MJ's eyes light up in shock. "Thelma mailed the packages. She might have remembered where the packages were sent."

Fernando whispers, "You mentioned her fear of water. Who else knew about it?"

MJ tries to brush off his question. "I've solved that murder. Renquist admitted to it."

"Did he know?"

"Yes. Everyone in the office did."

Fernando grows solemn. "I looked into Thelma's death. I found signs of torture."

MJ sobs, imaging Thelma's dying moments.

The agents keep their distance this time.

Fernando whispers, "Conventional torture must not have been enough. Thelma was found on the beach with seaweed between her toes. They used her phobia of moving water to torture her further. She was immersed in the ocean, until her heart gave out, literally scared to death."

Thelma suffered through her worst fears because of her.

MJ taps out.

"I need a moment." She sprawls out on the mat and covers her eyes with her hands. The guilt pins MJ to the mat, but her emotions aren't something she can tap out of. If only she could wish this away. Did she really have to make those damn recordings?

When she uncovers her eyes, Fernando is there to help her to her feet.

Fernando asks, "Ready to go?"

Zelda. Thelma. Gerald. Renquist's threat. How much more can she endure in a single day? She isn't ready. MJ nods anyway.

She doesn't have the energy to fake a defense. MJ wilts onto the mat, pinned under Fernando.

He whispers, "You're not paranoid. They'll come after you next. It could be someone in your life for a while, waiting until the other two were taken care of. Has anyone gotten close to you recently?"

"Besides you?"

"If I weren't on your side, this dojo would be a cemetery. Yes, besides me."

"Let's just say that None's girlfriend, Zelda, has been pretty friendly lately."

Fernando says, "Keep your eyes open for little slip-ups. It could be anyone."

MJ feels desperate. What if Renquist gets every copy and she dies for nothing. "We're not getting anywhere. Should I just release the info?"

"Without insurance, they'll stage your death in this dojo today. Hand out more copies." Fernando smiles at MJ. "I'll take one."

MJ sighs. "That keeps me alive, but Renquist needs to pay for what he's done."

Fernando considers options. "If you attack someone as powerful as Renquist directly, it always backfires. You've got two options: a powerful alliance, or, create powerful enemies for him."

MJ prefers a more direct approach. "You're Secretary of Defense. Don't you run the military?"

Fernando rolls MJ into a position where she can see his face. "On paper, but Renquist holds more sway. When Navy SEALs and special ops want to make real money, they retire to work for Renquist's security companies, like IJS agents. Renquist Aerospace has enough military contractors to fight wars on its own. If I oppose him openly or trust the wrong person, I'll have my own fatal weightlifting accident."

MJ feels the gravity of Fernando's words. "I used a black backpack for my go bag. It's almost ready. I bought a gun."

Fernando whispers, "I think your agents are catching on. I know another excuse so we can get close."

"Let me guess, pretend boyfriend?"

Fernando winks at her. "It doesn't have to be pretend."

She's drawn to his skill like a cat in heat, but he'll want things she can't give. MJ decides not to respond.

With no answer, Fernando switches subjects. "I'll investigate your agents to see who you can trust. What is on the tapes that's worth killing for?"

"Treason."

That word angers Fernando, a man who dedicated a chunk of his life to his country. "Get up. We have training to do."

When she sees Fernando's reaction, the importance of the recordings hits home. How could she even consider offering it to Renquist at the wedding to save Gerald and herself?

They get off the mat.

Fernando approaches her lead agent. "Agent Reynolds, I want to train Madame Vice President how to disarm a firearm."

Agent Reynolds squints at Fernando. "You'll have to get your own weapon, and I'll need to inspect and unload the weapon."

Fernando nods to Agent Reynolds. He turns back to MJ. "Did you bring your gun?"

MJ grabs a Glock 19 from her purse.

The agents immediately go on alert at the sight of the 9mm handgun.

MJ takes out the magazine and lets Agent Reynolds inspect her gun.

Fernando approaches MJ. "You kicked the weapon out of the gunman's hand, but with a more skilled opponent, you won't have that long. Let's go over some techniques from an Israeli martial art, Krav Maga. We'll practice weapon disarms."

He approaches MJ with his hands raised. "Keep your hands at the same level as the gun, never above the gun."

MJ points the gun at Fernando.

He tilts his body away from the gun barrel, grabs the gun, and rotates it around to point at her, all in a single second.

MJ gasps as how quick he took her gun. "Wow! Let's do that again."

Fernando continues the lesson, while he has MJ's attention. "If you have an active shooter in close proximity, close the gap and engage the hostile. It'll get them off balance. With a gun, the hostile will be overconfident, assuming they control the scene. If they engage at a distance, evade. Run away with no straight lines."

They practice the disarms until it's second nature for MJ.

Fernando follows her back to the Lincoln Bedroom.

Her detail stays outside, to give MJ and Fernando some privacy.

Fernando gets a copy of the Renquist recordings. He becomes her new dead man's switch.

MJ shows him the short sword and dagger with water serpent scabbards from her samurai sword collection. She frowns at the empty katana slot on her wall.

She looks at the ballgown that's draped over a chair. Fernando told her to look for slip-ups. MJ's situational awareness disappeared after the first drink with Zelda. Could that be the purpose of the bottle of champagne? Zelda rummaged through her cabinets on Air Force Two. Was she really looking for the recordings?

When MJ met her, she knew Zelda did her homework, with a steampunk outfit that would appeal to None. None told her Zelda tried too hard like he was her mission in life. What if he was her mission, a stepping stone to get closer to MJ and the package?

Zelda just happens to make a pass at MJ the day she learns both dead man's switches are dead?

Maybe you just hate men. Ha! None must have blabbed that MJ's a virgin. Zelda assumed MJ's a frustrated lesbian, stuck in the closet. She planned to get intimate to gain MJ's trust.

MJ whispers in Fernando's ear. "I don't have any definitive proof, but the pattern's there. Zelda's the spy."

Fernando whispers back into MJ's ear. "Don't risk approaching the president until you have evidence."

She quivers from the heat of his breath in her ear. Alone in her room, it might be overkill to whisper, but she enjoys the intimacy of being close with Fernando. Besides, anywhere might be bugged.

"I'll give the other copy to a friend I trust." Her body brushes up against his, while she whispers.

Fernando pulls away and shushes her. "It's better I don't know." He leaves without another word.

MJ lures Shen from the Oval Office to the Cabinet Room. She sits down at one of the dozens of chairs. MJ explains what a dead man's switch is and tells Shen about the two deaths.

Shen seems to weaken. He slumps into a chair two seats away from MJ. Shen nervously runs his fingers through his hair. "Why are you telling me this?"

MJ pulls a postmarked package from under the table. She slides the package towards Shen. "Just take the package. If they torture you, you'll have something you can give to them."

"Torture?" Shen jolts out of his chair. He stares at the package, terrified.

MJ continues calmly. "I mailed the package to myself. It's still sealed, proof you haven't seen the contents."

Shen retracts his hands into fists. "I don't want my fingerprints on that. Can't you just give the package back to Renquist?"

"I tried that. He threatened to bury me."

"You didn't hesitate to put my life in danger," Shen says. "You'd never do that to my brother."

If they suspect Shen knows anything, he might already be in danger. MJ better not point that out. She hears panic in Shen's breathing. How can she calm him enough to get his help? "He's president. It's our job to protect him from this. I ask because I trust you."

Shen shakes his head. "Find someone else to be your patsy."

MJ sighs. "I don't know who else to trust. Zelda's a spy. She's dating None to get close to me."

Shen scoffs. "Do you know how crazy you sound?" Shen approaches MJ and sniffs. "You're drunk. I can smell it on you."

MJ says, "I sobered up a while ago. I must have spilled champagne."

Shen asks, "What proof do you have of your allegations about Zelda?"

"Women are one of None's weaknesses, and Renquist exploited it," MJ says.

Shen loses his patience with her. "I know it's his weakness, one I've seen you exploit more times than I can count. Bring me proof, or don't waste my time."

66. **Retaliation**

Improvements None made to the educational system will be his permanent legacy. His tech boom is the last bright spot in the economy. The public forgets None's successes with their short attention span. His connection to Slam Dunk Jobs washes away all his achievements.

His best shot at restoring his legacy is regaining control of Slam Dunk Jobs. With Olivia Ford and Rick Slater, None and MJ began a hostile takeover of their own company.

The money for the purchase was supposed to be in the escrow account but never arrived. The lawyers for Olivia and Rick stopped responding. None, MJ, and Shen investigate from the Oval Office.

None charts the stock price for Slater Dynamics. The stock dropped under a dollar a share. How did it become a penny stock?

Shen calls the company's phone number. It keeps ringing but never picks up, not even with an automated message.

"Try Rick Slater's personal cell." MJ shows the number to None.

None dials Rick on his landline and puts him on speaker.

There is a deep breath, then Rick exhales. "Titanic, you're the last person I want to talk to."

"Rick, I can't buy Slam Dunk Jobs with an empty bank account," None says.

Rick laughs maniacally. "Renquist destroyed my business empire. You're the Titanic, and I'm collateral damage. Renquist won't rest until he sinks the lifeboats too."

None asks, "What happened?"

Rick says, "Slam Dunk Jobs tripled my labor costs. Things went downhill from there."

None remembers from Rick's corporate filing that he uses Slam Dunk Jobs at his company. Renquist must have a way to cheat and raise worker's salaries when he wants to. That can't be legal to rig his wage auctions. Then again, None knows Renquist made many laws toothless, with his endless lobbying.

Rick says, "Then major customers, like Galaxor Telecom, all canceled their purchases at the same time."

The Oval Office mood turns dismal. Galaxor Telecom is one of the many companies Renquist controls.

None asks, "Is that why your stock price is so low?"

Rick says, "Renquist spread rumors about the problems he caused. He got an army of short sellers attacking my stock. I'm selling my company to him for pennies on the dollar, while it's still worth pennies."

None doesn't know what to say. "I'm sorry that happened to you."

Rick snorts with contempt. "You're sorry? That's rich. Working with you cost me everything. I'll laugh when he destroys you. I'm out."

Click.

None stares at MJ with disbelief. "Renquist destroyed our ally…and made money doing it."

Olivia doesn't return calls. None should have talked to her about the escrow account at the wedding, but he had other things on his mind. None tries the next best thing. He gets Corella on the line. "Corella, what's going on with the deal to buy Slam Dunk Jobs?"

"How should I know? You didn't trust me to be involved, remember?" Corella sounds bitter, but she softens quickly. "Ask Olivia. She'll be at the big meeting tonight, an hour before we record the next episode of *The Internet President*."

None asks, "What's the meeting about?"

"I'll find out when you do."

None picks up worry in Corella's voice. He says his goodbyes and disconnects.

A highlight reel of the wedding took over most of the first episode in the new seasons of *The Internet President*. Trying not to give Corella embarrassing footage backfired. They made the narrative of the episode about Renquist instead, with the standoff with the Secret Service and Renquist's epic exit as the finale. Renquist outdid None on his own show.

Tonight, they film season two, episode two of *The Internet President*. None's deal with Corella was no surprises on air. To meet that, he gets the episode rundown an hour before the show. What preparations can he make in an hour?

Olivia scheduled her meeting at the same time he comes to pick up his rundown. She knows he'll be there. That can't be a coincidence. Does it still count as an ambush if you expect it?

In return for Zelda's free wedding ticket, he has to bring her to filming for today's episode. None hopes any ambush doesn't involve Zelda.

Apprehension sours his day. It's all he can think about until the motorcade arrives. None travels with MJ, Shen, and Zelda to the Reveal Channel studio. Shen brings his laptop to get work done.

Backstage near *The Internet President* set, security guards stop them.

Shen finds the nearest table to set up his laptop.

One of the guards hands None the episode rundown. "I'm sorry, Mr. President. The meeting is for Reveal Channel employees only."

They're using his auditorium. It takes 30 minutes to seat his audience, so the meeting can't last more than half an hour. None won't have to wait long. He scans the readout. Zelda's part is towards the end.

Shen checks the escrow account. It's still empty.

None slaps the paper with a loud thump. "This is not right. I need to see Corella. Now."

One of the guards leaves and brings her backstage.

Corella seems flustered. "What's wrong, Mr. President?"

He holds up the rundown. "It says five minutes in we'll have a surprise guest. Our deal was no surprises on air."

"I haven't gotten my rundown yet. Can I see yours?" Corella borrows None's papers. She shrugs. "Well, it's not a surprise if we tell you about it."

None shakes his hand in the air with frustration. "Aiya! How can a surprise guest not be a surprise?"

Corella half snarls, ignoring None. "Host two? What the hell does that mean?"

Bruce Cannon swings open the curtain and enters from the main stage. "You rang? I'm host two. I'm replacing host one."

None hasn't seen Bruce in person since the government shutdown confrontation.

Corella's mouth gapes open. It takes her a few seconds to remember to close it.

"There's real poison talked about you, Corella." Bruce doesn't veil his contempt for her. "I have pillows that are harder hitting than you. The owner brought me in to reign in the guests and add conflict to the show, something you've been unable to do. The amount of drama-worthy footage you got at the wedding? Unforgivable."

"They can't fire me. I have a contract."

Bruce wears an irritating smirk. "Good news, Corella. The network didn't fire you. They traded you to LDR news, for me. You can go back to the Word on the Street crew. I know how much you loved it. Then Tim Houser can be promoted back over you."

Corella's eyes light up in anger. "I risked my life getting a story! They can't demote me under Tim Houser! He's a coward."

Bruce smiles. "Don't worry. As the lead anchor, he's behind a desk, so he can't run away."

"I quit."

Bruce rubs his hands together with glee. "I'll let our lawyers know you plan to breach your contract."

The energy fades from Corella. "Wait…I'll do it."

None cringes at the humiliation he imagines Corella going through. She was a hero on election day. Now, it's for nothing.

A security guard comes to get Bruce. "They're about to introduce you."

Bruce returns to the stage with applause.

Corella glares at None and steps towards him. "This is what being nice gets you in this business. I had a promising career. It's ruined, because of you."

None says, "I'm—"

Corella holds out her hand to stop None. "Don't say another word, not even an apology. I have packing to do." She strides away with determination.

MJ exhales deeply. "This Bruce guy is worse than Corella."

None says, "I know. We'll have words with Olivia. This won't stand." None holds Zelda close. He needs the comfort.

Shen leaves his laptop on a nearby desk. "Could this be a negotiating tactic, to get better terms? Olivia and her family still haven't deposited into escrow."

None shakes his head. "No way. With the financials you showed me, this is the deal of the millennium. Anyone with the cash would want a piece of Slam Dunk Jobs."

None and the group discuss who the special guest might be. Their best guess is Astrid.

Olivia comes backstage after her meeting.

None blocks Olivia's path. "We need to talk."

"Yes, we do," Olivia says. "I'll send for you, after the episode."

None won't let her nudge past him. "Let's talk now."

"I can't." Olivia's words float out with a downbeat vibe. She seems more manager than TV personality.

"This is unacceptable." None stomps on the floor for dramatic effect.

"Throw as many temper tantrums as you want. A talk *after* the episode is the best I can do."

None lets Olivia pass.

Security guards let None and MJ know when it's time to come on for the show.

Bruce claps with the audience from the desk where Corella used to sit. "It's time for the crazy president show with, None of the Above, and Vice President More Jobs, welcome to the show."

None corrects him. "I'm not crazy. The name of the show is *The Internet President*."

None and MJ make their way to the host couch.

Bruce looks at None with another irritating smirk. "Your presidency is one big joke, and it's played out. Who can take you seriously with a name like None of the Above? Maybe we should go on online and find ourselves another president."

MJ gets up from the couch and puts her hands on the desk. "Talk with respect. That's the President of the United States."

Bruce says, "Sit down Miss Anger Issues. Lay a hand on me, and I'll sue you into the stone age. He's not a real president, more like comic relief."

IJS agents converge around Bruce's desk at the center of the stage. He isn't using the security pushovers Corella had.

MJ returns to her seat.

Seeing IJS agents puts None on edge. After the standoff, he certainly doesn't trust them. At least they aren't wearing Andrastes.

None sighs. "If I want my message heard, I have to deliver it with a certain entertainment value. I add humor to make the medicine go down easier, like Mary Poppins."

Bruce leaves his desk with an extra mic. "Let's get some audience reactions. Tell me if the president's policies have helped you or hurt you." Bruce walks down the middle aisle.

A middle-aged man in the front row stands and takes the mic. "This president pushes software as something that will solve all our problems, but his software is bankrupting my family." He hands the mic to Bruce and sits down.

The audience claps in agreement.

A white woman in a blue dress in the next row takes the mic. "My salary isn't some inefficiency to fix. My family depends on that money. Rent is so hard now."

Bruce points at None. "Is that man destroying your life?"

"Yes." She hands Bruce the mic and sits down.

None shakes his head at the leading question.

Bruce goes down the aisle and asks the same question. Yes. Yes. Yes.

None fumes. Bruce used his own shtick against him. He must remember being called out during the government shutdown confrontation.

Bruce walks back onstage towards None. "You ask America to find independent voices, but you treat your audience like sheep, expected to repeat your own words. We want more out of a president than a cheerleader."

None stands and moves from the guest couch to the front of the stage. He notices a better-dressed audience. Did they up the dress code? None addresses the audience. "Your anger is justified, but I don't control Slam Dunk Jobs anymore. Alvin Renquist does. He'll suck every penny from your pockets, into his. He's a machine that chews you up and poops out money. He's so ruthless and greedy that the devil sold *his* soul to Renquist. If shady were a person, it'd be him. Renquist uses wealth and power to crush anyone he opposes, but never in a fight fair. If he were here, I'd call him out."

Bruce snickers. "Then let's bring out our surprise guest, a minute early. Welcome the king of commerce, the sultan of stocks, the world's first trillionaire. Here he is, Alvin Renquist."

Renquist walks in to cheers, with a few boos along the center aisle. He moves towards the center stage, to get into None's comfort zone. Renquist points at None dismissively and laughs. "You forgot to say I eat puppy dogs and spit out their bones. I'm not some bogeyman. I'm the ultimate job creator."

None stares right back at him. "You lowered wages."

Renquist says, "Those jobs are competitive now. Unemployment is at new lows. I did that, not you, or some government policy. That's the open market at work."

None gets in Renquist's face, separated by inches. "I'm not sure I'd brag about trying to bring third world wages to America. You made it easy for employers to skirt regulations and skip benefits. You turned everyone into temps and part-timers. One job isn't good enough anymore. Yes, you did that."

Just the mention of regulations riles up Renquist. "It's my right as an American to make as much money as I can. Damn the consequences. It's an injustice when the government tells an owner what they can do with their hard-earned company. Regulations are just another excuse for big government to grab big taxes. Taxes are theft. That's my money they're confiscating. The only form of wealth redistribution I believe in is wages, and I'm not even fond of that one." Renquist laughs to downplay his wages comment, but it's clear he means it.

The crowd cheers Renquist on.

None shifts his focus to the crowd. How could they cheer for that? Renquist compared wages to wealth redistribution.

Bruce isn't the only one stealing None's moves. Renquist sidesteps None to go down the steps into the center aisle. "I'm a freedom fighter. I fight the PC police that tell me what to say. It isn't a crime to speak your mind. I have the god given right not to be a liberal. Those who judge me for judging others are hypocrites."

Everyone loves Renquist. Where did they find such people? None looks closely at their faces. He recognizes a few. They're from the wedding. Renquist stacked the audience with Renquist Aerospace shareholders. How did Renquist turn None's show into his turf?

None comes up behind Renquist. He shouts to get his attention. "Hey! It's not about liberal, or conservative. That's one of the ways they keep us apart."

Renquist turns around to face None. "So, what *is* it about?"

None says, "It's about the powerful and connected trampling on the rest of us. I was elected, not as a joke, but as a message from the American people, to the corporations and powerful."

"I've heard your message," Renquist says. "It goes like this. Blah blah blah software."

None shakes his head. "No, the message is: 'You have overstepped.' Corporations and the powerful took over the Republicans and Democrats. They've even corrupted the new parties. There's a battle going on at the grassroots level. I'm the warning shot. A global uprising is coming, against corporate greed, against oppressive governments, against every obstacle confronting the common man. It won't be communism, socialism, or even democracy. It will be something new."

Renquist can't help but laugh. "You're living in a dream world. Everything has been tried before. Governments have been around for thousands of years. What makes *you* think you can find something new?"

None swells with passion. He's a dreamer, and nothing will crush his imagination, not Washington, not Renquist, not anything. "The Internet disrupts everything it touches. Why would forms of governments be off limits? The powerful won't be able to keep us apart. The Internet will bring us together."

"See!" Renquist points at None. "There it is. Blah blah blah the Internet. You can't help but involve computers in everything. I'll tell you what brings people together. A paycheck." Renquist's eyes light up. He continues with the same passion None gets talking about computers. "Companies bring thousands of people together to achieve common goals, without the inefficiencies of government. There is not a single great thing you can do without a corporation. You want to rail against corporate America? We are America. American culture isn't baseball and apple pie anymore. It's Doritos and Budweiser. It's every corporation you put in your mouth."

None deflates and fills with angst. "I wish you were wrong. I wish corporations weren't taking over our culture. We used to have one day that was sacred, a day to spend with our families, away from our endless consumerism. It was a day to be thankful, but the engines of commerce stole Thanksgiving from us too. Now it's just another high holy shopping day, an appendix to Christmas. Tell your *wife*. We don't need a Second Christmas."

MJ claps her hands and laughs, alone on the host couch.

"You know what?" Renquist puts his hand over his heart. "I can't do this anymore."

None tenses up. Renquist puts his hand over his heart when he spins his biggest lies. What now?

Renquist says, "I know many employees have been hurt by Slam Dunk Jobs. I feel all of your pain. You see, the president isn't just the evil genius that founded Slam Dunk Jobs. The president set up this complicated

311

structure, so his company is always being sold, but the sale never goes through. It looks like I run the company, but he still runs it. He made me his patsy."

Gasps and murmurs spread through the audience.

None is exasperated. "Renquist set up the contract that way. I'm not in charge. That evil part is all him. Those were his ideas."

Renquist wears a hurt expression. "I won't lie for him anymore. The president tells me how to run Slam Dunk Jobs in secret meetings."

None's mouth gapes open, frozen in disbelief.

"The president forced me to have a secret meeting at my wedding. He made me...dance with him. The president turned off his mic, so his instructions couldn't be recorded." Renquist looks intently at None. "You're from L.A. Let me put this in terms you'll understand. Dancing with dudes isn't my style."

On the screen behind Bruce's desk, footage plays of None at the wedding. He ditches Zelda, then dances with Renquist. The parts with Renquist as aggressor are edited out.

Dancing with another man caught None off guard, but he's not so provincial as to be bothered by it. The inference that None forced himself on Renquist is another matter entirely. "That's not how it happened."

Bruce leaves his desk to join None and Renquist in the center aisle.

MJ shrugs and follows Bruce down the steps from the stage.

"We've got the video proof, right here." Renquist looks to Bruce. "He's trying to keep the loop going. Ask him if he's trying to buy his own company back, right now."

Bruce and MJ catch up to them in the center aisle.

Bruce says, "Is this true?"

None gets flustered. "What actually happened is—"

Bruce asks, "Are you buying back your own company, or not?"

"Yes." The truth sounds damning. None needs something better.

The crowd gets quiet for a moment.

Renquist took out Rick Slater to scuttle None's takeover. He might take out one billionaire, but he can't take them all out. Renquist isn't the only shark. None needs blood in the water to attract them.

"I'm sorry, Alvin," None says. He scribbles a note on a piece of paper.

Renquist acts self-conscious at being addressed by his first name. It implies a closer relationship they don't have.

None hands the note to MJ. "Make sure they broadcast that."

MJ goes backstage with the note.

None addresses Renquist. "I won't make you my patsy anymore. I'm selling Slam Dunk Jobs to the highest bidder. Anyone who's interested can contact me at the email that will be showing on your screen. Starting bid is $2 billion." Every greedy billionaire will want a cut of the most valuable company in the world. The low price ensures a bidding war.

Investors in the crowd pull out their phones, while they shove their way towards the aisle. It's a stampede out the door, as the frenzy begins to invest in the opportunity of a lifetime. Renquist's decision to fill the audience with wealthy investors just backfired.

Bruce, None, and Renquist run up the center aisle to the stage to avoid getting trampled.

Renquist holds out his arm and shouts, "Cut to commercial! Turn off the cameras!"

Bruce says, "Don't worry, sir. We'll edit that whole part out."

MJ returns to the main stage. "It's too late. I posted the sale on the Internet. It's trending."

Renquist storms out of the studio.

Bruce returns to his desk. He ignores the disappearing audience and pretends like nothing just happened. "Let's welcome our next guest, the first girlfriend, Zelda Remington."

None and MJ return to the host couch.

Zelda enters and plops down onto the host couch between None and MJ. She puts her arm around None. MJ edges away from Zelda.

Zelda says, "Thanks for having me, Bruce."

Wedding footage plays on the screen behind Bruce. Zelda leaves None. Then he hurries over to talk with Astrid. Without sound, his desperation to talk with Astrid is more pronounced.

Bruce says, "While you were going to the bathroom, the president was trying to chat up Astrid. Then from the footage we saw earlier in the show, he abandoned you for a secret meeting with Mr. Renquist. Zelda, how does it feel to be neglected by the President?"

Zelda pulls her arm from around None, stands, and moves away from him. "How could you do that to me? You said you were over her. Was it all a lie?"

None stands and grasps her hands. "Please, forgive me. I never meant to hurt you."

"I couldn't pee without you talking to that woman."

He gets down on one knee. "I'm begging you for one more chance. Astrid isn't real. You're real. With her kings and knights, Astrid's just a fairy tale."

Zelda stares into None's eyes. "I know how you are about oaths. Do you swear you will never talk to Astrid, or pursue her?"

None was a fool to go after Astrid in the first place. He stares at Zelda with repentant eyes. "I swear. I won't chase fairy tales unless it's with you."

His words move Zelda. She embraces None tightly. "I forgive you."

Bruce seems disappointed that they made up so quickly. He focuses the narrative about None and Zelda's romance.

None and Zelda go into nauseating details about their relationship for the rest of the show.

MJ acts increasingly uncomfortable as they share more details. When she can't take anymore, MJ moves to the guest couch.

By the end of the show, None and Zelda seem closer than ever.

Those left in the audiences get drawn in too. Their romance sounds like it's growing into a fairy tale of its own.

None, MJ, and Zelda go backstage after the show ends.

Shen greets them eagerly backstage. "I've got bids from billionaires, hedges funds, and institutional investors. Everyone wants to buy Slam Dunk Jobs."

MJ says, "With all our bidders, we need to break it to Olivia that she's out."

Shen checks his laptop. "Her funds never hit escrow. That's our justification."

A security guard approaches None's group. "Olivia will see you now."

Shen grabs his laptop, and they all follow into Olivia's office.

Secret Service agents sweep Olivia's office for electronics.

Agent Vincent says, "All clear, Mr. President."

Olivia asks for the agents to shut off their communications equipment again and for Zelda to wait outside. None meets her demands. Zelda rolls her eyes but obeys and waits outside. Agents put their earpieces in their pockets.

None says, "Renquist was running the show out there. It's unacceptable. I hate Bruce. I want Corella back."

Olivia sighs. "It's his show to run."

None can't believe his ears. "Aiya! I thought you owned the Reveal Channel."

"My major advertisers led a boycott over our Slam Dunk Jobs buyout. Galaxor Telecom pulled their ads first. I think we both know who was behind it." Olivia gives None a knowing look. "Without revenue, our network won't last long. I've sold the Reveal Channel to Renquist on the cheap, while I still can. I sign the paperwork tonight."

None shakes his head. "You have to stand up to him, Olivia."

"I can't. Renquist followed the money trail to my family. He knows who I am!" Olivia tears up. She acts overwhelmed and jittery. "Renquist threatened to reveal my secrets. That would ruin me."

None asks, "What about your family? You gave the impression they were powerful themselves."

"I'm afraid my family won't be investing. Renquist blackmailed them too. They're in the food business. If my family invests, USDA inspectors will magically find horse meat hidden in their supply chain. My family would lose

314

their business. I didn't take him seriously until I found a severed horse's head in my refrigerator."

None never forgave Renquist about the ants. Killing a horse to stoke fear takes his depravity up a few notches. If enough money were on the line, Renquist probably *would* eat a puppy and spit out its bones. Is there a line Renquist wouldn't cross?

Olivia says, "I never cared about the money. My dream was to have my own show. Renquist agreed to let me keep hosting the Daily Reveal, so I'll be alright. You should warn whoever your other partner is."

None sighs. "It's too late for that." Renquist took out both partners and made money doing it. Renquist only had to scare off one partner to scuttle the deal, but he destroyed them both.

67. **Winning is Losing**

None, MJ and Shen go through the competing bids in the Oval Office. Renquist took out their first wave of investors. Only a couple days after None's encounter with Renquist, they have dozens of viable options. They have to vet the bids to see which options are most palatable. They implied it was the highest bidder, but it matters who buys Slam Dunk Jobs.

Agent Vincent says, "Your next appointment is here, Alvin Renquist."

MJ asks, "He's here in the White House?"

Agent Vincent nods.

None shares a horrified look with MJ. He looks at Shen. "Did you put him on my schedule?"

Shen throws up his hands in surprise. "I'll get to the bottom of this." Shen leaves the Oval Office.

"We'll hide the details on the investors and see what he wants," None says.

MJ closes Shen's laptop.

None hides folders on the candidates in his desk. He motions to Agent Vincent to let Renquist in.

Renquist looks around the Oval Office. "How do you get work done in such a tiny office?"

None says, "What do you want? I'm busy."

Renquist smiles at None. "I'm delivering good news. When I had you dragged off, you said I'd pay, and I just did. In a few weeks, we'll complete my original $1.2 billion purchase of Slam Dunk Jobs, and you'll be one of those billionaires you rail against so often."

Getting his money should feel like a victory, but it tastes like ash in his mouth. None shakes his head. "I'd rather keep the company. I have a new set of investors to outbid you."

"It's too late for that. I deposited the money in escrow yesterday." Renquist revels in None's disappointment.

None thinks about the trillions of dollars in salaries Renquist gets a share of. "On my show, you talked about how much you hate taxes. When you take your share of all those salaries, you *are* a tax."

"The Renquist tax. I love it." Renquist tries not to laugh. He looks at None intently. "I'm surprised at you. I never saw you tell a straight lie before. You told the TV audience those ideas were mine."

Nothing incenses None like an accusation of dishonesty. None jolts to his feet, enraged. "What? It's the truth."

Renquist's brain wheels run in overdrive. He stares at None, confused. A look of recognition comes over Renquist's face. "You really don't remember. You must have blocked it out somehow." Renquist gasps. "That's why you acted so strange when you offered to sell the company."

MJ hangs on every word Renquist says.

None trembles and tries to hide his fear. Please, let Renquist be lying. Could he be talking about the missing memories? "What are you talking about?" A part of None hopes Renquist never answers.

For the moment, he doesn't answer. Renquist muses on his own question. "You know what the difference is between us?"

None says, "I'm not an ant killing, horse beheading, economy crashing psychopath?"

Renquist laughs off the insult. "The difference is one word: hypothetical. You could destroy the world if you believed it was a thought exercise. I asked you to come up with the best way to lower wages."

"What did I tell you?" None worries, afraid to hear the answer.

When Renquist reflects, he smiles fondly. "You said you wouldn't do it."

The answer relieves None. He knew he'd never come up with all those greedy ideas.

Renquist grins menacingly. "Then I asked you again, with an extra word, hypothetically. You said it would only happen if people had to bid on their own jobs. That was your billion-dollar idea. That's why I bought your company."

The news blindsides MJ.

None sees the disappointment in MJ's eyes. Did None really set off the chain reaction devastating the economy? None has to remember. He doesn't want to remember, but he must know the truth. Focus. Focus. That day was the day. That day. Bring it back. Remember!

To Renquist, these memories are a treasure chest of joy. "I played you better than an orchestra. You laid out my entire plan for world domination in an hour of brainstorming. Did you think I wouldn't take such a gift, just because you asked me not to?"

The memories seep back. The suffering of millions started with an offhand comment about bidding on existing jobs. When he realized what he unleashed, None compartmentalized, unable to accept the truth. None puts his head in his hands, devastated. He marinates in a broth of shame and guilt.

Shen comes back inside the Oval Office. He stares at Renquist and remains silent.

Renquist edges towards the door. "I'm the kingmaker, and you're less than a pawn. Every move you made has failed. This chess game we've been playing won't last much longer." He laughs on the way out. "Congrats again on your billion. The company makes that every few days."

Shen waits until Renquist is gone to speak. "I discovered another one of his pawns. I fired the staffer who put Renquist on the schedule. What happened here?"

MJ fills Shen in on the encounter with Renquist.

Shen confirms what Renquist said. With the money in escrow, it's too late for the hostile takeover option.

None blamed everything on Renquist, but he knows better now. None thought that robots would crash the global economy, but all it took was an app. His app.

68. **Kiss**

Twenty minutes after Renquist leaves, None remains at the Resolute Desk in a catatonic stupor.

MJ worries about him. None isn't the best at dealing with disappointments. He just sits in his chair, like he's part of the furniture.

Shen strokes his hands through his hair. He looks burnt out by the events of the day. "Can we speak privately?"

MJ nods. She follows Shen outside the Oval Office, down the hallway, into the Cabinet Room.

Shen nods a few times, summoning courage. "I've decided I'll be your dead man."

"It's dead man's switch." The way things have gone lately, MJ knows the difference might be academic. "What made you change your mind?"

He opens his mouth a few times before words come out. "Our last option to stop Renquist failed. We're out of safe options. I've got to kick fear in the balls like you do. The stakes are bigger than my life."

MJ knows Shen is usually a cautious person. She'd seen a glimpse of him overcoming his fears when they defaced the signs at Garfield headquarters. Taking the package is another sign Shen could be capable of more. He could be a valuable ally, especially if Zelda tries to freeze her out of None's life, as Astrid did. MJ smiles back in gratitude. "Thank you."

Shen fidgets with his hair. "Now what?"

MJ pulls a package from her purse. "You take my copy, and I'll retrieve the copy in my stash."

He hesitates, then takes the package. Shen handles it as if it were a bomb. "The man who makes my morning coffee was a mole for Renquist. He did what he could to get close to me. It reminds me of what you said about Zelda. I have to warn you. My brother bought a ring."

A rush of adrenaline. Quickened heart rate. MJ's body is ready for battle, but this isn't the sort of conflict she can win with fists. "Do you mean what I think you mean?"

Shen gulps, then nods. "If my brother's about to marry a spy, you better find proof quick."

MJ rushes out of the room without a goodbye. MJ marches down the hallway, faster and faster. Run. She has to run. She runs with desperation towards the Lincoln Bedroom.

Agent Reynolds and her detail pick up the pace to keep up with her. "Is everything alright, Madam Vice President?"

MJ wants to punch something. She needs to get centered and calm down so she can think. "I need a workout."

Her Secret Service agents follow her to the Lincoln Bedroom. After she changes into a gi, they follow her to the dojo.

MJ practices Karate katas in her dojo. An October chill keeps her muscles from overheating as she practices relentlessly.

Her mind is the weapon she must wield. MJ replays everything she knows about Zelda. She thinks back to the plane with Zelda. That two-faced spy lowered MJ's guard. It won't happen again.

Zelda started rapport building with some faked forgiveness about knocking her down on TV. What was she even talking about? MJ has all the episodes of *The Internet President* recorded. MJ doesn't have to replay it in her head. She can watch it for real.

MJ leaves the dojo and returns to her room. She fast-forwards through the "Jury of Lovers" episode to where MJ barreled through the rows of chairs to chase Corella. The video shows MJ knocking over Zelda hard, but that's not how she remembers it. MJ remembers barely tapping her.

This time, MJ plays the video frame by frame. Slowed down, Zelda leans in towards MJ. It looks like stunt work. This was a staged pratfall.

MJ repeats Zelda's words. "This is the last time I date a guy with female friends." She made MJ look crazier to scare off the other women. Zelda played hurt and then lead them off the stage. The act was all about removing the competition.

It's suspicious, but not enough to consider proof.

If MJ can't convince None to dump Zelda, maybe she can find him a better option. MJ needs a woman from None's past that he might still be hung up on. Even Astrid is the lesser evil now. Astrid might have worked before the wedding, but not now. MJ watches the entire "Jury of Lovers" episode again. MJ finds what she's looking for. Sarah.

Sarah gives None a chance to pick her, but he hesitates, and she walks away. MJ can read the disappointment. He wanted to choose her right there.

MJ still has Sarah's number. Maybe woman to woman, she can convince Sarah to give None another chance. MJ dials.

Sarah picks up. "Hello?"

"I don't know if you remember me. I'm MJ, the president's friend."

"I know who you are." Her words sound cold and artificial. This might be a hard sell.

MJ says, "I'm not sure the president has told you, or not, but I know he still misses you."

No response.

The silent treatment won't stop MJ. "This is difficult for me. The idea of anyone being more important in None's life scares me. Sarah, you had something special with him. I know it didn't last long because of the craziness in his life, but if you give him another chance…" MJ trails off. She chokes up. It's so hard to ask another woman into his life, regardless of how evil she thinks Zelda is.

"That is a path I choose not to follow. I wish you well on your journeys without me. Namaste." The words don't sound like the Sarah she remembers, almost lifeless.

Sarah hangs up.

MJ holds her thumb over the button to redial. Even if Sarah says yes, MJ can't bring yet another woman into None's life. There must be another way.

In espionage, the standard way to stop a spy is a counterspy. Who would Zelda let get close to her? Just None and MJ. She can't ask None without proof, or he might think she's crazy. That leaves MJ. She has to become the counterspy herself.

The easiest way to get close to Zelda is to let Zelda come after her. Zelda assumed MJ was a closeted lesbian, so that's what she'll pretend to be. She'll lure Zelda in and find proof.

From Mata Hari to Anna Chapman, sex has always been the weapon of the female spy. How's that supposed to work for MJ? Even the idea of sex gets her queasy. MJ will have to woman up and soldier on. She's run out of options.

MJ stops by the Queen's Bedroom, where Zelda sleeps.

Zelda acts surprised to see her. "I'm sorry about the other day. I guess I was picking up on the wrong vibe." Zelda doesn't stroke her hand today.

"I can't stop thinking about what you said." MJ sways back and forth with a bashful grin. "I've never kissed a girl before. Will I like it? I want to find out."

Zelda probably thinks MJ is walking into her trap, but it's the other way around. Zelda leans in closer to MJ, intrigued.

MJ wears a smile to put her at ease. "I've been doing a lot of thinking about us."

She takes the bait. "Us?" Zelda touches her hand with a predatory smile.

Zelda mocks MJ's Karate kiss-block move. "Let's try the kiss without the Kung Fu this time."

Her form was flawless. It hints at training. MJ's pulse quickens. Every moment she spends with Zelda, she'll be in danger. Who knows what other hidden skills this spy has.

Zelda's hands do a soft lingering touch as she pulls away. "It's your first time. I want to make it special. Give me an hour to freshen up."

MJ returns across the hall to the Lincoln Bedroom. She faces danger in Zelda's room that could strike at any time.

She takes off her clothes and jumps in the shower. MJ imagines all the attack vectors Zelda might try.

Poison. Drugs. She won't drink anything Zelda gives her.

Gun. Knife. MJ trained for that.

MJ dries off and begins to dress. She picks boots first. They have to be sexy and combat ready. MJ chooses knee-high boots with sturdy heels that won't break off, like when she attacked Renquist.

She remembers how naked and vulnerable she felt on election day without her boots. No matter what, she's not taking them off. Zelda will expect her to take off her clothes. MJ needs clothes that are easy to take off without removing her boots. She picks out a black mini-skirt and matching halter top.

Her goth steampunk seven-buckle arm warmer gloves are next. They extend from her knuckles to her elbows to cover her scars, but they also might provide some protection in a knife fight.

Is the lipstick Astrid gave her really battle worthy? MJ is ready to find out. She pulls out the bottle of Invincible Lips and applies it with deliberation. With her war paint on, it's go time. She returns across the hall to the Queen's Bedroom and knocks on the door.

"Just a minute." A few seconds later Zelda opens the door. She wears a lacy lingerie corset with straps that come down to her stockings. A bikini bottom covers her privates and not much else. Zelda pulls MJ inside and slams the door closed before an agent can see how she's dressed.

When MJ sees the rope on the chair, it seems more nefarious this time. If MJ stayed longer last time, what might have happened to her? Perhaps Zelda intended to tie her up and interrogate her to find the package. MJ would be too drunk to notice Zelda slip drugs into a second bottle of champagne.

Zelda notices her staring at the chair. "Have you ever been tied up?"

No way. MJ stuffs her apprehension deep inside her. MJ isn't ending up like Thelma, or Gerald. She'll die fighting, not tied down and tortured. "If I wanted to be tied up, I'd be married."

"I'll take that as a no." Zelda giggles and pulls MJ towards the bed.

Being on high alert heightens all MJ's sensations. She sits on the bed with Zelda. MJ notices every breath Zelda makes.

Zelda strokes her hands gently, up to the edge of the seven-buckle arm warmer gloves, right above the knuckle. She traces the edge of the gloves. "Can I take these off? You might be more comfortable."

MJ tenses up. She's self-conscious about her burn scars. MJ holds her breath a few seconds, then nods.

Zelda undoes the gloves and glides her hands up MJ's arms with a whisper of a touch.

Tingles follow Zelda's fingers. MJ never let anyone touch her so sensually, much less a spy who might become her assassin.

"These are the scars of a hero. That's very attractive." Zelda traces the scars up her arms with delicate hands.

The words are meant to gain her trust, but MJ enjoys hearing them. She smiles back at Zelda. So sensitive, MJ's hair bristles at the slightest tactile sensations.

Zelda's damp mouth and probing tongue engulf MJ's fingers.

The thought of Zelda's saliva slathered on her fingers makes MJ uneasy. She rationalizes: The fingers are just appendages, not really the main her. MJ lets herself enjoy the hot, moist feelings.

"You can touch me, too." Zelda slides the lacy corset down enough to reveal her bosom. She places MJ's hand on her left breast.

MJ gropes the breast firmly.

"Softer, you're not a man." Zelda coos when MJ touches her more sensually.

Touched. Touching. It's all so new and shiny. MJ is a new car, about to be driven for the first time.

Zelda goes in for a kiss.

MJ cringes. She tightens her lips. She feels queasy thinking about Zelda's tongue invading her mouth. She rationalizes: Her mouth is shut so tight that Zelda can't penetrate.

Zelda awkwardly kisses MJ's closed mouth. She tries to pry open MJ's mouth with her tongue, but MJ's clenched jaws thwart every attempt. Her hand strokes MJ's back, fingernails first. Zelda undoes the drawstrings on MJ's halter top.

The halter top drops to reveal MJ's firm breasts and well-toned torso, a dividend of her aggressive training. MJ retracts her arms to cover her bosom.

The flash of her perky breast brings hunger to Zelda's eyes. She breathes hot on MJ's neck and moves down.

MJ's willing to do anything to stop the ongoing siege of kisses before Zelda penetrates her. She rationalizes: Her breasts are outside her body. They could distract Zelda from her mouth.

She drops her hands to leave her breasts unprotected. MJ lets herself enjoy Zelda's small kisses descending from her neck to her right breast.

Zelda suckles on her nipples.

Quivering with halting breaths.

A crescendo of delight.

Wet.

This must be what arousal feels like.

Amazing. With the adrenaline of danger and the thrill of a new experience, MJ throbs with life.

MJ opens her mouth a tiny bit, forgetting herself in a rapturous moan.

Zelda seizes on the opening and moves back to kiss her mouth. She holds MJ's head with both hands and plants a hard kiss.

Tongue.

MJ can't rationalize away Zelda's intruding tongue, or the sickening ooze from her lips. Alien fluids in her mouth. Vile. Contemptible. Filth.

It feels like drowning ants, scurrying down her throat, alive.

MJ faces her worst fear of foreign objects invading her body. The fear wins.

Nausea.

Vomit surges up her throat into Zelda's trespassing mouth.

Zelda's eyes bulge in a mix of rage and shock. She covers her mouth and races to the bathroom.

"I'm sorry. It must have been bad fish at dinner." MJ's anxiety diminishes. She'd rather have her own vomit in her mouth, than Zelda's tongue. She grabs a tissue from the bedside. MJ wipes away vomit chunks from her lips and reties her halter top back on.

The bathroom door slams. MJ hears the toilet seat go up. It sounds like Zelda might be retching herself.

This is MJ's only shot. Zelda will never make out with her again. She has to find the evidence, while Zelda cleans up in the bathroom.

MJ rummages through drawers of the dresser and nightstand. There's nothing helpful in Zelda's purse. She didn't bring a screwdriver, so she can't check the vent on the wall. She looks under the bed. There's something underneath on the other side.

The bathroom faucet comes on. Zelda screams in the bathroom and curses MJ's name. She could come out any second.

It's a race against time. MJ races around to the other side. She drops to her knees to see better.

Her missing water serpent katana is under the bed. How did Zelda get it? What was she going to do with it? To smuggle that in, Zelda must be working with agents. How many accomplices does she have?

MJ grins. This is the evidence she needs. None knows they took her katana when Renquist's dirty cop arrested her. This katana ties Zelda to Renquist. If she shows it to None, Zelda's done.

Should she take the katana with her? Rushing towards the president with a sword is a great way to get shot. She'd better leave it here. Besides, it'll be more damning to find it in Zelda's room.

MJ rushes across the hall to her bathroom. She gargles with mouthwash and sprints to the Oval Office with her security detail in tow.

She barges into the Oval Office.

None stops whatever he was doing. "Good, you're here. You freaked out the Secret Service. First swords, then a weaponized shovel?" He brings up the website for the Decapitator Z9000. "Do I really need to approve a shovel that's sharp enough to cuts off a zombie's head? Last time I checked, we weren't facing a Zombie Apocalypse."

MJ says, "It's for my go bag. I'm worried about a Renquist Apocalypse, with assassins paid to silence me. I found another spy for Renquist. It's Zelda."

He almost falls out of his chair in shock. "You sound paranoid. I finally have an MJ friendly girlfriend, and you still go off. You've interfered with every relationship. I'm not dying single because you can't stop meddling."

MJ says, "I was trying to protect you."

"Protect me from what? Happiness?"

"No. Protect you from the wrong girl." Everything MJ did was for him. Can't he see that?

None only sees her meddling. "You don't decide that for me. MJ, you made all my relationships about you."

How dare he say that to her. "Your company. Your presidency. My entire life revolves around you. Our relationship is important to me. It's just…I can't give you what you want."

A part of her wants to be with him, but it would never work. Sex is the rift between them that she can't cross.

None says, "You act jealous, but you don't want me. I have to accept some of the blame. I can't believe I was hung up on you all those years like I was destined to be with you." He snorts in contempt. "I should have quit chasing you a long time ago. You can't keep me all to yourself anymore."

"That's not what this is about." MJ needs to steer the conversation back to why she came.

None blushes a little bit. "I feel embarrassed asking, but did you ask Zelda to take off your dress?"

It's pretty hard to explain away. He'd understand if he was wearing that damn ballroom gown. "Yeah, but—"

None says, "Zelda told me how you tried to kiss her after you downed a bottle of champagne."

He stopped listening when MJ admitted asking Zelda to take off her dress. It's obvious he'll take her side, just like he did with Astrid. "Zelda tried to kiss me."

None seems skeptical that Zelda would do that. "It's OK. I'm not judging you. If you like girls, it would explain a lot."

Accusing her of being lesbian is just an attempt to protect his fragile ego. MJ is offended he wouldn't believe her. Zelda manipulated him. "I have proof. Do you remember my katana, that Renquist's goons stole?"

None raises his eyebrow. "Of course."

"Zelda has it." MJ revels in playing her trump card. None won't have a choice, but to believe her.

He bolts towards the door. "Let's go. Show me."

None, MJ and their agents march towards Zelda's room.

Agent Vincent knocks on her door.

Zelda swings the door open, dressed in a conservative pantsuit. She goes to hug None, but he's distant.

None points at MJ. "Where is it?"

"Under the bed." MJ smiles in anticipation.

Agents look under the bed. "There's nothing there."

MJ goes to look for herself. It's gone. "It was right there. I swear!" MJ looks around. The rope is gone too. Zelda covered her tracks.

Zelda pretends to be confused. "What's going on Cuddle Muffin?"

None looks anguished. He turns to MJ. "Why can't you let me be happy? Why do you have to make things up?"

If the katana had been there, Zelda would be gone. Now that she put None on the spot, he'll have to choose between them to save face. MJ doesn't like her odds. She pleads with him. "Zelda's a spy for Renquist. Listen to me this one time, and I swear I'll never meddle in your love life again."

None doesn't take long to decide. "No. You aren't going to destroy another relationship for me."

MJ moves closer to None to make her case, but agents block her path. "Bro?"

None fights to get the words out. "We aren't bros anymore. From now on, you can address me as Mr. President."

She can feel him slipping away. Maybe he could convince him in private. "We shouldn't be having this conversation in front of her."

Zelda's smile gives away how much she enjoys the precarious position MJ's in. She knows she's won.

"Oh, I think she's the perfect person to hear this. I want to show her I'm putting her first." None gets down on one knee and pulls out a jewelry box.

Zelda doesn't wait for words. She puts out her hand. "Yes."

None puts a ring on her finger and rises. He takes Zelda in his arms and kisses her.

MJ's heart breaks. She can't imagine her life without him. Zelda didn't have to tie her up to torture her.

None holds Zelda in his arms, as he addresses MJ. "She's more than a bro. She's going to be my wife. I want you out of the White House. I'll have you moved to the Naval Observatory, where vice presidents are supposed to live."

69. **Overdrawn**

With their salaries squeezed, workers don't buy anything beyond the essential. The drop in sales across the economy drops corporate profits across the board. The only labor costs left to cut are managers.

All the red buttons get pushed.

Profits turn into losses, leaving few corporate winners.

Global stock markets drop almost every day for two weeks. Big drops on the stock exchanges trigger breakers that halt trading. Being unable to trade spreads panic further.

Wall Street deploys its most effective weapon, lobbyists. Industries come to Washington asking for bailouts.

Banks borrow money from the Federal Reserve to stay open. The conventional solutions seem to do little more than buy time.

Congress can't achieve anything beyond finger-pointing.

None and MJ attend one emergency meeting after another, but they communicate only through intermediaries.

MJ may only be feet away, but None misses her. The two of them have never been as emotionally distant as the last two weeks. Their friendship falls apart, just like the country.

Civil unrest spreads through America, with protests and riots.

Fear of the federal government has grown for a long time. A summer 2015 multi-state military training exercise known as Jade Helm 15 sparked unfounded concern that the military would invade Texas. While the exercise ended without issue, it was taken seriously enough that Texas Governor Greg Abbott instructed the Texas State Guard to monitor the operation.

Talk show demagogues spread a theory that Jade Helm was a dry run. It's going to happen for real this time. Conspiracy theories seem more plausible in a world that's falling apart. Fringe is the new mainstream. Many worry that the federal government will use the civil unrest as a pretext to declare martial law and confiscate weapons.

The paranoia centers in Texas, a state with a proud heritage of rebellion. For a decade, The Republic of Texas was a sovereign nation with its own currency. It's a place that clings to God and guns, guns they won't willingly give up.

Fear prompts the Texas legislature to pass the Don't Mess With Texas Act, which rescinds even the most minor gun regulations. It couldn't be easier to buy a weapon there.

Gun shops temporarily run out of inventory.

A few of the smaller towns go from open carry to required carry. Everyone in town has to own a gun and train to use it.

Gun control isn't about mass shootings anymore. It's about mass uprisings.

In the Oval Office, None and Shen plan for the international economic summit that starts in two days. His national emergency meetings will be replaced by international emergency meetings. With the speed things are unraveling, he hopes he has a country to come home to.

With all the recent changes, upheaval plagues None's personal life as well. MJ has her martial arts to center her. What centers None? Was it his relationship with MJ? Everything feels so out of whack. None can't afford to be distracted right now. He looks to his brother. "Am I moving too fast with Zelda?"

Shen laughs. "Maybe you should have asked yourself that before you proposed."

Did the proposal timing have more to do with MJ? Maybe he overcompensated, to make sure she didn't chase away yet another girlfriend. "I better introduce Zelda to Ma. Don't tell Ma I waited weeks to tell her."

"I'll call to let her know we're coming to L.A.," Shen says.

None signals him not to. "No. I don't want Ma to work so hard cleaning for our visit. We'll make it a surprise, so there's no fuss."

Shen says, "Surprise visit is the right call. Otherwise, the media will say the president ran home to mommy, in the middle of a crisis."

The president sighs. "They wouldn't be completely wrong. I need to clear my head."

"I'm glad we're going home. I need some alone time with Lan."

None, Shen, and Zelda leave for L.A., aboard Air Force One. It feels strange not to have MJ going.

Zelda doesn't seem nervous at all to meet None's mother.

He's the one who's nervous, about meeting his mother's boyfriend, her so-called roommate. It bothers him how her boyfriend took over the living room. His mother has made a point not to say a word about the relationship.

Lan meets them at Air Force One when they arrive in L.A. She seems a bit off. Lan doesn't act like her usual game show hostess upbeat self.

None introduces Lan to Zelda. "Hey, Lan. No greeting card today?"

She shakes her head. "I don't make those anymore."

After some small talk, Shen and Lan get ready to leave.

None smiles at them. They've been apart a while. None's glad they'll get a conjugal visit. None waves at them. "You two have fun."

"Oh, yeah. We'll have lots of fun," Lan says with a sarcastic sneer. She takes off with Shen.

None and Zelda travel to his mother's house via motorcade.

Inside, his shocked mother greets None and Zelda when they arrive.

None and Zelda field questions from Mrs. Wong about how they met, their wedding plans, and how Zelda feels about children. During the Q&A, None has something else on his mind.

His mother's two-bedroom house used to feel cramped, before her roommate took over the living room with his things. None remembers how the living room used to be. The missing dining room table and chairs. The missing TV. The dead stump that used to be a money tree. If it weren't for the plastic covered couch and tattered Bible, he wouldn't know it was her house. "You shouldn't let your roommate take over the living room. There's barely anything of you left in this room."

Mrs. Wong looks at None oddly. "No roommate forced me to do that. My mortgage did. I sold things to make my payments, but it wasn't always enough."

None jumped to conclusions. Perhaps he misjudged her boyfriend. He didn't force his mother to sell her bed. "Don't sell any more things. I'm going to be rich soon."

She smiles at the thought. "I'll let you buy me a bed. I'm tired of sleeping on the couch."

None cocks an eyebrow. "Your boyfriend makes you sleep on the couch?"

Mrs. Wong stares at him like he's an idiot. "What boyfriend? I live with Marjorie."

When all the furniture went missing his brain filled in the blanks with the wrong answers.

A shared bed. He assumed that meant sex.

None got it wrong when he woke up in MJ's bed. He got it wrong again about his mother.

The last thing None needs is imaginary problems. He already has enough real problems.

None remembers Marjorie was his mother's best friend at work. Slam Dunk Jobs gutted the HR industry. With the oversupply of skilled HR people, a call center job is the best Marjorie could hope for in her field.

Mrs. Wong says, "Marjorie spent her whole life in HR until your company automated away her job. When she lost her house, I had to take her in." Thoughts about what happened to her friend chill his mother's mood.

330

The impact of his creation extends here, in his mother's home. None can tell Marjorie's situation weighs on her, but his mother has nothing to feel guilty for.

The impact on the nation itself is more extensive. The origin of the national debt problems started decades before he got to Washington, but he lit a match.

How do you slow down an explosion?

He's buckling under the strain. He'll have gray hairs before he hits 40. "This country's overdrawn and so am I. I've given everything I have. It's not enough."

Mrs. Wong looks around, then back at None. "Where's Maria?"

Zelda gets uncomfortable at the mention of her name.

None can't change his mother. Like MJ's parents, she still calls MJ by her old name. "We had a big fight." None sighs. "I'm not sure if we're friends anymore."

His mother looks aghast. "She brings out the strength in you. How can you win without Maria by your side? This is not the time to be without your best friend."

"I'm his best friend now," Zelda smirks.

Mrs. Wong shifts her sympathies to her son. "With all the problems you face, you need more than a best friend. You need faith in something bigger than yourself." She grabs her tattered Bible and leads None and Zelda in a short prayer.

She hands the Bible to her son. "I have faith. You need faith right now. Son, I want you to take it. Reflect on its wisdom."

None tries to hand it back to his mother. "That's not really me. I need reason, not faith, to get me through my struggles."

"Maybe you need both." Mrs. Wong closes None's hands on the tattered Bible. "Promise me you'll read it."

He's never been much for organized religion. Too often it's used to manipulate the masses and turn off their brains. Give someone enough blind faith and you can make them believe anything. Religious wars. The Inquisition. Radical Islam. Jonestown. Don't drink the Kool-Aid. Like any tool, it's used for good and evil.

Promises haven't worked out well for None lately, but he wants to make his mother happy. He enjoyed Bible stories as a child and appreciated the poetry and parables as a young adult. Why not? He's been inspired by less before. "I promise."

None holds the tattered Bible with both hands. It must be vitally important to her, to entrust him with her most treasured possession.

The focus shifts back to meeting Zelda. They talk and eat for a few hours before it's time to leave.

Between spending time with his mother and a home cooked meal, None feels calmer and more energized. She's right about MJ, though. He needs her by his side. It's been much harder without her.

None and Zelda say their goodbyes and head back to Air Force One.

Shen greets them, as they enter the plane.

None winks at him. "How was your alone time?"

Shen won't dignify his wink with a response.

None asks, "Hey, what was up with Lan? She acted a little strange when we arrived. I was looking forward to one of her funky greeting cards." None chuckles.

Being peppered with personal questions irritates Shen. "I don't want to burden you with my problems. Let's leave it at that."

None puts his arms on his hips. Why can't his mother and brother be more open? "We're a family. Don't hide your problems, like Ma did with her mortgage."

Shen glares back. "You really want to know?" He winks back in a forced and aggressive matter. "I spent my alone time in bankruptcy court, holding Lan in my arms, as her real estate dreams died. You can't call her Monopoly anymore. You'll need a new nickname. The game Monopoly ends after you go bankrupt."

"I'm sorry."

He snaps back. "Private matters are supposed to stay private. Why do you always have to push?"

Agent Vincent gets None's attention. "The vice president is on the line for you."

None is relieved for the excuse to leave. He goes to his office on the plane and puts the phone on speaker.

Zelda follows him and listens as inconspicuously as possible.

On the phone, MJ is all about business. She still has a job to do. "Mr. President, the Texas National Guard refused to clamp down on the uprising. The situation in Texas could turn into an armed rebellion if we don't do something decisive. Armed civilians are massing at a gun rally in Houston. Whatever you plan to do, you need to hurry."

Hearing MJ address him formally feels awkward. He holds out the phone. "Agent Vincent, redirect the plane to Houston."

None gets back on with MJ. "I'll make things right between us. Just give me time. I need you." None catches himself. "I need you in Texas."

70. Defend Texas

The roots of civil unrest in Texas are easy to trace. Lower wages mean lower taxes. Small towns can't afford police and firefighters from their reduced tax base. The money that would pay for these things went to Renquist instead.

Making matters worse, fear of bank failures convinces people to stash cash in their homes, which leads, in turn, to an increase in home invasions. An awful time to be without police, small towns revive an Old West remedy to replace the missing law enforcement. Town sheriffs post sign-up sheets looking for volunteers to join posses.

Armed citizens keeping the peace and defending their towns become a movement in itself.

Today that movement unites. Militias and armed citizens descend on Houston for a convention called Defend Texas. The convention name was meant to stir controversy and unite the true believers. These armed citizens will defend their towns, whether against criminals, or encroachment by the federal government that they fear is coming. Demagogues continue to feed that fear.

Air Force One heads towards the epicenter of the uprising.

None sit next to Zelda on a couch in the presidential suite. He holds Zelda's hand tightly and looks into her eyes. "Houston could get dangerous. Can you promise me you'll stay on Air Force One, so I don't worry about you?"

"So, I have to worry about you, instead?" Zelda looks at None, sizing him up. "What aren't you telling me?"

"I may have to do something reckless at the gun rally." None gulps hard. "I need you to wait in the presidential suite. If I see your worried face, I might not be able to go through with it."

Zelda puts on a brave face and attempts to be cheerful. With her free hand, she runs her fingers down his back. "Let me go, and we'll join the mile-high club."

The offer of sex tempts him. None has to resist. He pulls away from her. "I'm sorry. I have to go." None makes a quick escape before he loses his willpower. He goes upstairs to his Oval Office in the sky.

Shen and Agent Vincent wait for him.

Agent Vincent says, "I've alerted the Secret Service Houston field office we're coming. I need to know where we'll be traveling in Houston so that I can send advance teams."

"The Defend Texas convention."

Agent Vincent tenses up when he hears the answer. "Mr. President, that location is filled with firearms and ammunition."

None sees only one path forward, and it leads to that gun rally. "I know that."

Shen loses patience with his brother. "Do you have a death wish? I've been watching these crazy people on the news. They hate the federal government. You want to waltz in there representing that very same government? I guess they aren't the only ones that are crazy."

Agent Vincent nods. "Understood, Mr. President. We'll remove all weapons in a section of the convention center so that you can speak."

None freaks. "Aiya! You can't take away their guns. That'll confirm their worst fears. Leave them armed, while I speak."

Anger, shock, and disbelief wash over Shen. He steadies himself. "You want them to have guns? Wow. Call me when your sanity returns. I can't listen to my brother throw his life away." Shen storms out.

Armed civilians in the same room with the president is the Secret Service's worst nightmare. Agent Vincent turns pale. "Mr. President, you're putting us in an untenable situation. We'll be forced to fire on anyone who points their weapon in your direction."

None looks at him with a steely stare. "No. Do not fire on the civilians unless they fire first. Look, if you create martyrs at the convention, we might have a civil war on our hands. Some of them think we're the bad guys. Don't make them right."

Agent Vincent says, "Mr. President, given your current situational parameters, I see only one way to secure the site. Do you promise to do everything I ask of you?"

"I promise." It occurs to None after the words leave his mouth, that he doesn't know what he just agreed to do. He needs to be more careful when giving his word.

The color returns to Agent Vincent's face. "Thank you. Good news, Mr. President. After the incident at the wedding reception, the Secret Service requisitioned a complement of ten Andrastes, for you, the vice president, and eight agents on your security detail."

None's gasps. "You want me to address the gun rally in battle armor?"

Agent Vincent nods. Confidence returns to his voice. "It's the only way to keep you safe. Showing superior force will send a message and keep them in line."

He looks back at the agent with crazy eyes. None appreciates they want to protect him, but Agent Vincent doesn't see the bigger picture. "Remember the Alamo? Standing up against superior forces is a Texas tradition. Just one martyr, one, could channel outrage against the federal government. Is that what you want?"

Only one thing matters to Agent Vincent. "I just want to bring everyone home safe."

"I'll wear the damn thing, as promised, but I can't allow my detail to be geared up for war at a gun rally." The armor sends the wrong message. There must be a way for None to fix it.

Agent Vincent updates the Houston advance teams on the rules of engagement. "I've got confirmation that Madame Vice President will also wear an Andraste."

None isn't surprised to hear she'll wear one. MJ's been jonesing to try one out since she saw one at the stockholders meeting.

The motorcade travels to the convention center in Houston. The Secret Service gets the convention to schedule None and MJ as surprise guest speakers. Their official topic is the Don't Mess With Texas Act, the state law that gutted all existing gun regulations. On the way to the convention center, in two converging motorcades, None and MJ stop at sporting goods stores. They both test the new law by buying guns.

None meets MJ backstage at the auditorium where they will speak. Lights and scaffolding extend overhead, as well as a few layers of curtains. All the curtains are pulled up.

The presence of firearms at the president's talk unnerves the Secret Service enough that they keep multiple ambulances on standby.

MJ confers with the Houston advance team about escape routes and whether the backstage can handle weapons fire without ricochets. Metallic plates cover the very back of the stage to catch any potential bullets.

Secret Service agents put rows of bullseyes up for target practice.

Before they let people in, MJ has her agents clear the stage and conduct live-fire tests to make sure the bullet traps work correctly. MJ must expect something to go down.

The back curtains come down to conceal the metallic plates. The next outer row of curtains comes down to hide None, MJ, and their security details.

Off in the wings are two Andrastes on pallets.

MJ steps into the battle armor. It encloses around her. She does a karate chop. "Mr. President, do you need a lift?"

She picks up the pallet with his Andraste on it, like it's a dumbbell. MJ does several reps. The lifting capabilities of her new toy delight her. "I'm invincible!"

None rolls his eyes. He points at his battle armor. "How am I supposed to make them laugh in that?"

MJ sets down the pallet with his Andraste. "Making a fool of yourself isn't in the presidential job description. Can't you let the ham sandwich win?"

"I have to make Americans laugh and remember who they are. I don't want the America I love to disappear while we're at the economic summit." None heaves a heavy sigh. He made a promise. None puts on his Andraste.

MJ stares back at None. "Do you think that's really possible?"

None can hear the audience being led into the auditorium. It won't be long now. "We're about to face a mob with guns. What could go wrong? No matter what happens, keep your Andraste on. You're plan B."

MJ nods. She practices karate katas with a battle yell. The Andraste responds immediately to her movements.

The sound of her battle yell disturbs None. What will the audience think? He shushes MJ, "don't scream, or you might startle the gun-toting hombres out there."

She quiets down but continues to practice martial arts in her Andraste during the half hour keynote speech before them.

None watches MJ as he contemplates what to say.

After the warm-up speech ends, an announcer comes on. "The Don't Mess With Texas Act passed with an overwhelming majority. Here to discuss it, give a Texas welcome to our surprise guests. From Washington—"

The crowd doesn't let the announcer continue. They respond to the mention of Washington with boos and catcalls.

None's curtain raises enough for the audience to see two sets of heavily armored legs.

At the sight of the Andraste legs, the crowd goes silent.

The curtain exposes the ominous torsos of their military grade hardware.

The silence is broken when a man cries out. "It's a trap. They're here to take our guns."

There's a rustling in the crowd.

Another man says, "They're coming for us."

Cocked guns. Inserted ammo clips. Dropped bullets. None hears the sounds of impending warfare. He realizes the audience can't see their transparent faceplates yet. For all they know, None and MJ could be killer robots. He ducks under the curtain with MJ in tow and comes out quickly, so the spectators can see their faces.

The chips are down. MJ's the sort of person to take the chips, stack them, and play poker with them. The danger doesn't faze her at all in her invincible suit.

Rifles and pistols in the first few rows point at None and MJ.

The crowd acts jittery around the Secret Service agents posted in the auditorium. Each agent attracts a group of armed citizens poised to strike.

None can only imagine how unsettled the agents must be, outgunned and outnumbered by a hostile army of citizens. His rules of engagement put their lives in danger, but None saw no other options. The feds can't be the ones to fire first. They can't provoke the crowd.

He raises his hands. None has to calm the audience. "I'm not here to take your guns. Not on my watch. I'm on your side. Powerful forces work against us. Those powers only respect the common man with a gun in his hands. That is why I support the second amendment."

Those aiming at None and MJ lower their weapons.

A heckler in a baseball cap screams. "You're a liar! You're not on our side."

Secret Service agents move to take the heckler down.

None motions for the agents to stop. "It's rude to interrupt. Listen to me, and I'll listen to you. I promise. After my speech and the Q&A, I'll leave five minutes for heckling."

The audience laughs. Some people glare at the heckler, and he sits down. The Secret Service agents back off.

"You see me as an outsider, but recently, I became a Texan. This place becomes part of you. I appreciate the people and food of Texas!" None fist pumps. Mostly he appreciates MJ's family and the food they brought home from their restaurant.

The pro-Texas comments get some applause. Some of the more trusting in the audience turn on their safeties and holster their weapons.

None gets somber and remorseful. "Unintended consequences always get you. I can't fix the past, but it's not too late to salvage the future. Tonight, I leave to meet with world leaders at an economic summit, to solve this crisis. I'm going up against Big Money people."

Big Money people. The crowd hates them as much as Washington.

None says, "Who do you think they'll save? Main Street, or Wall Street?"

The question riles up the crowd. They know the answer, and they don't like it. Too many of them feel left behind, struggling to make ends meet.

"I will be there to fight for you," None says. "This downturn has affected my family too. My brother lost his business. My mother nearly lost her home. I know what's at stake and I won't give up." None hopes his family forgives sharing those details.

Hearing about his family's troubles resonates with the crowd. They're going through similar problems themselves. Billionaire, or not, it makes him seem more like one of them.

None makes a dramatic pause, then continues his speech. "We endure difficult times. This isn't the Old West anymore. Mobs with guns make bad decisions, so we have to keep our heads. To survive and thrive, we must turn fear into laughter. Otherwise, we stand to lose everything that made America special."

The heckler shouts again, "It's easy to be brave behind bulletproof armor!"

None laughs. "You're right. This armor is what happens when we give in to fear."

He promised Agent Vincent he'd wear his Andraste. He didn't promise to keep it on.

"I'm going to take off this armor and let Texas prove itself to the world. Are you the gun nuts they think you are, or the misunderstood patriots I know you to be? I put my life in your hands."

The Secret Service agents panic. They rush the stage from all angles. Their president just went rogue.

None uses the enhanced strength of the Andraste to resist efforts to keep him pinned inside. He tries to exit his armor, but the agents reinitiate the Andraste close sequence. "Let me go. I can't wear this symbol of fear any longer."

Scattered audience members call out, "Let him go."

Agent Vincent leads the agents trying to clamp down the armor on None. "It's for your protection."

None could throw the agents off with the power of the suit, but he doesn't want to hurt them. Why can't they understand the grand gesture he's making? Can't they take some risks just once? "If I want their trust, I have to give them mine."

The audience chants. "Let him go! Let him go!" None becomes a symbol of resisting the very government he represents. It's clear to them now. He does fight for them, even against his own government.

Agent Vincent sees the situation spiraling out of control. He lets None exit the armor.

The spectators applaud. He got the feds to back down. It gives them hope.

The agents form a human shield around the president.

None forces his way through the agents to the front of the stage. He can sense the crowd is with him. He goes into rabble-rouser mode. "I believe in you. I believe in this country. I don't believe in letting fear control us. We stand together. We choose what kind of world we want to live in. We choose!"

The crowd gives None a standing ovation. He's their champion to fight the Big Money people.

MJ comes towards the front of the stage, still wearing her armor. She points at her Andraste. "Can I keep it?"

None says, "Only if you show us some tricks."

She does cartwheels and lifts herself up with one arm. She carries None's Andraste off the stage to show the lifting capability.

None asks, "Anything else to add, MJ?"

She leaves his question hanging, grabs a duffel bag from offstage, and returns. "This is a gun rally. I didn't come here to talk. I came here to shoot."

On her cue, the back curtain raises to reveal bullseyes near the back.

MJ smiles at None, as she sets down the bag. "With the new law, it's never been easier and more convenient to buy firearms. So, the president and I stopped and bought guns on the way here."

None reaches into the duffel bag and puts on a gun belt with dual holsters. He holds a pistol in each hand.

MJ pulls another gun belt from the duffel bag. In her Andraste, she's a bit big for human-sized belts. MJ hams it up. She pretends to be embarrassed and pats her tummy as if she put on too much weight. She shrugs and puts the gun belt over her shoulder. The audience smiles at her antics.

None lifts his guns up in the air. "Can you say open carry President? I've never held a gun before. I feel powerful. I'm president plus plus." He fist pumps with the guns, Yosemite Sam without the hat.

The sight of a gun toting president shocks and delights the audience. They act like they won a game show prize. The president got them to forget their fears for a few moments, even if it is to laugh at him.

None gestures wildly with the guns in his hands as he talks. "Bruce Cannon of the Reveal Channel says I'm crazy. Am I crazy? It doesn't matter. No more mental health checks to buy a weapon."

Secret service agents look like they're being tortured. Some turn their heads away and grimace. The first accidental death of a president could be on their watch.

None points the guns towards the audience. He lifts guns up and down like he's weighing them. "Feels great in my hand."

Spectators in his line of fire crouch down to take cover behind the seats in front of them.

He notices the crowd reaction and points one of the guns at the ceiling. "Don't worry. It's not loaded." He fires the gun.

Bang. Smoke. Ceiling tile nuggets drop to the stage.

Secret Service agents converge around the president to protect him. They look for the shooter. A couple of seconds later, they realize the president was the shooter.

None shrugs, still holding his pistols. "Maybe I should have taken that optional gun safety course."

Nervous laughter comes from the crowd.

MJ holsters her weapons. "I know my way around a gun range. I'll keep him out of trouble."

She quickdraws and hits all the targets with marksman level accuracy.

The targets are reset. It's None's turn to shoot.

MJ gives him pointers.

The recoil startles None at first. He steadily improves, but still has trouble hitting targets until he moves closer. Given it was his first time shooting, he did OK.

After the shooting exhibition, None addresses the crowd. "We may face another Katrina, or 9/11, or worse. I hope that if such an event happens, I can count on the patriots of Texas."

He receives another standing ovation.

After the Q&A, true to his word, None ends, leaving five minutes for heckling.

71. **Visiting Armageddon**

This economic crisis could turn into a global conflict if something isn't done. Empty stomachs breed war. It won't take much to ignite the next global flashpoint.

None, Shen, and Zelda travel on Air Force One to the G20 summit in Europe. MJ follows them on Air Force Two.

By the time they enter European airspace, gun safety courses become mandatory in Texas. After the mockery that None and MJ made of Texas gun safety laws, the state legislature reinstates them. It's a good sign. Things haven't gone off the rails back home. At least, not yet.

The city of Rome, in Italy, hosts the emergency G20 economic summit. G20 member nations account for the majority of the world's population. Heads of state and central bankers meet to address the global economic collapse.

None, MJ, and Shen attend meetings with world leaders and their bureaucrats. Because of his antics on TV, international leaders don't take None seriously. They can't openly disrespect him, though. He is president, after all.

The warring bureaucrats each have separate conflicting agendas. How can None get them to agree on anything, much less come up with a solution? He has yet to hear a single good idea from anyone.

If a dysfunctional summit wasn't bad enough, None finds out that Renquist leads a second American delegation. A dozen rogue Congress members help him push immigration reform and new trade deals. Renquist can already dropship thousands of workers on short notice in America. He's trying to take Slam Dunk Jobs global. The very idea makes None shudder.

Big Money people care most about bailing out banks and Wall Street. They assume bailouts are a given. To pay for the bailouts, world leaders talk about governments borrowing longer term, with 100-year bonds.

"Paying off the same debts for 100 years is not a solution." None can't stomach any more of this idiocy. He gives up on the conference and retreats into Zelda's arms at the hotel. None cuddles with her on the king size bed. He gets ten minutes of comfort before a knock on his hotel door.

Agent Vincent talks through the door. "Mr. President, your vice president and chief of staff wish to speak with you."

None wishes he could hit a snooze alarm on the door, but it could be important. He pulls away from Zelda and stands next to the bed. "Let them in."

MJ rushes in his room with her go bag slung over one shoulder.

Shen follows her in.

When MJ sees Zelda, she tightens her grip on her go bag. MJ looks at None intently. "Everything depends on the emergency session. They said you walked out."

None throws up his hands in frustration. "All they seem to care about is saving the banks and the stock market. Wages. Income inequality. That's what needs to be fixed. The elites have no idea what it's like for the rest of us."

Hearing None paint himself as an everyday man irks Shen. "The rest of us? President and future billionaire, how aren't you an elite?"

None will always feel middle class. Money and power won't change him. Why can't his brother see that? "Shen, you've seen what I went through to get here. The results are very different, but we both lost our businesses."

Shen doesn't take well to having their situations compared. "My business is none of your business. It's certainly not a talking point to share with the world."

Maybe None did play up Shen's business failure for sympathy, but it was necessary. Too much hung in the balance to worry about his family's privacy. "I'm sorry."

"Parrots keep secrets better than you." Shen shakes his head. "What's your point about elites?"

Does Shen forgive him? Whatever the answer, Shen doesn't want to talk about it. None picks up where he left off. "For the Big Money people, this catastrophe is a line item on a balance sheet. For the rest of us, it's financial Armageddon."

MJ scrunches her nose. "Sometimes people won't believe something unless they're made to see it, even if the truth stares them in the face."

In his gut, None feels there's some sort of subtext he's missing. Oh, well. If it's important she'll eventually be direct. At least MJ didn't give him another "be an adult" lectures for storming out of the emergency session.

"It's too bad Armageddon isn't a place we can visit, so it's real to them," MJ says.

A light bulb goes off in None's head. "MJ, that's brilliant. We'll visit Armageddon."

"Here comes the crazy train." MJ rolls her eyes. "What are you talking about?"

None rushes to a desk in the corner of the room. He waves over MJ and Shen to follow. None talks while he searches for images of Chernobyl on his laptop. "World leaders need to see what the aftermath of Armageddon looks like, so they internalize the stakes. We'll take them to an eerie, abandoned place, marked by doom."

He looks through pictures of the Chernobyl Exclusion Zone, the 30-kilometer area around one of the world's worst nuclear disaster. Soviet-era concrete buildings and a derelict Ferris wheel stand above the encroaching forest.

Rusted bed frames. Piles of gas masks and radioactive clothes. The disintegrating ruins seem like the perfect spot to get his point across.

MJ grimaces at pictures of creepy deformed dolls. She points at a photo of 450-foot tall metal mesh that stretches a half mile across. "What's that?"

None looks up details. "Duga-3. It's an early missile warning system the Soviets used, during the Cold War. Its signals were so strong that it disrupted global radio communications with a tapping sound. It was nicknamed the Russian Woodpecker." He's intrigued. Now he really wants to go.

More pictures. Outside town, it looks like a wildlife preserve. A catastrophe for humanity is a boon for nature. Wild boar, deer, and wolves roam freely.

MJ says, "So, after Armageddon, the planet becomes an animal paradise, without us? I'm not sure that's the message you want to send. Everyone's afraid of radiation there. It's abandoned for a reason. Besides, there isn't a place to land Air Force One."

The lush greens in the wildlife photos make None want to go camping. It's too nice. He needs a more ominous location, one with a place to land. "OK. Let's find some other post-apocalyptic playground to take them to."

None searches for abandoned airports and finds what he's looking for. In 1974, Turkey invaded Cyprus, taking over a third of the Island. Heavy fighting took place at Nicosia International Airport. After over 40 years, the deteriorating airport remains at the center of the U.N. controlled Cyrus Buffer Zone. Frozen in time, since 1974, a gutted Cyrus Airways plane waits on a runway for a departure that will never come. Inside the terminal, instead of passengers, sedimentary layers of pigeon poop and debris cover the rotting seats. It stands as a monument for the inability of warring peoples to resolve their differences, even generations later.

Shen asks, "How are we going to get heads of state, with egos even bigger than yours, to meet at a broken-down airport?"

MJ cuts off None before he can answer. "You're a straight shooter. You come at everything directly. Try a little mystery. Don't tell them where they're going, for starters."

None snaps his fingers with a smile. "If we could get the Russians to come, no one will want to be left on the sidelines."

Shen says, "Based on my Russian security briefing on President Boris Orlov, I say, we appeal to his vanity. I don't think a Russian leader has been on Air Force One since Nixon flew with Brezhnev. He'll want to make history."

None can feel the plan coming together. "We'll use the mystique of the most famous plane in the world. Now, all I've got to do is figure out what solution to give them." He turns to MJ. "MJ, if anything happens to Air Force One while it's full of foreign dignitaries, it'll be World War III. Take Shen with you and oversee the arrangements personally."

MJ nods with a smile and leaves to make it so.

Shen follows right behind her, on the way out.

None has one day to come up with a solution. He paces across the hotel room.

Zelda gets bored waiting for him to come back to bed. She takes a nap, leaving None alone with his thoughts.

None returns to the laptop to research the banking system. The details on how banks really work shock him. With fractional reserve banking, the whole system builds on stacks of promises, powered by debt, to create money from nothing.

Since only a fraction of a bank's money is available for withdrawal, the system is inherently unstable. To keep the banking system functioning during a crisis, it requires lending from central banks, FDIC insurance to protect depositors, and ultimately, bailouts.

The debts used to create money come attached with interest due. Interest on the debt compounds and grows. The debt feeds on itself until there's enough debt to destroy the global economy. Compound interest is a force powerful enough to bring civilizations to their knees.

Boom and bust cycles constitute a natural part of capitalism, like ocean currents that ebb and flow. Real problems arise when wealth gets concentrated in too few hands. Pile on enough greed, and you can break capitalism. Violent overthrows often follow, like the French Revolution, or the rise of communism. This crisis could end that way if a solution doesn't work quickly enough.

None wonders if there's something better than a band-aid to solve this crisis. Could there be a long-term fix to the fundamental flaws in capitalism? The cracks are showing.

Every year, technological innovation and automation increases productivity and propels us closer to a post-scarcity society. What happens when manufacturing capabilities outstrip global demand, and there's a glut of everything? What happens when robotic factories don't need workers?

Capitalism was designed for the world of scarcity. What happens in a world of abundance?

As advancements like self-driving cars raise technological unemployment and take away millions of jobs, most futurists agree that universal basic income is the most viable solution. Countries like Finland have pilot programs to test it out. How will governments pay for it, when they're already deep in debt? Tax the robots? Maybe that's not such a bad idea, but robots are tomorrow's crisis.

A more long-term fix might be a next generation capitalism replacement. Forward thinkers, like The Zeitgeist Movement and the Venus Project, tinker with designs for resource-based, next-generation economic systems, that consider sustainability and environmental impacts.

Even for a better new system, None can't afford the inevitable learning curve. These ideas need testing. None doesn't have time to test.

The cracks may show, but capitalism isn't broken enough yet. It can still be salvaged. He needs something simpler to buy time.

There was some talk at the conference about new digital currencies, like bitcoin, to replace cash. Mostly it was pushed to get rid of the cash economy and crack down on tax dodgers. Hackable money that needs tech support, there's a nightmare. Just, no. Not yet.

None's mind wanders. He looks at Zelda sleeping peacefully. It's been a while since he slept so soundly. None looks at his mother's tattered Bible on the nightstand, next to his side of the bed. He planned to read it before he went to sleep, to keep the promise he made to his mother.

He's out of ideas. None needs a break. He walks over to the nightstand and grabs the tattered Bible. None takes it back to the desk and flips through it. How can he find a new idea, reading such an old book?

The Ten Commandments isn't just a famous Bible story. The Ten Commandments are laws. The entire Book of Leviticus details extensive laws and regulations. Some rules seem weird. You can't eat pigs, but you can eat grasshoppers and locusts. Society changed a lot since the biblical times.

Could the Bible still be relevant? These are modern problems. Fractional reserve banking didn't exist back then. Then again, they had money. They had debt. Could this be an ancient problem too?

Unsustainable debt is the crux of the issue, both now and then. Once the wealth gets too concentrated and debts too large, interest compounds the problems, and a society's financial ship falls over and capsizes. How did they solve it in biblical times before central banks?

None doesn't have to wait long before he finds his answer. Leviticus 25 talks about the year of Jubilee, where every 50 years all debts were canceled.

Jubilee.

Who says he needs a new idea when an old idea might work?

72. **The Money Game**

In his office on Air Force One, Agent Vincent lets None know Air Force Two has taken off for the rendezvous at Nicosia International Airport.

MJ calls None from her airplane. "Mr. President, why did you mobilize the sixth fleet?"

Fernando must have told her. None says, "I want military backup. If anything happens to Air Force One, it's World War III. Do you trust me?"

MJ hisses air. "I'm still on your crazy train. What do you think?"

"MJ, you're a decoy." None pauses to let the words sink in.

The phone is silent for a while. "So, where are you going?"

"It's a secret. I'm not even telling the pilot where we're really going, until right before we reach the airport." None can sense her displeasure from the way MJ breathes into the phone. "MJ, you told me to be mysterious."

MJ scoffs. "I didn't mean to me."

None feels the plane taxi towards the runway. "I've got to go. We're about to leave."

"It's OK...It's OK." She seems to be reassuring herself, more than talking to None. MJ disconnects.

MJ's a trooper. None has every confidence in her. She'll be OK.

An Air Force C-17 military transport takes off next. It carries The Beast and other vehicles for the presidential motorcade. Usually, the transport would arrive well in advance of the president, but that might give away their true destination.

None shifts his focus to the precious cargo in the level below, the leaders of the G20. It's risky putting the principal leaders of the world on a single plane.

Nothing can happen to Air Force One. None arranged for a fighter escort and sent in Navy SEAL teams to covertly secure their final destination.

Air Force One takes off. The flight from Rome, Italy, to the island nation of Cyprus, takes a couple of hours. They head towards the abandoned Nicosia airport in the middle of the country.

Air Force One is always closely monitored since any damage to it on foreign soil would result in a declaration of war. Today is no different. A flight plan into U.N. controlled space was registered in advance. They're expected.

The cockpit is right in front of None's office. He walks over there with a folder and knocks on the cockpit door.

The pilot opens the door. He's surprised to see None. "What can I do for you, Mr. President?"

None pulls a map from the folder. A brown X marks a location by the ocean. "I want you to set a new course."

Concern creeps across the pilot's face. "This will take us across Turkish occupied territory."

"Then it's a good thing we have the president of Turkey on board." None smiles back at him.

The pilot gives an odd look at the map. "I don't see any airports. Where did you want us to land?"

None pulls out satellite images of the area. He points to a section of road that travels parallel to the beach. "Use that road to land on. This section doesn't have any buildings."

"Yes, Mr. President." The pilot isn't at ease with the instructions, but he's trained to follow orders.

None returns to his office, where Shen waits for him.

Shen asks, "Do you know what you're going to tell the G20 leaders? How are you going to fix things?"

With a wink, None reaches into his pocket. "I'm going to use the 100 trillion dollars I have in my pocket."

"Tell the joke right," Shen says. "Pretend the money is in your other pants."

Shen made that joke when they went over escape clauses in the Slam Dunk Jobs contract. None wouldn't steal his joke. It wasn't even funny. None shakes his head and pulls his hands out of his pocket.

Agent Vincent announces MJ landed in Air Force Two at Nicosia Airport. Air Force One has almost reached the Turkish controlled section of Cyprus.

"Send up the Turkish President." None grins broadly. It's lucky that Turkey is a member of the G20. He's invading Turkish airspace with their leader on board. No matter how angry their president gets, Turkey wouldn't dare shoot down Air Force One. Then again, they shot down a Russian fighter that invaded their airspace, resulting in Russian sanctions.

Shen says, "I hope you read the profile on Turkey. President Yavuz used civil unrest in his country to violently seize power. His vision for a new Ottoman Empire is dangerous and power mad."

None says, "He's the reason I needed MJ as a decoy. Restoring ancient empires always ends in disaster, like with Hitler and Mussolini." It only took a few such megalomaniacs to start World War II. None knows empty stomachs will thrust a wave of dangerous men into power if something isn't done soon. President Yavuz is merely the crest of that wave.

Secret Service agents accompany Turkish President Yusuf Yavuz and his translator into None's office.

President Yavuz doesn't wait for introductions. He quickly uses his translator to communicate back and forth. President Yavuz's voice grumbles with a low base, as his translator speaks. "Mr. President, my military identified American fighters entering our airspace, followed by unidentified radar contacts. This violates our sovereign territory. I demand an explanation."

None points out the window and smiles. "You detected our escort to Varosha."

At the mention of Varosha, President Yavuz gives None the evil eye.

The translator tenses up. "You can't bring fighter jets there. It's in a U.N. buffer zone. I'll file a protest with the U.N. Security Council."

The response disappoints None. "File a protest but tell your fighters to withdraw. I don't want to shoot them down."

"They warned me you were unpredictable." President Yavuz and his translator shake their heads.

None asks, "How long have our countries been friends?"

President Yavuz turns frigid. His translator says, "That friendship may end if it continues to be abused."

None gets somber. He knows the Turkish invasion of Cyprus is a touchy subject. "I understand why you don't want that particular history dredged up, but I have my reasons to go there. America let you invade Cyprus because the Nixon Administration valued Turkey's friendship. An American diplomat was assassinated because we put Turkey first. Do this favor for me. I could be a powerful friend."

After some reflection, President Yavuz nods.

The translator seems relieved. "I will recall my planes. Someday I will come to you, and I will be the one to ask how long our countries have been friends."

President Yavuz agrees not to tell the other G20 leaders. With his business concluded, President Yavuz and his translator contact their government, then return to the passenger section.

When it's almost time to land, None gets into a seat in the cockpit for a better view.

Air Force One flies past a high rise with one side of the building blown off. The pilot angles the plane so the passengers can't easily see the damage. The plane heads out over the beautiful turquoise waters of the Mediterranean

Sea and circles back for an approach. The pilot identifies a barren section along the road where the buildings were bombed out and never rebuilt.

The plane lands on the empty road, John F. Kennedy Avenue. This war-torn ruin still bears the name of a beloved American president. Two American presidential administrations after JFK, Secretary of State Kissinger let the Turks invade Cyprus. A single U.S. president can change the course of nations, a power None now holds.

None comes downstairs to the main cabin, excited to finally reveal his plan. He looks at the expectant faces of the G20 leaders. "The world has not faced such dangerous consequences since World War II. Then, Allied leaders met secretly at Yalta, to hammer out agreements on how to end the war. I brought you all here, without distractions, to do something similar. Before we leave, I expect us to find a solution to our shared crisis."

His words remind the G20 leaders of the gravity of their situation. They grow quiet.

None, Shen, and the G20 leaders disembark from Air Force One, onto the beach. A motorcade awaits, with Secret Service agents and translators. It's an unusually cold day for Cyprus, with temperatures in the mid-50s. Storm clouds in the distance explain the chill in the air.

Manly and rugged, Russian President Orlov wears a short sleeve shirt that shows off his muscles. He projects an aura of confidence and power. "Yalta was also a beach resort, a Russian beach resort." He surveys the barren beach. President Orlov cocks a suspicious eyebrow. "It's very nice, but where are all the people?"

He can't share the whole truth, but he can share a misleading truth and hope Orlov doesn't inquire further. "Only certain people are allowed here."

President Orlov squints at None. "So, it's a very exclusive resort." He smiles when he says the word exclusive.

The stares from President Orlov make None uncomfortable. None feels pressure to say something more to explain away the empty beach. "It's October. Maybe the water's cold."

President Orlov responds to None's words like a challenge. "Ha! In Russia, I swim with ice." He takes a few steps towards the beach and tugs at his shirt.

None holds out his hand like a stop sign. "Keep your shirt on."

The other G20 leaders laugh.

The laughs turn President Orlov's display of strength into a moment of embarrassment. The Russian president doesn't take it well. He tucks his shirt back in and strides towards one of the many black SUVs with tinted windows in the motorcade. "Let's go."

Secret Service agents block President Orlov from entering the black SUV.

None stands quietly, while the other G20 leaders wait to see what happens.

The tension builds. The other leaders look between None and President Orlov, but they remain standing next to None.

A smile spreads across None's face. The others leaders don't know why the Secret Service agents stopped President Orlov. The SUV he went to isn't for passengers. Inside is a rotating turret with a 6,000 round minigun that pops up from a concealed section of the roof. It's part of the heavy armament that protects the motorcade.

The unintentional power play reveals that None has their support, at least for the moment. None nods to break the tension. "Yes. Let's go."

Secret Service agents lead President Orlov to another vehicle.

President Orlov sneers and gets inside.

None enters his limo, The Beast, while Secret Service agents lead the G20 leaders into various vehicles in the motorcade. None has Secret Service agents seat the Turkish president with him in The Beast.

The motorcade drives along John F. Kennedy Avenue into Varosha, a beach resort abandoned in the middle of a no man's land. The area has been left derelict, while it's been used as a bargaining chip for the decades-long negotiations between Cyprus and Turkey.

Cyprus was a British colony until 1960, so British Prime Minister Scott Glendorn quickly realizes where the motorcade heads.

The other leaders may not know where they are, but once they pass barbed wire and rusted-out cars from the 1970s, they know something is wrong. Varosha is a short drive from Air Force One. Before the panicked G20 leaders have time to react, the motorcade arrives.

A ghost town overlooks what should be a beautiful beach resort. A rusting construction crane lies dormant, next to the steel beams of a hotel that will never be completed. Armageddon has come to paradise.

Crumbling walls. Broken windows. Curling paint. Decades of neglect have taken their toll. Dust obscures the labels on antique soda bottles, discarded in the street.

Rows of decaying high-rise hotels line the beach side of John F. Kennedy Avenue. The view from some of these once famous hotels must be spectacular if you can get to the top without falling through a rotting floor. Much smaller buildings dot the other side of the road.

The Secret Service fields questions from scared G20 leaders who won't leave the safety of their vehicles. Agents get out and hold the SUV doors open for their VIP guests to get out.

Several leaders pull their doors shut. Others stare at the open door with fearful eyes.

None gets out of the Beast, with Shen and the Turkish president in tow. He approaches President Orlov's SUV.

President Orlov stares at his door with confusion, not fear.

None says, "You are a man of action, President Orlov. Come join me on this adventure."

President Orlov steps out of his SUV, the first G20 leader to emerge. He looks around with a grin. "In Russia, we have such places as this, to explore."

None can't help but ask. "Do you mean Chernobyl?"

"Yes. I used to hunt deer there." President Orlov seems lost in thought for a moment, reflecting on a happy memory.

His words tell None that President Orlov still considers former Soviet bloc countries as part of Russia. Chernobyl is in Ukraine, not Russia.

President Orlov seems to remember that inconvenient truth as well. "Perhaps you could arrange a trip with the Ukrainian President. We're no longer close."

"I'll see what I can do." None smiles back. Because of tensions with Russia, there are places, like Chernobyl, that western tourists can go, that President Orlov can't.

The other leaders reluctantly come outside. Many of them are as curious as they are fearful. Only British Prime Minister Glendorn still hides in the safety of his SUV.

Chinese President Liu Yong examines the area with a calculating eye. He keeps his emotions in check, which makes him hard to read.

None remembers that President Liu was a nuclear engineer before he joined the government. In China, most leaders have engineering degrees, instead of law degrees. None hopes their shared technical backgrounds will give him a better chance at building a rapport.

American politicians couldn't be more different. They proudly trumpet their ignorance, by saying they're not scientists, before parroting the slanted positions of their corporate sponsors. Would it change American politics if more American leaders had technical backgrounds?

Presidents, prime ministers, and translators of the G20 surround None, wanting answers.

President Liu speaks to None in Mandarin.

None understands Mandarin but doesn't want President Liu to hear his strong American accent. He doesn't respond.

President Liu scoffs and switches to English. "You're Chinese. Why don't you speak Mandarin?"

"I'm American."

"I expected more from the man who wrote Adaptonium. Why are we here?"

None is president, yet the Chinese leader seems more preoccupied with the educational software he wrote. Given the high value placed on education in Chinese culture, it makes sense that he would fixate on that. The Chinese must be adopting Adaptonium. None basks in the glory of his achievement for a few moments.

None has to focus on the present. He still has to answer President Liu's question. "This is Varosha, a casualty of the 1974 war between Cyprus and Turkey. It was a top tourist destination in the 1970s, attracting celebrities of the era, like Richard Burton, Elizabeth Taylor, and Brigitte Bardot."

President Liu loses patience. "That answers where we are, not why."

None says, "It's too easy to forget the dangers we face in air-conditioned conference rooms. We face bigger challenges than the squabbles between nations. This beach has been left desolate for decades because two countries couldn't agree. We'll need almost two hundred countries to agree to solve this crisis. Look around. Is this the future you want? If Armageddon can come to paradise, which of your countries is next?"

A smile creeps across President Liu's face. "My father was part of the ruling elite. When I was young, he sent me to live with farmers, so that I would understand the rural part of China. I understand your purpose. I will listen to what you have to say." He stretches his hands outward. "This, I would expect from the man who wrote Adaptonium."

At least someone understands. None wears a bigger smile than he should, given the circumstance. "Welcome to the forbidden zone. Explore this time capsule and think about what's at stake. Then we'll reconvene to brainstorm a solution in 45 minutes."

The Turkish president Yavuz gives None a dirty look. He effectively welcomed everyone to Turkish occupied space. It's not None's place to welcome people. It's his.

The G20 leaders speak among themselves about the situation. They split into small groups to explore the surrounding history lesson. Everyone is careful not to stray too far from the motorcade.

The overall reaction is calmer than None expected. The G20 leaders probably assume he arranged permission to be there well in advance. It's a good thing None told President Yavuz he was going to Varosha. Wait. He never got explicit permission to be there. President Yavuz only agreed to turn his planes around.

None better keep the Turkish president in sight. He looks around. Shen and the President Yavuz are both gone. None has to find them.

He follows a group past chipped concrete pillars, down into a subterranean parking structure. Light seeps in through holes in the ceiling near the back wall. Hoods on the antique cars are up, their engines looted. Thick layers of dirt obscure the windows.

President Orlov struggles to force a driver's side door open. He braces his feet against the car for leverage. It still won't budge. It must be rusted in place.

None keeps a calm face, but he's worried about finding President Yavuz.

President Orlov climbs into the car through an open window for a closer look and dusts himself off. He looks at the odometer. He waves for None to come look. "Ha! Zero miles. It's a new car." President Orlov lets out a boisterous laugh at his discovery and talks to himself in Russian.

A new car. The idea of a rusted-out heap with four flat tires being "new" is indeed laughable. This must have been a car dealership. None would like to explore too, but he has to find President Yavuz. None rushes back up to the surface to find the next group.

Agent Vincent and his detail have to pick up the pace to keep up with him.

None bounces from one group to the next but doesn't see President Yavuz, or his brother.

He'll need help with his search. None leads his detail down the street, out of earshot from the primary group. He addresses Agent Vincent. "I want all agents on the lookout for President Yavuz and my brother. Put SEAL teams on alert."

None squints. Down the road, it looks like Shen. He's running towards them.

Secret Service agents draw their weapons until they can assess the potential threat. When Shen arrives, they put away their guns, but remain on edge.

None scolds his brother. "It's dangerous to wander off. Why didn't you stay with the main group?"

Shen pants as he tries to catch his breath. He struggles to speak in between breaths. "Yavuz…I warned you…can't trust…soldiers."

His brother warned him about President Yavuz. Shen must have followed him. "Is he coming here with soldiers?"

Shen nods vigorously.

It's None's worst nightmare. His plan is unraveling. "Agent Vincent, get everyone back to the motorcade."

Agent Vincent scrambles his agents. They accompany None and Shen back to the motorcade.

Most of the G20 leaders stand near the SUVs, confused about the abrupt end to their exploration. This area isn't too defensible. A couple decaying hotels loom on the beach side with a clearing on either side. The sections between the building on the non-beach side are overgrown with plants. A sneak attack would probably come from that direction.

Shen is too tired to go on. He sits down on the pavement. Shen explains how President Yavuz searched until he found a patrol with six Turkish soldiers. They chased after Shen and got on their radios. He lost them just before reaching the motorcade. That patrol probably isn't far and likely called for backup.

None sees President Orlov and President Liu examining a hotel entrance on the beach side of the road. He runs over to them. "Sorry to cut it short, but we need to head back to Air Force One."

President Orlov shakes his head. "This place is older than Chernobyl. I want to see more."

On a pockmarked wall, at a hotel entrance, President Liu touches small holes. "Some of these bullet holes seem more recent. Does that have anything to do with our sudden departure?"

None hurries them along. "I'll explain everything on Air Force One."

President Liu and President Orlov accompany None towards The Beast, while most of the G20 leaders make their way towards the rear of the motorcade.

"Ha! Honest Guy, this is a war zone, not the exclusive resort you promised." President Orlov looks smug.

It's disputed territory, not a war zone. None decides not to correct President Orlov. He smiles back wryly. "It's so exclusive that only soldiers can vacation here."

President Liu looks at None suspiciously. "The soldiers I saw weren't vacationing."

"Soldiers?" None's eyes widen. His smile disappears. Oh, no. Is it too late? "Where?"

"On the signs, for the forbidden zone. Moments like these are why I learned English." President Liu says.

Those red signs are pretty memorable. They have forbidden zone translated into five languages in white lettering. No trespassing signs don't typically have a drawing of a soldier carrying an assault rifle. None almost laughs in relief, when he realizes President Liu means drawings, not actual soldiers.

None gave the G20 leaders a misleading impression of the forbidden zone with his gleeful welcome. None couldn't help it. He's fascinated with abandoned places. At least President Orlov shares his interest.

They reach The Beast. Secret Service agents open the door. Shen waits inside.

President Liu looks at None intently. He raises his voice. "Why are there soldiers on those signs?"

The voice of the Turkish translator comes from the far side of the road, hidden in trees. "Because my men have orders to shoot trespassers on sight."

It must be President Yavuz and his patrol.

Turrets with 6,000 round miniguns pop up on the roofs of three SUVs. A semicircle of metal armor on each side provides flank protection to the gunners.

"We're not trespassers." President Orlov turns to None. "Are we?"

Secret Service agents pull their weapons. A few agents try to push None into The Beast.

None resists and shoves his way to the front of his limousine, where he is clearly visible. He raises his left fist.

President Yavuz continues in a grumbling tone, as his translator shouts. "You invade my country and dare to raise your fist at me?!"

None glances at his raised fist. "It's the only thing keeping you and your men alive. It's my signal not to engage."

He notices something in his peripheral vision. None glances to his right. A convoy of vehicles zips down the road towards them at high speed. That must be Turkish soldiers for backup.

Secret Service agents watch tensely. None's rules of engagement have tied their hands again. Agents evacuate G20 leaders into their vehicles, while they wait for None's fist to drop.

President Liu and President Orlov hurry inside The Beast. President Liu tugs at the limousine door.

"This door doesn't close until our president is inside." Agent Vincent blocks the car door from closing. He points 20 and 50 feet away. "If you want a closed door, your vehicles are over there."

Neither of the foreign leaders seems interested in running exposed to their SUVs.

A few automatic weapons jut out of the foliage.

The translator speaks again from his hiding spot. "You speak to me like a little man. I'm president of Turkey. I have power over you here. Only the U.N. is allowed in this area."

None glances over his shoulder at the escaping G20 leaders and shouts back. "Look around. India. Russia. China. Europe. We represent many nations here. We are the U.N."

About a mile down the road, the convoy speeds towards them. In the distance, the convoy looks like a long worm slithering down the road. Turkey has 6,000 troops in the area. He hopes they're not all coming there.

President Orlov seems to realize his survival depends on None's. He exits The Beast and crouches next to None, using the car for cover. "Guns aim at my head. Lower your fist."

None's keeps his fist raised. "There are already guns at our heads. These are just the ones you can see."

President Liu follows President Orlov and crouches behind him. "Spare us your metaphors. You put us in danger to make a point. I assumed you had a plan. You don't."

None cringes, his left arm still up. "I wanted the G20 leaders to feel the same sheer terror their citizens feel right now."

"There's a word for people who spread terror." President Orlov looks at None in disbelief. "You're a terrorist."

With a heavy sigh, None shakes his head. "It's a poor choice of words. It's not terror I want. It's empathy."

"We get it. Now lower your fist, so your men defend us." President Orlov stops crouching. He tries to tackle None.

None won't let President Orlov force his hand down. He runs in front of The Beast, without protection. His hand doesn't waiver.

"You're crazy!" President Orlov ducks down behind The Beast and cusses out None in Russian.

At half a mile, None sees the individual jeeps and personnel carriers in the approaching juggernaut of death. He could end this now with some airstrikes and snipers, but None didn't come here to start a war.

None calls out to President Yavuz from the middle of the street. "I came here to fix a crisis, not start a new one. I apologize for overstepping. I thought we were friends. I want to talk before my fighter jets and snipers do the talking for me."

Voices scream back and forth from the tree line.

The convoy halts. The troops abandon their vehicles and flee from the road with whatever weapons they can carry. They must have realized how vulnerable they were to air strikes. They're still coming, but the troops will be much slower on foot.

The thought of a sniper seems to change President Yavuz's attitude. He and his translator walk into view. He looks up at the abandoned hotels. The hotels might be disintegrating, but they still could hold a sniper nest, and he'd never see the bullet. President Yavuz lifts his left fist in the air like None.

The translator says, "You came here to talk, then talk. I will raise my fist too. May our arms not tire."

President Liu yells out to None from behind The Beast. "No more theatrics. I want solutions, before your arm cramps up."

Great. Now, None won't be able to stop thinking about arm cramps.

"I'm listening, too." President Orlov stays put behind The Beast, next to President Liu.

None could still salvage the plan. If the other superpowers, Russia and China, agree. They could influence the G20, which in turn could sway the U.N., assuming they don't get shot first.

None looks back at Shen. "When everyone goes bankrupt in Monopoly, you end the game, but we're still playing. Let's start a new game with a reboot. We'll reboot the economy, just like a computer."

President Yavuz shakes his fist with anger. For a moment, his fist almost drops. His translator says, "Do you expect us to step aside for new leaders, in this reboot? I fought too hard to give up power."

None looks back for responses from the other leaders.

President Liu and President Orlov peek their heads over the limousine, to express their dislike of a reboot. Existing governments will fight to stay in power, at any cost. At least everyone agrees on something.

"What I'm really talking about is debt forgiveness. How about a global Jubilee? They canceled debts during a Jubilee year, in the Bible. That would help sell the idea in some of the more religious cultures."

President Liu stands up behind The Beast. He raises his eyebrow at None. "I didn't take you for a religious man."

"Anyone can find wisdom in ancient books," None says.

President Orlov jolts to his feet and stretches his hands onto the hood of The Beast. "No Jubilee! Honest Guy, debts must be paid."

President Liu says, "Interesting. If borrowers expect debts to be forgiven, it leads to future bad behavior."

None shakes his head. "Come on guys. It's every 50 years. How old will you be in 50 years? For most of us, it'll be a once in a lifetime event."

President Orlov pounds his fist on the hood. "No! If debts are forgiven, powerful men lose money."

"China has hundreds of billionaires now," President Liu says. "Interesting."

President Orlov glares at President Liu. "Not interesting. Dangerous. I take money from Russian oligarchs. I get Russian civil war."

None knows he's not exaggerating. President Orlov built his power base on unsteady alliances. He rose to power by bringing peace among the oligarchs. There's still the occasional assassination, but nothing on the scale it was.

Shen peaks his head out of The Beast. "We have the safest car in the world. Why is everyone outside? Soldiers are coming. We've only got five to ten minutes before they get here."

President Liu and President Orlov look at Shen and the open door to safety.

What if this is None's only chance to convince the other superpowers? None's voice grows louder with conviction. "When debts grow too large, they're unpayable, whether you choose to accept the fact, or not. In the end, math always wins."

President Liu looks back at None. "We spent trillions trying to control prices in our stock markets. It worked for a while, but they still crashed." He hangs his head. "Mean reversion. Math always wins."

"I don't understand your words. Are they English?" A conversation about math leaves President Orlov left out.

"The economy is math. Why do you think we're here?" None resists the urge to throw up both hands. With the awkward position of his left arm, None feels the strain. Did he clench his fist too tightly? His muscles ache.

Unseen soldiers charge towards them. It all seems hopeless, but None won't give up. He has to make them see. Everything depends on it.

None says, "Money was invented to replace barter. Commodities like gold and silver were used as the first money, real money that was worth something. Then came paper money and fractional reserve banking. We've layered on all this complexity: economics, stock markets, central banks, derivatives. We're left with a growing disconnect between the real economy and the financial alchemy of Wall Street."

President Liu looks at his watch. Perspiration beads on his forehead. "Paper money started in China. I'm quite aware of this history. Your point?"

"We created the economy. The economy didn't create us. Money only has value because we say it does. What we value as money is arbitrary."

None points to antique discarded bottles, blackened by the dust of time. "Bottle caps could be money if we all agreed. Money's an illusion, a game we play with ourselves. If the game stops working, change the rules."

President Orlov pulls a wad of cash from his pocket. He storms out into the open, towards None. He raises a clenched fistful of money. "Crazy guy, cash is real. I hold it."

"Do you know how many currencies have failed? America didn't start with dollars." None awkwardly fishes in his left pocket with his right hand. He pulls folded money out and hands it to President Orlov. "If you think cash is real, here's a Zimbabwe 100 trillion-dollar bill. Shove that in your wallet."

President Orlov drops the fistful of cash. He stares at the zeros with excitement. He counts 14. He counts them again. "I've never seen this much money."

None smirks. Any second now, President Orlov will get his point. "Hyperinflation in Zimbabwe made their currency almost worthless. If that were in U.S. dollars, a couple of those pieces of paper could pay off the world's debt."

The joy drains from President Orlov's face. "It's not real." He watches with disillusionment, as his fistful of cash blows away in the winds that warn of the coming storm. His cash becomes part of Varosha, lost in time with everything else.

Shen runs to None from the safety of The Beast. "We're almost in grenade launcher range. We have to leave now. One itchy trigger finger could ignite World War III."

"My men would not risk explosions near me, but your time runs short," says the Turkish translator.

None looks at Shen. "My arm is cramping, but I need more time." If None's hand falls, it's war and death.

Shen nudges his shoulder under None's left arm to help support it. "MJ isn't the only one on your crazy train. Say what you came to say."

The brothers face Armageddon together, side by side. In his quiet way, it's moments like these that Shen shows his love. It's the moment between them that means everything, without words. Stronger together, None feels a renewed strength.

President Yavuz looks at his translator and taps his left elbow. The translator takes the hint and puts his shoulder under President Yavuz's fist arm.

With his brother's support and a broad smile, None finally suggests his solution. "If we can't do debt forgiveness, we'll give out money to pay the debts. Money is artificial, so we'll create more. We'll hand every government in the world 300% of their GDP. That's like three years of salary. It should be enough to solve the debt crisis and leave a bit more for stimulus."

Confusion and shock scroll across President Liu's face. "If you hand out that much money, everyone will realize the truth of money. It could end confidence in fiat currencies. We'll be thrown back to the dark ages of barter."

None smiles back at President Liu. "People think they understand banks. So, every 50 years we'll create the Bank of Jubilee to transfer the money. You could even bail out normal banks if you want to."

"Interesting." President Liu nods several times. "I agree to your plan."

The conversation has President Yavuz's undivided attention now. He smiles like he's spending the money in his head. The translator asks, "What about countries under sanctions?"

Turkey has lived through multiple sanctions, so the question isn't surprising. None grins. "We need the world to come together, even the bad guys."

Russia suffered through sanctions too, when it took Crimea from Ukraine. President Orlov scowls at the comment. "I'm not a bad guy. I just don't like Europe eating pieces of Russia."

Dozens of Turkish troops stream across the nearby clearing. They'll be surrounded in no time.

"If we all leave safely, I'll owe you two favors." None extends his right hand to President Yavuz.

President Yavuz looks out at his approaching troops. He weighs his options. With a smile, he shakes None's hand. His translator says, "Two favors." President Yavuz sends the six troops with him, from the patrol, to intercept the oncoming Turkish soldiers.

"Thank you. Your guns helped us focus only on what was important," None says.

The translator says, "When you invaded my territory, I mistook you for an imperialist. You see the world...oddly. You wanted us to see what you see."

"I guess we're done here." None sighs relief, as he and President Yavuz slowly lower their fists.

"No." President Orlov waits until he has everyone's attention before continuing. "I hold my tongue no longer. I cannot trust those with such friends."

None's jaw drops. "I know we act foolish on TV, but Shen and MJ are good people."

President Orlov shakes his head. "No, the friend you dance with at the wedding."

"Renquist?" None snickers. "He's my sworn enemy."

Recognition lights up President Orlov's face. "Ah! I'm surrounded by such friends."

None shrugs. "Keep your friends close and your enemies closer."

President Orlov pretends to slow dance. "Maybe enemies shouldn't be quite so close."

With a laugh, None uses the only Russian word he remembers. "Da."

"Da." President Orlov returns the laughter.

President Liu acts insulted. "You speak Russian, but not a word of Mandarin?!"

None shakes his head in frustration. "Aiya!"

President Liu calms down. "I know 7 languages, but I can curse in 12."

President Orlov curses in Russian. President Liu cusses back in Russian. They all laugh.

"I thought negotiating under the gun was a figure of speech," President Liu says.

President Orlov pounds his chest. "In Russia, when I negotiate under the gun, I hold the gun."

None lets out an uncomfortable laugh. He can imagine President Orlov doing that.

The superpower super friends bond over their near-death experience.

None addresses President Yavuz. "Varosha served its purpose. Give our people time to leave, and your two favors are earned."

President Yavuz looks like he feels left out. He turns to leave with his troops.

None waves him over. "Come on. You're with us. We have to go save the world."

73. **Doomsday Clock**

MJ enjoys adventures but doesn't have the same fascination with abandoned places that None has. Stepping in 40 plus years of pigeon poop in the airport terminal gets old fast. MJ can't wait to leave. Once Air Force One takes off from Varosha, MJ leaves Nicosia Airport.

On the way to Rome, None and Shen conference in MJ. None gives a rundown of what he wants in the agreement. He promises MJ he'll explain what happened in Varosha, but not yet.

Once Air Force One and Two land, it's a mad dash to the G20 summit.

Each country strategizes how to get the most out of the Jubilee Agreement. Support spreads quickly among the G20 leaders that were on Air Force One, and beyond. It doesn't take much to convince nations that are inches away from bankruptcy. The others nations fear being left out and not getting their share.

Fierce negotiations take place around a giant circular table where the G20 leaders sit at the summit.

MJ and Shen stand behind None, with the bureaucrats and bodyguards surrounding the table.

MJ hears conflicting rumors about what happened in Cyprus. Everyone assumes Turkish President Yavuz was in on what happened. MJ knows better.

The strangest rumor she hears is that None was going to personally pay off half the world's debt with the money in his wallet. What? How? None's almost broke, until the Slam Dunk Jobs deal finalizes.

None fights hard to keep the agreement simple, but the leaders haggle over the smallest details for hours before reaching an agreement to submit to the U.N. General Assembly.

When they're done, None signs the agreement and stands up. "Sound the trumpets. Let's call it Jubilee."

The G20 leaders rise and give None a standing ovation. He gained their respect for his bold approach. They congratulate each other and shake hands.

MJ and Shen follow None from the conference room. Their American delegation is the first to leave. None strides into the hallway with a spring in his step. He halts when he sees Renquist and his dozen rogue Congress members, including Senator Garfield, 50 feet down the hall.

Because of the sudden stop, MJ rams into None. The impact shoves him forward.

When she sees Renquist, high alert mode kicks in for MJ. She clutches her go bag tightly. MJ takes it everywhere. She feels like going someplace safe. No place is safe. It's a matter of time before Renquist has all the packages.

Renquist leaves his congressional minions behind and strides to None.

"What do you want?" None looks Renquist up and down.

The smile on Renquist's face can only mean bad news. "I heard about your secret meeting. I wanted to thank you."

None isn't happy being thanked. "For what?"

Renquist says, "I used derivatives and credit default swaps to make money on banks, as the market crashed. Because of your Jubilee idea, I'll make even more money on the rebound when you bail out the banks."

The SEC suspected Renquist of insider trading, but his cronies in Congress slashed funding for the SEC. They don't have enough staff left for an extensive investigation.

People like Renquist complain about the inadequacies of government. Then they sabotage government agencies they hate until they make it the truth. MJ clenches her fist. She'd like to wipe the smirk off his face.

"Don't thank me." None thinks for a moment, then smiles. "I'm not bailing out the banks."

Renquist stands motionless, with a horrified look.

It takes a lot to shock Renquist. None takes satisfaction in that, as he strides triumphantly past his stunned nemesis.

MJ and Shen follow closely behind.

Once they're beyond earshot, MJ laughs. "He bought it. Of course, you're bailing out the banks. We all saw what happened when Lehman Brothers failed."

None turns towards her. "Not you too, MJ. The Federal Reserve was set up in a secret meeting for a reason. Private bankers control the dollar. The dollar is their religion, and I'm ready to commit heresy."

Has he thought this through? MJ hopes None didn't choose this course of action just to upset Renquist. "A big move like that will disrupt the status quo and create powerful enemies. Renquist won't be the only one sending minions after us."

He gets adamant. "Let the banks fail. We'll nationalize them. The Federal Reserve isn't part of the U.S. government, but I plan to create banks that are."

"We're riding the crazy train again." MJ rolls her eyes.

All None can do is grin. "I'm not crazy, but I am going postal. I'll bring postal banking back to America."

MJ's eyes squint in confusion. "Aren't there enough banks? Why post offices?"

None says, "The poor are more likely to have a phone than a bank account. Banks began abandoning low-income neighborhoods in the 70s. Without banks, payday lenders and check cashing businesses prey on poor people who have no other options, but those neighborhoods have post offices."

The idea doesn't seem that radical after all. MJ looks for holes in his plan. "Are post offices equipped to handle money?"

None says, "They already handle money orders. Post offices will handle the basics: banking, checking, and microloans. Then we'll tie those postal bank accounts into the mobile payments and websites of the failed banks."

"The banking lobby will stop you," MJ says.

None seems pleased with himself. "The global Jubilee money will be the opening balance. Over 130 countries have postal banking. I was able to sneak in a provision that any nation that had postal banking would use that for their opening balance. Since America *had* postal banking…"

He laughs like an evil mastermind. Congress won't stand in the way of a payout to every American, no matter how many lobbyists the banks send.

MJ says, "I thought you were trying to keep the agreement simple."

None smiles back. "It's nice to be on the winning side of a sneaky contract clause."

The G20 leaders finish congratulating each other and head down the corridor with their entourages.

The window for any privacy at the G20 summit just vanished. None and MJ say their goodbyes. They get into their separate airplanes and head for their separate homes.

On the long flight back, MJ waits for a phone call from None. She expects him to share tales about his adventures in Varosha. Instead, silence greets her. He's probably doing who knows what with that spy, Zelda, in the presidential suite on Air Force One.

MJ dozes off here and there but doesn't get much sleep.

Back in D.C., MJ returns home to the expansive three-story vice-presidential residency at the Naval Observatory, officially known as Number One Observatory Circle.

By the time she gets home, it's late at night. She heads straight to her bedroom. Motion controlled lights turn on when she enters.

Just a room. Just a bed. Just some furniture. It's not the Lincoln Bedroom. It's not the White House. It's not with None. It's 9,000 plus square feet of loneliness.

MJ sets her go bag on a chair and slumps onto her bed.

In the White House, she noticed items in her room moved when she wasn't there. The Secret Service blamed the cleaning staff, but she knows better. They're still looking for her packages.

She installed motion-controlled lights so they couldn't sneak up on her and plugged it into a UPS. The UPS isn't just for backup power. If they cut her power to keep the lights off, the UPS will beep.

Because of her shot nerves, MJ started a new routine: She takes anxiety medication with a warm cup of milk before bed. Now that Zelda isn't across the hall, MJ should feel safer, but she feels Renquist lurking in the shadows. MJ looks around. Everything seems to be undisturbed. Maybe she's just paranoid.

MJ doesn't have time for fear, or self-pity. She gets up. MJ grabs the Glock 19 from her go bag and lays it under her pillow.

With MJ out of the country, today would have been the perfect opportunity to search through her things. If they didn't do it today, then it's all in her head, and she's worried about nothing. MJ pulls a ruler out of the nightstand next to the bed. She takes photos with the camera on her phone to measure the location of her items.

She compares before and after photos. The measurements differ, but never by more than half an inch. Housekeeping staff wouldn't be so precise. To keep it that close, Renquist's henchmen must have taken photos too, to put everything back in the right spots.

MJ sends for a warm cup of milk.

An agent lets in a man from the kitchen staff with the milk.

After he leaves, MJ locks her door. The security detail outside doesn't make her feel safer. Renquist could get to anyone. She puts on a nightgown and takes her anxiety medication with the milk.

MJ throws open the covers.

A blue envelope contrasts with the white sheets in her bed.

Who was in her room? Are they still there? Her pulse races. MJ grabs her gun from under the pillow. She surveys her bedroom and checks the bathroom for intruders.

MJ grabs her phone and calls Fernando. "Come right over. There's something in my bed I want to show you."

"I'm on my way." Fernando hangs up. He seems to be in a hurry to get there. Fernando must have picked up on the worry in her voice.

What's in the envelope? Could it be anthrax or a nerve toxin that kills on contact? She certainly won't touch it with her hand.

MJ puts on daytime clothes and a pair of gloves. She flips the blue envelop over with the ruler. It's completely unmarked. She decides to wait for Fernando before she proceeds.

Fernando arrives with a sly grin.

MJ lets him in and locks the door behind him.

He winks at MJ. "I'm here. Show me to your bed."

She hurries to the bed.

Fernando follows her eagerly. His smile disappears when he sees the blue envelope. "So, that something in your bed you wanted to show me, isn't you? Of course, it isn't."

MJ gives him a puzzled look. "Why would I ask you to come look at me? You know what I look like."

He takes a deep sigh. "You wanted me here. I'm here. What's going on?"

Before MJ can finish her explanation, Fernando grabs the envelope and rips it open. "You haven't even opened it yet."

MJ cringes. She hopes it's nothing after all. Maybe the housekeeping staff left a letter, asking for tips, like on a cruise ship, or a fancy hotel. That would be worse than anthrax. She'd look like a silly little girl afraid of nothing.

Fernando shows MJ the cryptic note. "James Danger" is scribbled at the top of a map with Washington, D.C. streets. An arrow points to a location on the map with a caption that says "Doomsday Clock. Meet me hair." The note is signed 'Digger.'

He says, "It should say 'meet me here.' Anyone who can't spell here is clearly uneducated. The Doomsday Clock must be a landmark, near that location. It's addressed to James Danger, probably an alias, like Carlos Danger."

MJ takes the note. She keeps her gloves on as a precaution. "The president used to be named James. James is in danger."

Fernando asks, "How do we know when to meet them?"

"The Doomsday Clock. It represents how close the world is to a global catastrophe. It always points to a few minutes before midnight. That's when we meet them." MJ looks at her phone. "We have almost an hour to get there."

Fernando shakes his head. "You told me Renquist threatened to bury you and you want to meet some guy named Digger? Send agents."

MJ says, "I have a short list of people I trust: you, me, Shen, and the President. Agents aren't on that list."

James. None doesn't use that name anymore. Only someone who's known him for a while would use that name. What if hair isn't a mistake? It's a clue. Gold Digger. "I know who this is."

"You're going to meet him?"

"Her." MJ gives Fernando the crazy sort of smile that None gives her. "Looks like you're going for a romantic walk with your pretend girlfriend."

74. **Rendezvous**

MJ has the security detail drop her off, with Fernando, a mile from their true destination, where they plan to meet Astrid.

The pretend couple walk hand in hand down the streets of D.C.

Secret Service agents stay tight on them, in spite of MJ's request for privacy.

MJ and Fernando close in towards the spot from the map. How can they ditch the agents, if they're followed so closely? MJ whispers in Fernando's ear to make it look real.

Fernando nibbles on her ear.

The public display of affection makes the agents take their request for privacy more seriously. They give the pretend couple a wide berth.

When agents aren't looking, MJ wipes her ear off with an anti-bacterial hand wipe from her go bag.

Fernando watches with disappointment.

They continue to a set of back to back wooden park benches, with only a few minutes to spare.

A sketchy looking homeless guy in a hoodie hangs out nearby.

MJ gets a creepy vibe. She's being watched. MJ leads Fernando to sit down at one of the park benches.

The homeless man walks towards them and sits on the adjoining park bench with his back to them.

Astrid will be here any minute. MJ worries the creepy homeless man will scare her off. She holds her nose at his urine scent. "Can you hang out somewhere else?"

"Don't look at me." The homeless man sounds crazed. He keeps his back to MJ and Fernando.

MJ clenches her fist. If Astrid doesn't show up because of him, she'll do a lot more than look at him.

The homeless man attracts the attention of MJ's detail. They walk towards him calmly from a distance.

Something pokes MJ in the back. Is that homeless man touching her?

The homeless man whisper screams, "Take it, so that I can get out of these clothes."

The idea of this homeless guy taking off his clothes horrifies MJ. She looks down. It's a phone. MJ shoves the phone in her pocket.

The homeless man mutters to himself. "Why couldn't she get Steel to do it? He looks like this anyway."

Secret Service agents shoo the homeless man away.

MJ smiles and thanks the agents. Steel. She doesn't know any Steel, but the name sounds familiar. Renquist isn't the only one with people. The homeless man must be a messenger from Astrid. MJ just has to wait for the agents to leave, so that she can check the phone.

The phone vibrates. The agents are still close enough to see the phone if MJ answers it.

"Kiss my ear," MJ says.

Fernando glares at MJ with a skeptical eye. It's clear he hasn't forgotten how she wiped off her ear.

MJ can't answer the phone, or the nearby agents will see it. Each vibration makes her more desperate. She shouts, "Kiss me!"

The agents cringe and step back a few feet.

Excitement spreads across Fernando's face. His skepticism seems to disappear. He leans in to kiss her ear.

At the last moment, MJ sneaks the phone over her ear.

Fernando kisses the phone with shock and irritation. He clues into the ruse quickly. Fernando cups his hand around the phone.

MJ answers the phone, concealed from the agents. "Hello?"

The mysterious caller uses a voice modulator so they can't be recognized. "Don't say names. Computers monitor every phone call."

MJ asks, "What do you want?"

"I want to keep your best friend alive, but I need protection."

If None's in danger, MJ will do anything to keep him safe. "If your information checks out, you'll have it."

The garbled voice says, "There's a meeting at a warehouse, tomorrow at midnight. 'A' plans to put a hit out on 'J.'"

An assassination. Wow! The note referred to James Danger, so "J" means James. MJ needs a better clue than just "A." "Who?"

The modulated voice sighs. "Your favorite two people, with and without sarcasm. Find 'A.' Watch him. He'll lead you where you need to go."

Favorite? None. Favorite with sarcasm? Renquist. His name doesn't start with "A." Oh, Alvin Renquist. It must be Astrid. No one else calls them both by their first name. MJ asks, "Why contact me?"

"You're being watched. You're in danger because you watched 'A' before. This time will be more dangerous."

MJ knew they were watching her. She can't believe things could get worse. "More dangerous? Two people died."

"Move quickly, or they won't be the last. No more questions. A smart girl knows when to cash in her chips. Destroy the phone." The line goes dead with a crunching sound. The mysterious caller must have taken her own advice.

Astrid warned MJ of computers monitoring calls. Renquist controls Galaxor Telecom, one of the major telecommunications companies. The NSA can find cell phones, even when turned off. Maybe Renquist can too.

MJ needs to break the phone, without tipping off whoever is watching her. She put her hands over Fernando's hands. She moves his hands and the phone near her left breast.

Fernando lights up at the prospect of touching her. He takes a deep breath and stares deep into her eyes. He shifts closer to MJ on the park bench.

With the phone over her breast, MJ leads Fernando's hands to push in on the ends of the phone. The phone bends but doesn't break.

From a distance, it looks like Fernando gropes her chest. The phone cops a better feel than Fernando does. His fingers get a glimpse of side boob. Fernando leans in to whisper in MJ's ear. "Kinky. I've never had phone sex like this."

She whispers back, "I need to destroy the phone."

Her focus is on the phone, not Fernando. MJ sees disappointment on his face.

Fernando pulls her hands and the phone down to the park bench. He positions the middle of the phone over the edge of the park bench.

MJ and Fernando push down on opposite ends with all their weight. They lean into kissing range, but Fernando doesn't go for the kiss.

He says, "You're breaking my heart with all these games."

The phone splits in two. Fernando and MJ each put half of the phone in their pockets.

MJ doesn't have time to deal with Fernando's feelings. She explains the phone call in hurried whispers.

"Madame Vice President, I'll run surveillance on Renquist and loop you in, when I know something." Fernando pulls away.

He holds her hand on the way home for appearances, but Fernando seems distant.

With None barely talking to MJ, she has to stop the assassination on her own. She'll need Fernando's help. MJ worries that she just pushed him away

75. **Through the Looking Glasses**

It's 10 am. That leaves less than 14 hours to find Renquist's meeting location. Fernando can use NSA resources, but Renquist has many ties in the intelligence community. It's not paranoid to consider that any government agency could be compromised.

Secret Service agents lead Fernando to MJ's bedroom in Number One Observatory Circle.

MJ comes to the door and gives Fernando a short hug in front of the agents. She leads him in and locks the door behind him.

Fernando wears wraparound sunglasses and carries a paper bag. "I'm not playing games anymore. Here's a parting gift." He hands MJ a box from his paper bag.

Parting gift? "Fernando, don't abandon me when I need your help the most." MJ opens the box, to find another pair of wraparound sunglasses like his. "It's Winter. Why do I need sunglasses?"

Fernando says, "These are data glasses. You control them with eye movements, voice commands, or you can link it to a secure mobile device. I whitelisted your glasses so they can connect to my satellite uplink."

MJ asks, "We can use these glasses to communicate? Why didn't you give me data glasses earlier?"

"No one was plotting to kill the president." Fernando pauses in reflection. He shakes his head. "The truth is, I used to want to be close to you."

"You don't now?"

Fernando looks away. "I thought this was going somewhere, but you wiped me off your ear like I was filth."

He tosses the data glasses manual on the bed. "Hurry over. There's something on your bed to look at." Fernando flashes a sarcastic smile.

He moves to the door and unlocks it. "With all your mixed messages, it's too painful to be in the same room." Fernando blows her a kiss on his way out. "I'll call you."

He was a pretend boyfriend. Why does MJ feel dumped?

MJ puts on the data glasses and inserts the connected earpiece. Darker than sunglasses, the data glasses block outside light to keep classified information private. Miniature built-in cameras display what's ahead since they're too dark to see through.

She reads the manual and tests out the various ways to control the data glasses. MJ focuses on the eye movement controls since she might be in a situation where she can't talk or use her hands.

A video conference request from Fernando comes up on her glasses. She accepts. Fernando said he would call her. She assumed he meant on the phone.

Fernando checks in with MJ throughout the day. Presidential assassinations should be Secret Service jurisdiction, but they could be compromised. Fernando uses only his most trusted NSA staff on the op. He doesn't access Renquist's NSA file. That might tip off his allies.

Fernando's team locates Renquist 40 miles away, at the Galaxor Telecom headquarters in Baltimore. They monitor all his communications.

MJ remembers the Galaxor Towers building from the strike.

Satellite feeds, and phone transcripts come up on MJ's glasses. She sifts through the transcripts. Nothing is damning, but multiple calls reference a meeting tonight. Astrid's information checks out.

NSA agents assess the most likely warehouses for the meeting. They take over Wi-Fi networks and hack nearby devices: laptops, phones, DVRs, smart TVs, smart watches, thermostats, anything with networking capability.

An hour before midnight, Renquist leaves Galaxor Towers in a limousine. Fernando's team tracks him with a combination of satellite and traffic cameras. He arrives at a Galaxor Telecom warehouse filled with cell phones on conveyor belts. Almost a dozen limousines disappear into the warehouse. Those must be Renquist's co-conspirators.

All hacked devices in range of the warehouse have their camera, webcam, or microphone turned on. Every voice-activated device with Internet capabilities has a microphone and listening capabilities. All those gadgets can spy on their users. Today they do.

MJ watches feeds from the warehouse security cameras on her glasses.

Renquist and the co-conspirators walk upstairs to a conference room on the second floor of the warehouse. It's hidden from the security cameras.

The only networked device in the room is a TV. It's a good thing Renquist's people didn't update the firmware on their off-brand Korean smart TV. Fernando's techs exploit a security vulnerability to turn its microphone into a listening device.

Renquist's voice comes in clear over the TV microphone. "Even without me, 50 billion dollars sits in this room. The President wants to let the banks fail and give the money that could save them to the people. He's a damn communist."

A second voice speaks up. "If the president wants to call us banksters, we'll show him banksters. Kennedy pissed off the wrong people and got whacked. I say it's this guy's turn."

NSA servers try to match the video from the warehouse security cameras and facial recognition data so they can identify the co-conspirators.

Another voice says, "I'm not letting this man destroy everything I worked to build. I'm in." The voices chatter among themselves.

Everyone goes silent, as Renquist speaks up. "Only we can save America from disaster. I don't want him unelected. I want him dead."

Many voices chime in with agreement. After the decision to assassinate None, the conversation changes. It feels more like they're eavesdropping on a social club, than a plot to kill the sitting president.

"We should arrest them right now," MJ says.

Fernando sighs deeply. "No. We have to find out how far this conspiracy goes. We'll watch all of them and collect more evidence until we know who we can trust."

He sees and hears what she does. MJ knows Fernando is right, but she has to tell None. She also has to keep her promise to Astrid. "Fernando, find Astrid before Renquist's goons do."

Facial and voice identification matches trickle in over the next several hours. The second voice matches Jonathan Maluki, CEO of Cragmire Investments. Most of the co-conspirators control large financial institutions or defense contractors.

It's all too much. There's not much MJ can do at four in the morning. She has a glass of milk sent up from the kitchen. Scared to miss any updates, she gets into bed with her data glasses on. MJ takes her chill pills and gives in to sleep.

Fernando's wakes up MJ on her glasses. "My team has been tracking a group of rogue intelligence assets. Originally, we thought it might be a terrorist cell. They've murdered some important people, including both your friends."

Guilt churns MJ's stomach at the mention of her two dead man's switches. She can feel more bad news coming from the heaviness in Fernando's voice.

"SIGINT identified the most likely places they might use as body dumps to dispose of their next victims." Fernando pauses for a few seconds with a sigh. "One of the teams found a body. It may be too late for your friend, Astrid."

Astrid would still be alive if MJ hadn't thrown her into Renquist's arms. Why did she have to meddle in None's love life? Guilt throws her insides into a blender. MJ can't bear to hear anything more.

MJ turns off her data glasses and throws on daytime clothes. She has to share the burden of this news with None. She gets her security detail to drive her to the White House.

On the ride over, MJ puts her data glasses back on. Video streams of a decapitated corpse in workout clothes with the top digits of all its fingers sliced off. Bloody orange foam padding wraps the badly decomposed body. Police and CSI technicians roam through sections of a landfill looking for evidence.

Fernando says, "They brought in cadaver dogs to search for the head."

Dogs can be trained to search for different types of scents. Cadaver dogs are trained to search for decomposing flesh.

A professional apparently didn't want their victim identified. When unraveled, the foam padding around the body extends over six feet.

It's a yoga mat. Yoga. Could the corpse be Sarah?

A preliminary time of death matches the filming day for the "Jury of Lovers" episode of *The Internet President*.

It can't be Sarah. MJ talked to her on the phone weeks after that. She shudders, as she remembers asking another woman to come into None's life again. That thought troubles her until she reaches the White House.

What is she missing? MJ feels unsettled. On the way to the Oval Office, she escapes into the bathroom for privacy, away from her agents.

She's going to call Sarah and remove any doubts. She keeps the data glasses on, so that Fernando can hear. She asks Fernando, "Do you believe in ghosts?"

Fernando says, "I don't know about ghosts, but my past haunts me. MJ, there's nothing you can do about the two friends you lost."

"I thought Sarah North died, but I talked to her. I have to call her again, just to be sure."

He doesn't know what to say, so he just listens in.

MJ dials Sarah with her phone.

Sarah picks up. "Hello?"

It's so such a relief to hear her voice. "Great! You're alive. It's MJ, the president's friend."

"I know who you are." Her words sound cold and artificial, like a voice from beyond the grave.

MJ feels the hand of death. She's talking to a dead girl.

Sometimes MJ asks her phone questions. She's wondered about the person the phone's voice comes from. Even after the original person dies, their voice could be immortal, living on as a ghost in the machine.

If it's a computer behind the voice, there's one question they probably didn't think to script for. "Are you human?"

After a long pause, the voice answers. "That's an interesting question. I need to reflect on that. I'll talk to you later. Namaste." The words sound almost lifeless.

The phone line goes dead.

MJ redials. The phone number has been disconnected. Did her question trigger a kill switch? "This smells like Renquist. What cell company did Sarah use?" MJ gives Fernando the phone number.

It takes Fernando several minutes to find out. "Galaxor Telecom. I'm guessing you already knew that."

Who would want to kill Sarah? She didn't seem like the kind of person that makes enemies.

"I have to tell the President." It's not Astrid, but this news will still hit None pretty hard. MJ pulls herself together and leaves the bathroom. She continues to the Oval Office.

Agent Vincent and some other agents stand watch in the Oval Office. Shen and a handful of support staff fill up both couches. They read papers or stare at laptops set up on the coffee table between the couches.

None grimaces at his laptop on the Resolute Desk. He cheers up when he sees MJ. "I'm glad you're here, MJ."

"Mr. President, I have some bad news." MJ's lips curl in sadness.

None says, "I know. The bank lobbyists fight dirty. I'm losing support for the Jubilee deal."

MJ takes off her data glasses and stares into None's eyes. "You're going to want to clear the room."

He waves her off. "No, it's fine. I'll bypass Congress, go straight to the people. I could use some help on my Jubilee speech, though." None has tunnel vision. He can only see the problem he's focused on.

She needs his attention. MJ thrusts her data glasses over None's eyes.

Agent Vincent and another agent pull MJ away from him.

"Clear the room." None stands up at his desk. He rips the glasses off and points at MJ. "She stays."

None fumes, while he waits for everyone else to leave.

Once only None and MJ remain, his voice rumbles with anger. "Gross. I'm not into these gory movies."

"It's no movie. That dead girl is Sarah."

None scoffs. "Why would Sarah wear a waitress uniform?"

What is he talking about? MJ puts the glasses and earpiece back on. The headless corpse with its digits removed does wears a waitress uniform. Where are the workout clothes? "The dead don't change clothes. Fernando, what happened?"

"The cadaver dogs found another headless corpse."

MJ asks, "Wasn't there a waitress in the Jury of Lovers?"

None nods with a smile, then a scowl. It must be a bittersweet memory. "Yeah, Betty. After what happened on TV, she disappeared."

"No," MJ says. "She was murdered, along with Sarah."

The news stuns None. He braces himself against the Resolute Desk and slumps back into his chair.

MJ tries to remember. "I was wearing my sexy power suit that day."

None says, "Your power suit got everyone's attention. I remember."

It wasn't just horny stares. MJ remembers feeling watched. None talked about both the dead girls that day. "What did Betty and Sarah have in common?"

"I made a list of who I could see spending a life with. Then Corella blew everything up, and Lashay showed her true colors. So, I crossed them both off my list."

MJ asks, "Who else was on that list?"

"Zelda."

The neck hairs bristle on MJ's neck. "Zelda was there that day, watching. She took your list of potential mates and turned it into a hit list. Zelda killed Betty and Sarah, to make sure you picked her. I told you she was a spy. There's your proof, rotting in a landfill."

None calls his security detail back in. "Arrest Zelda Remington."

Agents look for Zelda, but it's too late. Somebody tipped her off and helped her escape.

None cringes. "At least, I made a new mistake. I'm sorry I didn't believe you, MJ. If you forgive me, I'd like you to move back into the White House."

It's hard to forgive his lack of trust in her, but she misses him too much to say no. Without Zelda's hooks in him, MJ hopes things can go back to normal. "I'd like that, Mr. President."

"You're my best friend, MJ. You don't need to call me that."

76. **Murder Room**

None looks worried when MJ asks him to keep the Oval Office clear. With Zelda's escape and the murder of Betty and Sarah, he's barely holding on. He asks for quiet time to reflect until Shen can get there.

When Shen arrives, MJ shares everything she knows about the assassination plot.

None doesn't speak a word.

MJ calls Fernando for a status update on her data glasses. She puts it on speaker so None can hear. When Fernando answers, talk radio plays over a blank screen. It's clear he's on speaker too.

"There's something wrong with your feed. I can't see anything," MJ says.

Fernando says, "Guess who followed me home. I'm taking Astrid and a couple of her associates to a safe house. I'll turn video back on when I get there. We have a problem. How can we hide a famous person? Everyone knows what Astrid looks like."

Astrid says, "You have to fake my death."

At the sound of Astrid's voice, None's mouth gapes open. He signals MJ to end the call.

"I have to go." MJ takes off her data glasses.

None remains unsettled. "I'm sorry, I can't be a part of this."

"Are you saying we shouldn't do it?"

"I'm saying I shouldn't be part of this conversation." None says each word deliberately like there's a secret meaning.

She reads between the lines. None can't know anything about it, especially given how honest he is. Proof the president helped fake Astrid's death would be just the excuse Congress needs to impeach him. He has to maintain plausible deniability.

MJ puts the data glasses on and backs away. "I'm going to work in the other room."

Shen moves to follow MJ, but None holds him back.

None nods and smiles back at her. "Sounds good."

MJ walks to the nearby presidential study and makes sure she has privacy. She resumes the call with Fernando. "I assume we'll need a body to fake her death."

Fernando says, "In the old days, we could fake some dental records and swap in a body from the morgue. Now the DNA has to match."

"So, we'll cut off her foot, or something," MJ says.

Astrid shrieks.

Fernando waits for Astrid to calm down before he continues. "There's a top-secret defense project to clone organs for the battlefield. It's a complete failure, but for our purpose, we don't need living organs. We can't clone an entire body, so we need a cause of death that explains that away."

MJ asks, "Do we have the cause of death yet on Betty and Sarah?"

Gravel pings the underside of Fernando's vehicle. It sounds like they've gone off the main road.

"They were both stabbed." Fernando holds back uncomfortable words until he can't wait any longer. "The murder weapon was a long knife, possibly a sword."

"Zelda killed them with my katana. I can feel it." MJ imagines Sarah's ghostly voice whispering in her ear. Even if it was just some computer mimicking Sarah on the phone, the connection and loss MJ felt was real.

Fernando tries to downplay it. "That was my first guess too, but it's a preliminary report. We don't know for sure yet."

"I know she did. She could have shot them, strangled them, killed them a thousand different ways. Why did she have to use my katana? Was she planning to frame me for the murders?" MJ loved that katana. The sword's destiny was to kill in glorious battle, not murder innocents.

When Zelda used MJ's katana, she made it personal. What did MJ ever do to Zelda to be so hated? Puke on her? No. Zelda killed them long before that. MJ doesn't want to think about it anymore. She'll find Zelda and make her pay.

MJ downloads the investigation files onto her data glasses. The widths of the stab wounds on the autopsy match the width of her katana blade. She goes through the evidence multiple times until Fernando gets to the safe house.

Fernando turns on video and introduces MaxPlume and Craig Steel. The safe house living room is pure suburbia, with pictures of a pretend family and their poodle on the walls.

MJ watches through Fernando's eyes.

Steel wears a crumpled-up suit that looks slept in. He's a disheveled mess, but he beams with intense happiness.

MJ met MaxPlume at the wedding reception. He wears the blue, purple, and green feathered hair with puffy clothes he's famous for. Dressed like that, does he understand the meaning of hide?

Fernando looks Astrid up and down, focusing on her hair. "Your signature is your hair. If we want to sell your death, we should rip out your hair, to leave as evidence."

"Can't you just clone my hair too?"

He scoffs. "We don't need cloned hair on the battlefield. It was never attempted."

Astrid gasps with horror. "You're really going to shave my head?"

Fernando shakes his head. "No. We'll rip it out, to leave behind hair follicles. Otherwise, the CSI techs can only do mitochondrial DNA testing. The more conclusive the evidence of your death, the better. Besides, who would believe you'd rip out your own hair?"

Her eyes bulge out. Astrid didn't realize the price she'd have to pay. Risking her life was one thing, but that hair is part of who she is. "I want to say my goodbyes to the President, while I still look like me."

"OK. Hold on." MJ rushes into the Oval Office. She takes off her data glasses and holds them out to None. "Astrid wants to say goodbye."

None waves her off. "I can't."

After what she did for him? MJ shakes her head. "Forget plausible deniability. Spend five minutes with the woman who gave up everything, to save your life."

None says, "I swore to Zelda that I would never talk to Astrid again."

MJ stares at None in shock. "You're keeping that oath?"

None cringes with guilt but stands his ground. "I've never willingly broken an oath in my life."

MJ can't believe her ears. "She murdered Betty and Sarah!"

"I gave my word." None closes his eyes and bows his head. He doesn't like saying the words any more than MJ likes hearing them.

MJ takes a deep breath and shakes her head. Brick walls are more flexible than None and his oaths. MJ returns to the presidential study. She puts on her data glasses and delivers the bad news to Astrid.

Astrid narrows her eyes with a bitter smile. "This is his dream come true. He wanted me scalped and suffering, cast out of his life."

"None would never want that for you," MJ says.

"Then he shouldn't have said that to me on TV." Astrid leans in towards Fernando. "Don't make excuses for him."

Caught in the middle, Fernando sighs. He changes subjects before things heat up. "Let's plan out Astrid's death."

Astrid looks at Fernando with dead eyes. "Feed me to the pigs."

MaxPlume looks confused. "Pigs can eat people?"

Astrid says, "Pigs eat anything. They don't wait to gnaw on your face, like a dog, or cat. You're meat when you die." Astrid describes in gory details how the neighbor's pigs ate her pet calico. It's a gruesome reminder she grew up on a farm. "Don't I deserve a memorable death? Besides, you won't need a full corpse."

MJ shudders. She imagines the crunching sound of cracking bones when the pigs bit into Astrid's unfortunate cat. Crunchy and cat are two words she never wants to hear together again.

Astrid looks at Fernando. "What are your plans for my hair?"

Since MJ sees through Fernando's eyes, Astrid could be talking to either one of them. MJ waits to see if Fernando answers her.

Fernando says, "Before we left for the safe house, I gave my team Astrid's DNA sample. We need the hair in three hours to match the cloned blood and body parts. I'll make arrangements,"

Steel asks, "How can we help?"

Fernando looks through the safe house, then comes back to the living room. He addresses MaxPlume and Steel. "Can you two move all the furniture out of the master bedroom?"

They both nod and go into the master bedroom.

Fernando looks at Astrid. "It's too bad you're disappearing alone. I have a great cover for a married couple. It's OK. I'll find a way to make it work. I'll see you when I get back."

"Wait." Astrid waves her hand at Fernando to get his attention. "Before I go through this ordeal, can I talk to someone who isn't an ex-boyfriend I've publicly humiliated?"

He pulls out his smartphone to check the battery level. "I'll do things the old-fashioned way. MJ can keep you company." Fernando pulls off his data glasses and earpiece, then puts them on Astrid.

MJ watches Fernando leave through Astrid's eyes.

Astrid chokes up. "Renquist tracked my money. So, I left everything behind, but the diamonds. I'm glad I listened to my own advice." She tries to laugh.

"MJ, We're not frenemies anymore, right? I need a friend to help me through this."

MJ doesn't feel like much of a friend. She destroyed Astrid's life with her suggestion to date Renquist. "I'll be your friend, but I need to know why you warned None. You dumped him."

"Dumping him was your idea." The video feed shakes back and forth with Astrid's head. "You two have some warped thing I don't understand. You're the last person I should confide in."

"Do you want me as your friend, or not?" MJ asks.

Astrid pauses with a long sigh. "The President wouldn't even talk to me. What does it matter? Fine. I told Alvin I paused before burning the President as part of Alvin's plan to lure him into a confrontation. The truth is, I still love him."

It seems odd to MJ that such a materialistic girl would dump the wealthiest man in the world, the gold digger's holy grail. "Why didn't you stay with Renquist after you sent the warning?"

Astrid says, "Alvin loves contracts, so I wondered why he didn't insist on a prenup. One day, I asked him if he believed in marriage. Alvin said 'yes, especially the until death clause.'"

MJ doesn't see a problem. "Committed until the end, that doesn't sound bad."

"'Until death' means lifetime commitment, for normal people. To Alvin, it's just an escape clause out of a contract. I knew that day, if I ever crossed him, I'd pay with my life.

"I asked for a friend, not an interrogation. Don't you see?" Astrid runs her fingers through her hair. "They're about to rip my life out and discard it on the floor."

"It's just hair," MJ quips.

Sniffles betray Astrid's fear and sadness. "And you're just a nasty little voice in my ear. I'm better off without friends."

MJ says, "I'm sorry. It's better to lose your hair than your life."

"Like me, Marie Antoinette's signature was her hair. She was meant to be seen, but not heard. Her appearance was her only source of power. When the guillotines came, the last thing she clung to was her hair. They took that too. They chopped off her hair before they chopped off her head. My hair is who I am. Now, the guillotines come for me."

MJ doesn't have any comforting things to say.

Astrid fills the empty space with nervous chatter until Fernando returns.

Fernando comes back through the garage entrance. He goes to the master bedroom to make sure it's empty. It is. He thanks MaxPlume and Steel, and goes back into the garage.

Ten minutes later, Fernando comes back with four other men. They look like astronauts, dressed in yellow hazmat suits, gloves, and transparent masks. They carry rolls of plastic into the master bedroom.

MaxPlume sneers at their suits. "That shade of yellow is so last year."

Steel nudges him to be quiet

The noises of crinkling plastic and ripping tape come from the master bedroom.

In worried whispers, Steel asks questions about the hazmat men, but MJ doesn't have answers. Her best guess is the suits prevent DNA contamination.

Fernando and the hazmat men go back to the garage.

Astrid sneaks over to see what they did in the master bedroom. She whispers to herself loud enough for MJ to hear. "It's a murder room."

Plastic covers the floor and walls up to eye level. They could shoot Astrid and capture the blood splatter with the plastic. MJ worries. What if they decided to kill Astrid for real? She doesn't know anything about Fernando's men. Renquist can get to anyone. What if Renquist got to Fernando? MJ brushes the thought away. She trusts Fernando. She tells Astrid not to worry.

Astrid doesn't believe MJ. She rushes into the kitchen and searches through drawers. Astrid whispers, "How could a kitchen not have knives?"

She grabs the best weapon she could find, a rolling pin. Astrid hides it under her shirt.

The hazmat men carry panels of tempered glass and a chair into the master bedroom.

Fernando comes in from the garage, with a drill in his hand. "Alright Astrid, I'm ready for you."

Astrid edges towards Fernando, with hesitant movements.

MJ rethinks her trust in Fernando. SEALs are trained killers. MJ imagines him drilling through Astrid's eyes, or in through her temple. Should she tell Astrid to run? Fernando could butcher Astrid in the murder room, without leaving evidence.

Steel tries to go with Astrid, but hazmat men block him.

"Why do you need the drill?" Astrid gulps. She might not like the answer.

Fernando directs Astrid towards the master bedroom. He smiles. "I'll use it to finish your part in this."

Astrid won't budge.

He points towards the master bedroom. "Sit on the chair! We've got a deadline." Fernando's losing patience. His heavy breaths fog up his mask.

Astrid takes out the rolling pin and swings at Fernando.

He catches it with his free hand. Fernando gives Astrid a surprised look. "I'll take that."

Reluctantly, Astrid enters the master bedroom they turned into a murder room.

The tempered glass panels tile half the floor on top of the plastic. A metal folding chair sits in the middle of the room.

Fernando whacks the metal folding chair with the rolling pin. The rolling pin handle breaks off on the chair. "Sit."

Astrid looks around. The room is barren with no exits and no ready weapons. She sits down.

He braces the rolling pin against the floor and drills into the broken end.

It's not too late. MJ could tell Astrid to use the chair or broken glass as a weapon. No. Astrid doesn't have MJ's training. She'd never have a chance.

Fernando comes towards Astrid with the drill. He doesn't bother to remove the rolling pin. "Don't worry. I'll make this quick. This is going to hurt."

MJ feels helpless. What can she do, other than watch? Even if she knew where the safe house is, she'd never get help there in time.

About a foot from Astrid, Fernando spools her hair around the rolling pin. He leaves a bit of slack, then turns on the drill.

The drill spins the rolling pin. The hair coils up then pull taunt.

Astrid cries out.

A clump of hair flies off with a piece of her scalp.

She looks down to see her reflection in the glass. Astrid stares in horror where her hair used to be. A flap of skin hangs loose in the three-inch diameter bald spot.

Fernando turns off the drill and harvests the evidence.

MJ cringes at the sight of a layer of skin dangling towards the reflection. It's a relief Fernando didn't kill Astrid but tearing out 20 more sections of scalp is horrific on its own.

Astrid summons her strength and stands up in front of her chair. "I can't let you hurt me again. Hand me that drill."

"This op fails without your hair." Fernando seems disappointed.

"Hand me that drill." Astrid remains defiant.

Fernando hands Astrid the drill.

Astrid spools hair around the rolling pin. "I have to do this myself."

MJ speaks encouraging words in Astrid's ear, but she closes her eyes. She can't look.

She continues the cruel process until she's completely bald.

Astrid leaves behind her hair and old life when she exits the murder room.

Fernando and his hazmat men meticulously harvest every hair fiber and skin cell. They close the murder room after she leaves.

With Astrid's sadistic makeover complete, she stumbles into the living room.

MaxPlume gasps and covers his mouth. "They murdered your hair."

Steel looks sympathetically at her head, covered in scabs and unattached skin. He tries to comfort her with a hug.

Astrid turns to the wall to avoids sad looks. She stares into the pictures of the fake smiling family and catches her reflection off the glass. Without her golden mane, Astrid dissolves into the background. She sheds her regal persona, from lioness to kitten.

MJ has felt a lot of things about Astrid, but never pity.

"Who will love me, looking like this?" The question escapes Astrid's lips involuntarily.

MaxPlume nods with a solemn face like he's doing her a favor. "I will."

Steel says, "Astrid, I fell in love with you, not your hair. Until the wedding, I thought you'd forgotten me."

She eyes Steel with skepticism. "You didn't see your oversized marionette? I lit it on fire, kind of hard to miss."

"I was the tourist," Steel says.

Astrid gasps in shock and denial. "You can't be. I would have remembered a strong name like Craig Steel."

"You ditched me in a subway station bathroom." He speaks the words with decades of hurt that stretches to one single moment of abandonment.

Silence.

It takes time for Astrid to reflect on her younger self. She's been so many Astrids since she was that scared French farm girl. "After the excitement of leaving home wore off, I realized I didn't know you."

Steel ignores her excuse. "You were obsessed with movie stars, so I decided to become one. I became Craig Steel and changed my look. I fought my way through the Hollywood wannabes until I became an action star. The money, the fame, they were a means to an end." With each word, Steel seems bigger.

His gaze was meant for Astrid, but MJ sees the fire in Steel's eyes and feels his intensity. That drive to chase Astrid is what made him a star.

Steel flashes a cocky grin. "I waited until you hit it big so that I could find you again, and I made you mine."

It seems like Steel is winning Astrid over until he talks about her like a possession.

Astrid's voice sounds defiant and strong. "I'm not something that can be owned or collected. You turned your mansion into a cage."

The worst cages are the ones you can't see. None. Fernando. MJ kept everyone at a distance. She made her own cage, never willing to be vulnerable, or show any signs of weakness.

"I'm sorry! I couldn't bear to lose you again. I held on too tight. When I lost you, it destroyed me." Steel gets down on one knee. "I saw something in you when no one else did. Let me be your plus one. Let me be your husband in hiding."

MJ may be far away, but she can feel Steel's longing through the glasses. It reminds her of grappling with Fernando. It's intoxicating to be desired so deeply by someone, but MJ never gave in to it. From little noises, MJ feels Astrid drifting towards saying "yes" to Steel.

MaxPlume notices the shift too. He approaches Astrid, barely outside her comfort zone. "I never put you in a cage. When you need someone, you want me by your side. I wore piss clothes for you! Take me." His eyes avoid Astrid's scalp. MaxPlume can barely look at her.

Piss clothes? MaxPlume must have been the homeless man that delivered the phone to MJ. Clothes can make the man, but they can also unmake the man. It's hard for MJ to believe that disgusting creature is the same hunk she sees now. For someone used to elaborate outfits, the homeless disguise must have been torture.

Astrid takes MaxPlume by the hand.

MaxPlume smiles like he's won.

"I need you to stay behind," Astrid says.

He spits out words with anger. "Who else would wear piss clothes for you?"

She answers his question with a glance at Steel. "Our fans struggle to feel unique, to matter, in a world that views them as interchangeable cogs, replaceable by the lowest bidder. With me gone, my fans will latch onto you. Can you be there for them, when they mourn my loss?"

MaxPlume, placated by the role defined for him, nods and gives Astrid a goodbye hug.

Astrid shifts her focus to Steel. "Whenever I want to remember my old life, I'll look in your eyes."

He leans in to kiss Astrid.

Steel's face hurls at MJ.

He's kissing Astrid, not her. Her pulse quickens. It's too real. MJ rips off the data glasses before Steel's kiss connects.

Looking through Astrid's eyes, MJ saw her own cage. Astrid was smart enough to get out of her cage. Why can't MJ?

77. **Losing is Winning**

Fernando transports every precious piece of evidence between glass plates, wrapped in butcher's paper.

His team sneaks onto a pig farm in the middle of the night. They stage Astrid's death with the cloned body parts and Astrid's hair. In case the pigs overeat, they leave behind disturbing Polaroids, hair and a leg, outside the pig pen.

The pig pictures make their way onto social media. The whole world buzzes about Astrid's death.

Astrid's music tops the bestsellers list. Astrid wouldn't sign a prenup, so all that music money goes to Renquist.

Rumors swirl around who killed Astrid. Clips of Steel's breakdown at the wedding go viral. Some believe Steel stalked and murdered her because she burned his effigy. No one can find Steel to ask him.

Some suspect Renquist because of the $200 million life insurance policy he took out the day he married Astrid.

Conspiracy videos show None discussing pigs eating through bone as he mentions pork-barrel spending. It's an eerie coincidence, or was it? It's not hard for people to imagine None killing an ex-girlfriend who humiliated him multiple times on TV and married someone else.

MJ wonders if Astrid intended people to make that connection. Did Astrid choose death by pig to point the finger at None? It could be revenge for completely ignoring her and not saying goodbye.

Several days later, Fernando calls MJ on her data glasses. Renquist's assassins plan to kill None during his Jubilee speech. She rushes into the Oval Office.

None sees the concern on her face. He clears the Oval Office, leaving him alone with MJ.

With MJ's data glasses on speaker, Fernando explains details about the attack.

The response is obvious to MJ. "You have to cancel the speech."

None shakes his head. "I can't let the powerful steal Jubilee. This speech is too important."

Fernando says, "I agree, Mr. President. Do the speech to flush out the conspirators and wear an Andraste. It can survive sniper rifles, even heavy machine gun fire at close range. An Andraste only takes so many hits before the armor degrades. So, if you see the warning indicator, turn around and use your rear armor to buy time to escape."

Fernando gives MJ access to surveillance feeds over the Capitol, where tomorrow's speech will take place.

None thanks Fernando for his advice. He asks MJ to disconnect her data glasses.

The request troubles MJ, but she takes off the glasses anyway.

"I've got good news." None tries to sound upbeat but fails miserably. "I had another lawyer look over the contract. The death and incapacitation clause is still valid for a few more days."

"What are you saying?"

"If I'm killed, or incapacitated, it cancels the contract. Control of Slam Dunk Jobs passes to the next executive, you."

"I don't want to hear this." MJ feels her world falling apart.

None says, "I've tried every escape clause in the contract. I brought Slam Dunk Jobs into the world. To make things right, my only chess move left is a king sacrifice."

MJ says, "At heart, I'm a warrior. I'm meant to be sacrificed, not you. None, you're important."

"Am I important to you?" None's question falls like a dead weight.

MJ avoids his question. "Reality doesn't come with respawn points or extra lives. It's game over when you die."

None gasps in disbelief. "You think I don't know what death is? I've lost people I care about."

She holds her hands together, pleading with him. "I don't want you to die. Take an Andraste."

"Our Andrastes stay in the bunker. I can't be a symbol of fear. MJ, don't forget the big picture."

"Why should the economy overshadow our lives?" MJ wouldn't trade him for the whole damn global economy. Let the world fall.

MJ knows chess rules. The game ends, when the king falls. "A king sacrifice isn't a valid move."

None says, "I don't care about rules. Losing is the only way to win."

"There has to be another move," MJ says.

"What other move?" He eyes the floor and talks softly. "A queen sacrifice? I have no queen. Astrid married my enemy. Zelda was a spy. Betty and Sarah died, just for dating me. MJ, I'm going to die single. No matter what, I always end up alone."

"I'm right here." MJ throws up a guilty smile. How many relationships have been casualties of her meddling?

He pushes out bitter words, with tears in his eyes. "My love life has a body count."

Every time something, or someone, came between them, MJ knew in her gut it was temporary. She can't imagine a life without None. He needs to know how important he is to her.

She knows three words she can say to make him stay. Those are power words she's never spoken to someone that wasn't family. All she has to do is open her mouth and make the right sounds come out. Just say the words. Damn it! Why is it so hard?

MJ will have to work up to it, until then, she'll stall.

"Have you told Shen yet?"

None shakes his head. "Don't you dare tell him. I'm worried I won't go through with it."

MJ smiles in relief. He just gave her an easy way out. "I'm going to tell Shen. He'll back me up."

None says, "If something happens, promise me you'll make Shen file paperwork, before the sale to Renquist finalizes."

She knows None wouldn't ask for a promise if it wasn't important. "I'll promise if I get another shot to convince you to wear an Andraste."

"Agreed. I'll wait for you in the Oval Office before the speech." None shakes MJ's hand to seal the deal.

She trusts him at his word, but today MJ has to be damn sure. MJ makes None swear on his mother's Bible.

MJ leaves with her detail for the Lincoln Bedroom. She's glad to be back in the White House, but it doesn't help one bit with her anxiety.

The reality of what she faces hits hard, just before she gets to her room. An ounce of losing None shakes her to the core. MJ screams silently in her mind's eye. Her knees go weak. MJ braces herself against the hallway wall. MJ should turn around now and tell him how she feels.

Agent Reynolds looks concerned. "Are you alright, Madame Vice President?"

MJ struggles to hold back a flood of tears. A sniffle breaks free. She pulls a tissue from her pocket and blows her nose. "I must be coming down with something."

"Do you need a doctor?"

She needs time to think. "I need sleep. Make sure I'm not disturbed."

Agent Reynolds says, "I'll make sure the next shift knows, Madame Vice President."

MJ escapes into her room. If she says those three words, she'll have to face her biggest fears. So many times, she closed that door before she opened it. MJ panics at the thought of their relationship getting physical. The stakes are too high. She has to say the words.

She'll call Shen, and they'll come up with a battle plan. MJ feels like she's going to have a panic attack. She has a glass of milk sent over and takes her anti-anxiety medication.

MJ puts her Glock 19 under her pillow. She feels so tired.

78. **Rigged**

MJ wakes up in a panic, wearing yesterday's clothes. She doesn't remember going to sleep. What time is it? Oh, no. None's speech started ten minutes ago.

She has to stop the speech. MJ jolts out of bed. Her head feels dizzy from getting up too fast. Her world is spinning, a roller coaster she can't get off of.

MJ grabs her Glock 19 from under her pillow. She stumbles over to her go bag and shoves her gun inside. It's go time. She grabs her go bag, puts on her data glasses, and bursts into the hallway. MJ runs with wobbly legs. A surge of adrenaline wakes her up.

Her security detail chases after her.

"We're late for the speech. Why didn't you wake me?!" MJ screams at Agent Reynolds but doesn't waste time looking back.

"Madame President, you asked not to be disturbed." Agent Reynolds conserves breaths to pace himself.

She races through the White House. MJ doesn't stop until she's inside her limo.

Reynolds gets inside. They depart for the President's speech at the Capitol, two miles away. With crowds and road closures, it will be difficult to get there quickly.

The President's speech plays on the limo TV. He addresses the nation behind a podium with his presidential seal. A plastic shield protects him from chest level down along the stage. "Wall Street tells us open markets are efficient, open markets bring prosperity, open markets represent freedom. What they don't tell us is that markets can be rigged."

With somber eyes, MJ imagines None waiting for her in the Oval Office before the speech. Alone. How did he feel, when she didn't show up? Did he think she didn't care? MJ missed her last chance to talk None out of going unprotected, her last chance to tell him how she feels. The thought makes her nauseous. MJ's heart aches with despair.

None continues his speech on TV. "Every year, traders make hundreds of trillions in bets, called derivatives, on the LIBOR interest rate. Over a dozen of the world's biggest banks rigged the LIBOR for over a decade, to cheat on those bets."

MJ calls Fernando on her data glasses. No answer.

She skims through a bunch of status updates from Fernando. She stares in shock at the words, "multiple shooters." MJ was worried enough when she thought it was just one assassin.

Fire builds in None, as he speaks. He's powering up. "Foreign exchange markets trade different types of currency, like dollars and euros. Traders from most of the same banks manipulated exchange rates between currencies, rigging the value of money itself."

None pounds the podium each time he begins a sentence. "Again: *the biggest banks*. Again: for *over a decade*. Again: a market worth *hundreds of trillions*."

The crowds listen intently to his words, drawn to his intensity.

MJ pulls up surveillance feeds Fernando shared with her yesterday. There's too much to sift through. She needs Fernando's whole team working on this. Where's Fernando? He was her single point of contact for the team. Without him, the plan to stop the attack will fall apart.

On TV, None points to the audience. "You know what the traders said when they ripped everyone off? They said, 'If you ain't cheating, you ain't trying.' That's the gospel of Wall Street."

Another message brings MJ answers. "Trigger word…Jubilee." Somewhere the assassins wait for their trigger word, to synchronize attacks. When None says Jubilee, he'll be death marked, killed by a word. For once, MJ is thankful None likes to give long-winded speeches.

None says, "Let's recap how Wall Street stole billions in boring ways: they rigged it." None encourages the audience to repeat his words.

MJ calls Shen on her phone. "Shen, stop the speech. None can't say the word Jubilee."

Agent Reynolds raises his eyebrow and listens in closely.

Shen says, "How can he give a speech about the Jubilee deal, without saying Jubilee? He talks about it right after banks."

Anything she tells Shen will be overheard by Agent Reynolds. Does it matter, if she doesn't trust the Secret Service? They can't all work for Renquist.

None speaks with a rhythm. "LIBOR: they rigged it. ISDAfix: they rigged it."

Thousands chant, "They rigged it." The audience may not understand every detail of None's speech, but his words ring true, and they know how to repeat a chorus.

Deafening chants muffle Shen's words.

MJ's running out of time. None already talks about banks. Jubilee is next. She screams into the phone. "They're going to kill None. Stop the speech." MJ hangs up.

None recounts banking scandal after banking scandal. Municipal bonds. Foreign exchange markets. Silver markets.

After each one, waves of people chant louder, "They rigged it!" Their voices thunder across the Capitol.

On TV, police retreat towards the main stage. The crowd seems to be on the verge of a riot.

The chants grow more intense, as the limo approaches the Capitol.

Agent Reynolds moves to the edge of his seat. "Is there an imminent attack?"

MJ nods.

Agent Reynolds asks, "Who's your source?"

"It doesn't matter." MJ tries to deflect the question. She hears the surging power of the chants, coming from outside.

"I have to decide if it's a credible threat," Agent Reynolds says.

Why is he so interested in her source, instead of just warning the other agents? MJ won't tolerate his questions. "I don't care. Get the president to safety. Do your job."

Agent Reynolds' nostrils flare in resentment. He breathes heavy, then calls in the threat.

None shakes his head. "Those are just the times they got caught. Over $100 billion in fines hasn't changed anything. It's just the cost of doing business."

MJ feels helpless. She has to do something. She resumes reading Fernando's messages. She trembles at his last message.

911.

It's code for an emergency. Fernando was in such a hurry that he couldn't type out what was wrong. MJ tenses up. Dead man's switch. She rummages through her go bag, looking for her package.

It's gone.

Secret Service agents on TV scan the audience. Their heads swivel quickly.

None says, "Fraud. Money laundering. Bribing public officials. Why don't bankers go to jail?"

MJ tries to remember last night. Grogginess, nausea, dizziness, even her wobbly legs, they're symptoms. They drugged her milk last night so that they could take her package. Fernando's missing. Shen must still have his package since Renquist's people didn't kill MJ when they had the chance.

Agents run down the steps of the Capitol Building towards the president.

None stands fearless, even though he knows what's coming. Defiant and angry, he won't back down. "Now Wall Street wants to steal your bailout. No! Let the banks fail."

Agents pull None from the podium and surround him like a football huddle. The mass of agents pushes towards the Capitol Building steps.

"I'll bail out Main Street, not Wall Street." None struggles to move his head above the tide of the agents. He raises his fist. "Let there be Jubilee."

He said it.

Time stops.

MJ freezes mid-breath. Terror engulfs her soul.

The mass of agents pauses at the base of the steps. They pull None back down. His fist submerges out of sight.

Bullets fly through an angry sky.

Armor piercing rounds detonate the torsos of retreating agents. Three carcasses slump to the floor.

Piercing screams and panic ripple through the spectators.

Snipers shoot another volley. The huddle of agents turns into a mound of gore and carnage.

Weapons fire from far away rooftops. Counter-snipers take out their first assassin.

MJ stares at the TV, helpless.

The front of the huddle creeps up the Capitol steps in formation, with their heads up. It's harder to conceal the President going up stairs.

An exploding head propels bone shrapnel into what's left of the security detail.

The hand of death sends chills down MJ's back. Color vanishes from her sight. Emptiness and despair recolor her world in black and white.

Counter-snipers take down two more assassins. Are there any more?

Agent Vincent and another agent drag None up the Capitol steps by the arm. His feet flop along each step.

There's something wrong with None's head. Is that exposed brain?

Cold and numb, emotions overwhelm MJ. She blanks out.

The limo fishtails into an emergency U-turn.

Crisis center. Continuity of government. Agent Reynolds' words wash over her, unheard.

Her hands twitch. MJ remembers to breathe. She gasps for air.

MJ looks around, to regain her bearings. They've turned around. "Where are we going?"

Agent Reynolds says, "The White House, Madame Vice President."

"No. Take me to the hospital."

"Sorry Madame Vice President, we're routing everyone in the line of succession, down to Secretary of State, to secure locations." Agent Reynolds speaks with authority. He acts like he's the one in charge.

"Don't tell me what's safe. Hospital. Now!"

Agent Reynolds scowls at MJ, impatient with her answer. "With all due respect, a VP does one thing. Do your job. Go to the White House. Become president."

How dare Agent Reynolds speak to her that way and disregard her orders. He's fixated on going to the White House. MJ doesn't care what he wants. A president is never more than ten minutes from a trauma hospital. She has to act now.

MJ reaches into her go bag. She grabs her Glock 19 and points it at Agent Reynolds' head. "I'm not letting my best friend die alone."

Sweat beads on Agent Reynolds' forehead. He calls in for a new destination, the hospital.

She keeps the gun on Agent Reynolds.

Agent Reynolds smiles back at her.

His smile unnerves her. Does he not take her seriously, or does he know something? MJ's going to a hospital, but what if it's not the same one they're taking None to? MJ hopes she's paranoid.

After four minutes, they arrive at the ambulance hospital entrance. MJ points to the door with her gun.

Agent Reynolds gets out.

MJ gets out, her gun still drawn. She looks around. No motorcade. No best friend. She missed him. Tears well in her eyes. "Where's the President?"

Her security detail surrounds her. A few agents look oddly at her weapon but remain silent.

"He's on his way," Agent Reynolds says.

MJ wants to believe him, but she doesn't. There's something about Agent Reynolds she doesn't trust.

Sirens in the distance get louder. It must be None's motorcade.

She was paranoid. It was all in her head. MJ puts her gun away and wipes her eyes.

Secret Service agents move vehicles out of the way. None's coming in hot.

Agent Vincent and the surviving agents deploy in bloody suits. Skin and bone fragments litter their hair.

The doors of None's ambulance swing open.

White House Physician Indigo Parker jumps out. Shen jumps after her.

EMTs lower None's gurney to the ground and run towards the entrance.

MJ and Agent Vincent keep the hospital doors open.

Doctor Parker runs ahead, barking orders to EMTs and Secret Service alike. She sprints fast enough to leave the gurney behind.

Shen grabs None's left hand. He speaks to None in Mandarin. His words remain private, even while others listen in.

As the gurney enters the hospital, MJ grabs None's right hand. She joins the rush to the operating room.

More ambulance sirens approach behind them. The sound muffles, as the hospital doors close behind them.

MJ grasps None's hand tightly. "I don't want off the crazy train. Don't die on me. Don't make me president."

Nothing can stop the tears. Lines of mascara streak down her face, dripping below her data glasses.

"Fight your way back to me. I'm alone without you." Blurry eyes distort her vision.

They reach the operating room. Doctor Parker and another set of Secret Service agents wait inside, in medical garb.

Shen lets go of None's hand.

"No visitors in the OR." Doctor Parker motions for the gurney to stop.

MJ reluctantly lets go of None's hand. She drops to her knees sobbing.

None's gurney disappears into the operating room. He's gone, perhaps forever. What if MJ never sees him again?

MJ wallows in her own personal Armageddon. She's so angry at the world. She's a warrior at heart, but there's no way to fight back. She wants to cry and scream and run and run and run. What good do tears do? They just make it hard to see. All she can do is wait and worry.

She needs to focus elsewhere, or she'll be trapped in the pits of despair. MJ looks at Shen. Dead man's switch. She needs Shen's package to make more copies. She gets off the floor. MJ lifts up her data glasses to clean off tears and runny mascara, as she approaches Shen.

Shen runs his fingers through his hair. He fights his emotions, but Shen seems on the verge of ripping out his hair.

MJ says, "Where's your package? I need it."

Shen looks around, shocked and confused. "I had it, right before the shooting. I…I don't know. It all happened so fast."

Horror replaces sadness on MJ's face. "That was the last one. None of us are safe now."

Shen holds back tears. "They shot my brother, and all you're worried about is yourself."

MJ grabs both of Shen's hands. She peers deep into his eyes. "I need a favor. The incapacitation clause is valid. File the paperwork before Renquist can react and we'll finally take back Adaptive Unlimited."

He pulls back his hands. "No. I have to say final goodbyes."

MJ faces Agent Reynolds. She points at Shen. "As acting president, I'm relieving Shen Wong from duty as White House Chief of Staff. Please remove him from this facility."

Agent Reynolds winces. He orders two agents to escort Shen out.

Shen glares back at her on the way down the hall. "You don't own my brother."

"File the papers, and I'll let you come back," MJ says.

Shen's removal puts everyone on edge.

Agent Vincent approaches MJ with an unsteady gait. "I was with him to the end. I'm sorry for your loss, Madame Vice President."

MJ hardens at his words. The expressions on his face tell MJ that Agent Vincent wants to comfort her, to calm her down. She doesn't want to calm down. Her best friend lies dying behind those doors. "He's going to die because you didn't protect him. I'll bust you down to gate duty, so I never have to see you again. Get out of my sight."

With his head bowed, Agent Vincent disappears down the hallway.

Agent Reynolds runs after him.

The other agents try to avoid MJ's gaze. Anything could set her off.

Was MJ mad at Agent Vincent, or herself? What if she'd told Agent Reynolds about the attack when she woke up? Why couldn't she just trust him? None might still be alive. What if she'd told None she loved him? MJ catalogs every wrong move in her head.

Enough. MJ can't change the past.

Agent Reynolds returns with a black smartphone. "Madame Vice President, the 25th Amendment Section 4 has been invoked. A cabinet meeting is scheduled in three hours."

He hands the black phone to MJ. "This is the most secure phone ever made, the presidential cell phone. When you're ready to take calls, world leaders wait to share their condolences."

The phone means None isn't coming back. No. MJ won't accept that. She isn't ready to give up hope. MJ wants to throw the phone back at Agent Reynolds and tell him None will be fine.

She sighs. None isn't fine. MJ has lost Fernando, and after what she just did to him, Shen may never speak to her again. In one day, MJ has lost everyone she trusted. She'll have to start over.

She'll need powerful allies to fight Renquist. Talking with world leaders seems like a good first step. "I'll take the calls."

China. Russia. Germany. From major powers, down to the minor nations, the calls are all the same. The rulers express how sorry they are that None was attacked and how much they respected him. Why do people wait until you're on death's door to admire you?

Even her personal cell rings. MJ looks at the number. Unknown caller. Yeah, she's not answering that. MJ isn't in the mood for reporters, the vultures they are. Or maybe it's a computerized voice, waiting to sell her something.

MJ resumes the endless stream of sorry, sorry. It's like world leaders read off the same script.

Turkish President Yavuz doesn't follow the script. His translator asks, "How long have our countries been friends?"

Those words are None's words. MJ's heard None ask similar questions many times. "Why do you ask?"

The translator says, "Your president promised me two favors. I need a favor. Will you keep his word?"

MJ gasps. None is six inches from the grave and all President Yavuz cares about is favors, and she thought reporters were vultures. MJ does her best to mask her contempt. "What's the favor?"

"There are rebel encampments I'd like destroyed," the translator says.

"Don't you have your own bombs?"

President Yavuz yells in the background, as the translator speaks. "Of course, but I cannot be seen attacking these rebels myself. That's why it's a favor."

The unknown caller rings again. MJ greets any excuse to get off the phone.

"I have important matters to attend to. Send me details, and I'll see what I can do." MJ hangs up the presidential phone.

She answers her own cell phone. She hears car noises. "Hello?"

Fernando laughs with glee. "I never thought I'd hear your voice again. They took me and my package. I escaped. My package did not."

MJ whispers into the phone. "What happened?"

"They made the mistake of sending a soldier, that I trained, to interrogate me. He freed me. Then we neutralized my captors. I'm heading into the Situation Room with reinforcements. I've drafted every friend I trust. I'll take the fight to Renquist and his co-conspirators."

"Agreed." MJ has her first smile since the shooting.

Fernando ends the call.

She'll bring war to Renquist and his minions. She'll pay any price for her vengeance.

MJ takes more calls on the presidential cell. She hangs up abruptly on whoever she was talking to when she sees Doctor Parker. Hopefully, that wasn't an important country. "How is he doing, Doctor Parker?"

"I put him in a medically induced coma, to stop the brain swelling and give him time to heal. We're monitoring his vital signs closely."

"Will he live?"

Doctor Parker considers her words carefully. "It's premature to give a prognosis. The next few days will be critical."

Recovery isn't the only danger None faces. The hospital doesn't feel safe to MJ. "The shooting was a coordinated attack. There may be further attacks. Can we move him to a secure location?"

"Once he's stable, that might be an option." A hum comes from her pocket. Doctor Parker retreats into the operating room.

MJ stares at the closed doors like she's in a trance. She spaces out until she gets a call on her data glasses.

The call is Fernando. "I'm leaving for the cabinet meeting in ten minutes."

It's easy to forget Fernando is Secretary of Defense, with all the covert stuff they've done together. She hates to leave None's side, but the White House is minutes away. She can rush back if anything happens. MJ whispers. "What if Renquist tries something at the cabinet meeting?"

Fernando says, "With the heightened security levels, he won't be able to attack again today. See you in ten." Fernando sends MJ data links to his most promising leads and disconnects.

MJ and her security detail make the short trip back to the White House and find their way to the cabinet meeting.

Five cabinet members video conference, the rest sit at the gigantic table.

Fernando and MJ sit across from each other with matching data glasses.

MJ reads the 25th Amendment aloud. Section 4 explains how MJ ascends into the presidency if the cabinet votes that None can't do his job as president.

With the first vote, MJ votes aye. She calls out cabinet positions one by one.

Just before it's his turn, Fernando cracks a smile. His vote is an enthusiastic aye, that seems out of place on such a solemn occasion.

Is he happy to see None go? Does Fernando think she'll be his with None gone? MJ would give him a dirty look if the data glasses didn't hide her eyes. She turns her head and continues the vote.

Several minutes later, the vote is unanimous, declaring None unfit to carry out his duties as president. The results will be forwarded to Congress, and she'll be sworn in.

Fernando rushes out without saying hi to MJ.

MJ feels slighted until she sees a message from Fernando. "Meet me. Situation Room."

She thanks cabinet members as they leave, anxious to leave herself. As soon as she can break away, she heads for the Situation Room. MJ asks her security detail to wait outside.

At the door, Fernando greets MJ with excitement. "Jonathan Maluki, CEO of Cragmire Investments, chartered a private Gulfstream plane to leave the country, flight NQ836."

A soldier with a headset repeatedly hails flight NQ836, without response.

Fernando says, "You won't believe our luck. We've compared the flight manifest to the list of conspirators that we identified at the warehouse. They'll all on board, with—"

MJ shares his excitement. "Renquist is on board?"

His smile disappears. "*With* one exception. Renquist isn't on the plane. I've dispatched fighters. We'll escort them back to American airspace and force them to land."

"Shoot it down." A crazed smile spreads across MJ's face.

Soldiers in the situation room exchange glances.

Fernando's mouth gapes open. He pauses a few moments to get over his shock. "Madame Vice President, I'd feel more comfortable if you gave that order with full presidential authority."

Why is Fernando addressing her formally? MJ hopes it's for appearances. "I'll burn the world if I have to. They'll all pay."

Agent Reynolds knocks and talks through the door. "Madame Vice President, Chief Justice Dornan arrived. They're ready for you."

The interruption relieves Fernando. "We'll continue this discussion after your inauguration."

MJ taps her data glasses. "No. Send me the feed." She leaves the Situation Room and hurries down the hallway with her security detail.

Agent Reynolds asks, "Do you have a Bible for the ceremony?"

"It's in the Oval Office," MJ says.

On the way to the Oval Office, MJ passes a line of reporters waiting to enter the Cabinet Room.

MJ goes to the Resolute Desk. She opens the drawer where None keeps his mother's tattered Bible. She grabs it and follows her detail back to the Cabinet Room.

Secret Service agents accompany reporters inside the Cabinet Room.

A dozen reporters clap as MJ enters. TV camera operators swivel towards her.

Chief Justice Dornan waits in his Supreme Court robes, at the other end of the gigantic table.

A radar screen comes up on MJ's data glasses. Two radar contacts closely follow a third. Those must be the fighter jets. MJ whispers. "How long?"

"Three minutes to target," Fernando says.

Agent Reynolds whispers back. "They should be ready for you now."

MJ will have to be more careful what she says out loud. Agent Reynolds heard her. She rushes to Chief Justice Dornan. They exchange greetings.

Fernando calls MJ on the data glasses. "They may be treasonous swine on that plane, but they're still Americans. MJ, please don't do this."

Oh, so now she's MJ. MJ purses her lips. Nothing will change her mind.

Video streams from both jets come up on her data glasses.

"I ask you to reconsider. Let me abort," Fernando says.

MJ looks at Chief Justice Dornan. "We are a go."

Her word choice catches Chief Justice Dornan off guard, but he gets her meaning. He raises his voice. "Quite right. Let us begin."

The reporters quiet down.

Fernando says, "You are weapons free to engage,"

From the onboard camera, the first pilot says, "Fox three."

"Fox three," the second pilot says.

Weapons bays open on both fighters. Hydraulic arms push out an air-to-air missile from each jet. Two missile cams show on MJ's data glasses.

The Supreme Court Chief Justice raises his hand and begins the oath.

MJ raises her right hand, with her left hand on the tattered Bible. She repeats the oath, after Chief Justice Dornan. "I, More Jobs, do solemnly swear."

Only the blue on blue of the sky and ocean show on the missile cams. Twenty miles.

"That I will faithfully execute the office of President of the United States."

Twelve miles.

"And will to the best of my ability."

Eight miles.

"Preserve, protect and defend the Constitution of the United States."

The Gulfstream grows bigger as the missiles approach, from a pinpoint to a full-sized target. The first missile hits the main fuselage of flight NQ836. Missile cam one goes dark.

The second missile hits a chunk of flaming debris, as it falls towards the ocean. Missile cam two offline.

"So, help me god." MJ lowers her hand.

The first pilot cheers. "Good kill."

Reporters congratulate their new president.

79. **End Game**

In the White House Press Room, on a screen behind MJ, three bomb icons overlay a map of Turkey. The map dissolves into footage of explosions. Three Turkish rebel encampments disappear in billowing clouds of smoke.

MJ stands behind a podium. She addresses a packed room of reporters. "I have one thing to say, to the terrorists that blew up flight NQ836. You thought we would be vulnerable, in this time of mourning. You thought wrong."

The press jumps to their feet, to ask questions.

Bruce Cannon of the Reveal Channel says, "Turkish President Yavuz called this an unprovoked attack on his sovereign territory. What—"

MJ cuts Bruce Cannon off. "No questions." She retreats from the room, while cameras flash.

There's one known conspirator left alive. Three days after the attack, Renquist is still out there. She needs to kill him before he kills her.

MJ walks down the hallway, anything to get away from those reporters and their questions. Yavuz makes her the bad guy? He'd better not ask her for that second favor.

Fernando recorded damning evidence at the Galaxor Telecom warehouse, that proves Renquist plotted to assassinate None. If they use that evidence, any competent reporter would figure out what really happened to flight NQ836. MJ classified the warehouse recording Top Secret, to hide it under the veil of national security.

"I have Shen Wong to see you," Agent Reynolds says.

She stops dead in her tracks and nods to herself. "Bring Shen to the PEOC entrance."

The PEOC, the President's Emergency Operations Center, the White House Bunker, it's the safest place in the White House. It got safer when Fernando assigned his most trusted military personnel there. Until she takes down Renquist, it's MJ's base of operations.

Agent Reynolds says, "Shen Wong doesn't have clearance for that area. You had it revoked, Madame President."

"Then, get it unrevoked." MJ resumes her walk to the PEOC.

Agent Reynolds calls in the change of plans. "Shen Wong is on his way."

MJ and her security detail arrive in front of giant steel doors with an input screen in the wall next to it. She sends Fernando a message on her data glasses. "Meet me. PEOC."

Agent Reynolds types some of his access code into the input screen.

MJ grabs his hand and pulls it away. "Not yet."

A few minutes later, three agents arrive with Shen.

Shen carries a briefcase. He clenches his fist and screams at MJ. "You took my brother! Where is he?"

The agents with Shen restrain him from coming closer.

MJ stares back with an emotionless face. "Do you have something for me?"

"You greedy…" Shen gets so choked up he can't talk. "You were in such a rush to collect your money that you couldn't let me say goodbye to my brother. I bet you talked him into it."

Shen unlatches his briefcase and flings it to the side. Papers scatter everywhere. "There. Take it."

She deserves his anger, but MJ had no choice. At least, that's what she tells herself.

MJ picks up legal papers, while agents hold Shen back. She's not in the mood to order the pages and decipher the legalese. "What does it say?"

Shen says, "The emergency injunction worked. The contract is canceled. Adaptive Unlimited and all its companies are yours. I hope you're happy. You selfish…" Tears well in Shen's eyes. He tries to turn from her.

The agents release Shen.

He keeps his back to MJ. Shen runs his fingers through his hair.

"No. I'm not happy. Your brother made me swear. I'm sorry I had to hurt you, to keep my promise." MJ looks at Agent Reynolds. She points at the steel doors and nods.

Agent Reynolds types in his code and presses his thumb on the input screen.

Motors slowly open thick steel doors to the PEOC.

Inside the PEOC, a video screen, called the Big Board, dominates most of the back wall. The Big Board shows the overall status of the secret manhunt for Renquist, with hundreds of green dots, overlaid on a map of the East Coast.

A couple of dozen handpicked soldiers sit at rows of desks facing the Big Board. Some investigate leads on the search with their computers. Others coordinate with trusted counterparts in the NSA and other government agencies, on secure land lines.

Three guards stand watch at a security desk facing the entrance.

"If you want to see your brother again, you'll come with me." MJ marches past the security desk with her detail.

Shen wipes his eyes and turns around. He follows MJ inside with his escort.

The steel doors close behind them, with the thud of a bank vault closing.

No signal. MJ's data glasses disconnect. Without network access, her only video feed is what's in front of her.

MJ walks past the rows of desks in the operations section, to an entrance at the back wall. A locked storage locker stands on the side wall nearby. She checks to make sure Shen is behind her, then enters the Executive Briefing Room.

A conference table, just a bit shorter than the Cabinet Room table, could seat about 20 people if it weren't shoved up against the wall to make space. A handful of laptops sit idle on the table.

None lies in a hospital bed with a bandaged head and a breathing tube connected to a ventilator. Air hisses from the machine, as it breathes for None. The ventilator rhythm sounds like a heartbeat.

Doctor Parker and a nurse monitor a readout of brain activity on an EEG machine.

None moved two Andrastes and their charging stations to the back wall, after the Defend Texas gun rally. They remain there, untouched. If only MJ convinced None to use an Andraste during the speech.

Behind her, Shen comes to an abrupt halt, when he sees None. "Why is my brother here?"

MJ says, "I can't run this country from a hospital. I won't abandon him."

Shen shakes his head. "Yet, you forced me to?"

Nothing MJ says will get through to Shen. She turns her back to him. "Doctor Parker, how's he doing?"

"The intracranial pressure stabilized. If the swelling stays down, we can stop the IV and take him off the machines, to revive him."

Shen nudges past MJ, to take None's hand. "When can we wake him?"

What if he doesn't wake up? MJ won't acknowledge that fear out loud. Doubts are kindling wood in the fire of her mind.

Doctor Parker says, "I could wake him now, but I would feel more comfortable waiting. If we revive him prematurely, the swelling could return, resulting in permanent brain damage, or even death."

MJ senses jealousy from Shen that she was closer to None. She decides not to mention None gave her power of attorney for medical decisions. "What do you think we should do, Shen?"

"If Doctor Parker says wait, we wait." Shen appreciates the opportunity to feel heard.

MJ nods. "Is there brain damage? Will he be the same?"

Doctor Parker says, "There's still so much we don't know about the brain. If there's brain damage, recovery can take months. He'd have to relearn things, but a new treatment, called Adaptonium, has shown promising results."

"I've heard of it." MJ flashes a hopeful grin. His recovery might be helped by the very software he created.

Doctor Parker says, "I must warn you, even if the revival goes smoothly, the President will be disoriented and confused. He may not understand what happened to him. He could get frustrated, even aggressive."

Aggressive, or not, MJ wants him back. She latches onto any strand of hope. MJ wants to take None's hand, but she can barely hold back her sorrow. She can't afford to break down or rest until she stops Renquist. Shen deserves time alone with his brother, anyway.

MJ returns to the operations section of the PEOC. She's overdue for her security briefing. It'll help fill the time until Fernando arrives.

Her security adviser goes over a laundry list of bad news. Global banks continue attempts to steal the bailouts for themselves. Some countries add rules that will apply the Jubilee money to debts first, making it a massive payday for banks and credit card companies. Uncertainty and fear rule the day.

MJ feels so overwhelmed. She whispers a little too loud. "Is there such a thing as good news anymore?"

The security adviser smiles. "Yes, there is good news. Many of the passengers on flight NQ836 ran companies. Suspicious stock trades on their companies might lead us to the terrorists that shot down the plane."

Even her security adviser doesn't know MJ had the plane shot down. She smiles. "That is good news."

Her security adviser smiles back, then returns to more bad news. Russia seems on the verge of collapse, with rumors of an impending coup by the five richest oligarchs.

She stops the briefing the moment Fernando arrives. She'll listen to the details on Russia later.

MJ leads Fernando next door and clears the Executive Briefing Room, so that only None, MJ, Shen, and Fernando remain. "Everyone I trust is in this room. I want us—"

Shen twists around, to glare at MJ. "You expect my help, after what you did? I want no part of it."

MJ says, "You have no choice. You think this is over? None of us are safe until we take care of Renquist. Fernando has people, so he'll be safe. Shen, I need you to move into the White House and return as Chief of Staff."

Shen asks, "Why would I do that?"

"So, I can assign you a security detail. When None wakes up, don't make me explain how I got you killed."

"And stay where?"

"The presidential suite," MJ says.

Shen gulps. "Shouldn't you sleep there?"

"No. He's president." Sniffles betray MJ's feelings. "I won't give up. He's coming back if I have to rip him from the jaws of Heaven, or Hell." MJ closes her eyes and bows her head, as she struggles to compose herself.

MJ's pain hangs over the room like a dark cloud.

Shen never heard MJ use the word love, but he feels it. The power of it breaks through his own pain. Sympathy for MJ radiates through him. Shen reaches out his arms.

Tears stream down MJ's cheeks. "I'm so sorry, Shen. I promised. He made me promise. I couldn't say no. Not to him."

Two broken people hug and cry on each other's shoulders. They bond over the one man they both hold most dear.

Shen says, "You're right. This isn't over. I'm on your team. Let's take down Renquist."

Fernando watches awkwardly, unsure what to do. He waits, alone.

When the sniffles and tears die down, Fernando clears his throat to get their attention.

MJ and Shen pull apart. They wipe tears away and exchange understanding glances.

Fernando asks, "MJ, what was on these recordings worth killing over? Don't just tell me treason."

She stops to think about all she saw. "I recorded so much vile crap, but it's got to be the weapons deals."

"Renquist runs a defense contractor, that's not treason," Fernando says.

MJ shakes her head. "It is when you sell advanced weapons to the Russians."

Fernando gives MJ a skeptical stare. "Russia doesn't need weapons. The only country that exports more weapons than them is us."

How can he doubt her at a moment like this? "It's what I saw. Renquist's henchman made arrangements for Timoshenko, Kazakov, and some others I don't remember."

Surprise replaces doubt on Fernando's face. "Do you mean Kazankov?"

MJ nods.

Fernando lists other names: Uglitsky, Ivashov, Buryakov. She nods at all of them, except Ivashov. His face shifts back to doubt. "You recognized those names from the security briefing."

"I stopped my briefing when you arrived. Five names. Oh. Are those five Russian oligarchs planning the coup? That's where I left off." MJ covers her mouth and cringes.

Fernando freaks out. "I resisted my urge to watch the recordings because I thought I was protecting you. I was blinded by..." Fernando shakes his head and trails off. "If you had told me what you recorded, I could have done something."

Maybe it was a mistake to keep the details secret, but it was necessary. MJ needs Fernando to see that. "The dead man's switch worked. It slowed down Renquist until he could take us down simultaneously. Without it, he'd attack earlier."

Shen says, "MJ, here's what I didn't get about the recordings. Why have those secret meetings in *your* offices? Renquist owns big companies. He must have his own office."

MJ says, "What if Renquist knew someone was onto his secret? He didn't officially own Adaptive Unlimited, so our offices couldn't be traced to him. He probably went there to avoid snooping."

Fernando appreciates the irony with a smirk and a chuckle. "Renquist uses your offices to avoid being watched, and you watch him."

Shen asks, "Fernando, can you bring me up to speed on the security briefing? Maybe it'll remind MJ what was on those recordings."

With a nod, Fernando begins his recap. "There's a shadowy alliance of Russian oligarchs, and military leaders called Kombinat. They use discontent over the bad economy to gain supporters. Once they amass enough power, they'll topple Russian President Orlov."

MJ asks, "Renquist is selling weapons to Kombinat?"

"Unlikely." Fernando can't help but scoff at the question. "Weapons systems are expensive. It can cost a billion for a dozen planes. Even if the richest Russians pool their fortunes, it won't be enough money."

He should trust MJ enough to accept what she says on faith, even if it doesn't make sense. "Look, I heard Renquist plan weapons deals with his henchman. They named all those men, except Ivashov. Maybe *you* got the name wrong." MJ's words sound defensive.

"Unlikely." Fernando gives MJ a cocky smile.

That answer isn't good enough for MJ. "Because?"

"Ivashov was President Orlov's close friend, until they had a falling out. Membership of the Kombinat is secret, but it's known that Ivashov created it. When President Orlov found out, Ivashov fled to the United States."

MJ asks, "What made them enemies?"

Fernando quiets down to just above a whisper, like he's sharing a secret. "I heard a story, from an intelligence asset. One winter, Ivashov went to an exclusive Yalta beach resort with President Orlov's closest friends. President Orlov likes to display his manliness. He took off his clothes in chilly weather and jumped into freezing water. Then he dared all the men in his inner circle to jump in. Everyone did, but Ivashov, who stayed on the beach in his coat. President Orlov ridiculed Ivashov and had him thrown in the water with his coat still on. Ivashov founded Kombinat that day. He is never seen without a coat, even on the hottest days. They say Ivashov will take his coat off for no man."

MJ gasps. She covers her mouth with her hand. "None told me about a Russian who came to the police station. He wore a trench coat on a hot day."

Fernando asks, "What are you saying?"

"That Russian wasn't Renquist's henchman. I didn't hear the name Ivashov because people don't say their own name. Ivashov made weapons deals himself. He had as much to lose, as Renquist, if the recordings got out." MJ feels vindicated. It'll all starting to make sense.

"I hope you're wrong." Fernando goes to a laptop on the conference table and searches for Ivashov's NSA file.

Chills run down MJ's spine. She stares off into space. "It's you, from the police station."

Her odd comment unsettles Shen. It takes a lot to freak out MJ. That's enough to freak him out.

MJ switches into detective mode. "That's what None said during election results. It was Ivashov that called, to tell None he wouldn't be president."

Shen says, "I remember that. Lan and I had to get off the couch, so that they could search for more phones. Right after that, we lost California, and our electors defected to other candidates. It almost cost us the election."

"Ivashov tried to rig the election against None," MJ says.

Fernando chuckles. "Russians influencing our election? That's the most ridiculous thing I've ever heard."

MJ feels like she's onto something. "Shen, how could someone accomplish what happened?"

"If you bribed the California Secretary of State and intimidated our electors…maybe." Just the possibility disturbs Shen. "I don't want to believe that."

"Fernando, does Ivashov have a criminal past?" MJ asks.

"Russian capitalism began in the black market. Many early Russian oligarchs, like Ivashov, came from the criminal element."

"If he's a mobster type, bribery and intimidation are part of the job description." Whether Shen and Fernando want to believe it, or not, it seems plausible to MJ.

Fernando brings up pictures on the laptop from Ivashov's file to show MJ.

A blond male in his 50s wears a trench coat with wraparound sunglasses and a hat. MJ recognizes him from the recordings. There's something familiar about his sunglasses. "It's him. Are those data glasses?"

The realization stuns Fernando, but he quickly recovers. "Ivashov might use data glasses to run Kombinat. He could get satellite access from Renquist."

MJ says, "Fernando, these men are too dangerous to live. Put Ivashov and Renquist on the kill list."

"As president, you have that authority, but Renquist is an American citizen," Fernando says.

Shen waves his hand to get MJ's attention. "You can't just add someone to the kill list. They have to meet the legal criteria."

MJ rolls her eyes. "Which is?"

Shen looks at the ceiling, while he tries to remember the details. "One. They must pose an imminent threat to Americans. Two. Capture isn't possible. Three. It's consistent with the principles of the laws of war."

MJ says, "They tried to kill your brother, and they may try again. I'd call that an imminent threat."

Fernando checks off another criterion. "Kombinat isn't a nation. It's a non-state actor, so it won't violate the laws of war."

Shen says, "As long as you can capture them, you can't legally put them on the kill list. Also, the kill list doesn't target Americans on American soil."

MJ looks at None, lying there. She listens to the rhythm of his ventilator. A machine breathes for None. Renquist must pay for what he did. "Renquist will use any means necessary. Why won't we?"

"Renquist had me tortured, but he still has due process rights." Fernando sighs. "You gave an oath to uphold the Constitution."

It's the last thing MJ wants to hear.

"If we can't find Renquist, this is all moot anyway," Shen says.

Fernando summarizes the search efforts. They've put Renquist's known associates under surveillance, raided his businesses and properties, even boarded his yacht. "If he leaves the country he'll have to use a plane or a ship."

MJ mimics an explosion with her hands and grins. "I don't think he'll take a plane."

Shen looks at her hands with confusion.

Fernando says, "I'll focus our efforts on harbors and docks. If Renquist hides in a shipping container, he'll be hard to detect."

Shen says, "There's an option you forgot, luxury submarines. Besides beachfront property, a celebrity client of mine had me look into submarines. They're becoming more popular than yachts for the ultra-rich. There aren't any paparazzi underwater."

Fernando pulls out a drawer underneath the conference table. He calls the operations room from a secure phone inside the drawer and gives new orders to focus on submarines.

MJ hasn't slept much since the shooting. She pulls up a chair next to None to rest her feet.

The ventilator drones on with a steady beat. It comforts MJ to think of the sound as None's heartbeat. His body lies there like a corpse, but she knows he's still in there. MJ takes his hand to feel him closer, to sense him.

Heavy eyelids win against MJ's battle to stay awake. She dozes off in her chair.

Two hours in a chair, that's what passes for sleep these days. When MJ wakes, she lets go of None's hand and gets a status update on the search.

The entire operations room investigated every submarine. They found a division of Renquist Aerospace builds submarines, but all of them were accounted for.

"I'm sorry your idea didn't work." MJ gives Shen a sympathetic look.

Shen says, "Maybe we're looking for submarines in the wrong place. Remember how Renquist used your offices, in a company he didn't own yet, so that he couldn't be traced? What if he's using the same trick, a submarine he doesn't own yet, one under construction?"

Fernando holds out some papers. "We already have the list of submarines under construction, but all the information we could get is submarine model, port of origin and the name to paint on the side." Fernando flips through the ten-page list.

"Let me see it." MJ gets the list from Fernando. She reads through the names looking for something familiar. Her eyes bulge out on page six.

Shark Eater.

Renquist likes to brag he doesn't swim with sharks, he eats them. That's got to be his.

She reads the port of origin.

Baltimore.

That's the same city as Galaxor Telecom headquarters.

MJ grins. She's found him. "I want heavy surveillance on all docks in Baltimore. We're looking for a sub called Shark Eater."

Fernando gives new orders on a secure phone inside a drawer.

He takes back the list and brings up specs for that model of submarine on his laptop. At 220 feet long, it has room for all the amenities, from its wine cellar and Jacuzzi to its living quarters for the chef and maid. Fernando finds the features he cares about. "His sub has a 3,000-mile range and an airlock for a mini-sub. Once he gets inside, Renquist can go anywhere."

What if he already left the country? MJ hopes it isn't already too late.

Fernando looks intently at MJ. "I'm sorry I doubted your story. There's still one thing that doesn't add up. To buy weapons they'd need a lot more money."

MJ thinks back to the last time she saw Renquist. "At the G20 summit, Renquist said he'd make lots of money when the banks were bailed out. If he's heavily invested in banks, maybe he got Ivashov to invest too. In my security briefing, I was told about suspicious stock trades, related to the plane that we shot down. Can we detect suspicious trades in bank stocks?"

Shen says, "We shot down a plane? Never mind, I don't want to know." Shen searches financial news on his laptop. He points at the top financial news story. Banks: bailout, or bankruptcy?

Bank stocks and derivatives trading halted on all stock exchanges, after record trading volume. Trading will resume in a few hours, after Congress votes on two competing versions of the Jubilee bill.

Fernando nods. "If banks get their bailout, Renquist and Ivashov could make billions. That would be enough money for the weapons deal. We could face thugs with nuclear weapons. There have been rogue states before, but never a rogue superpower."

"What if I convince Congress not to bail out banks?" MJ asks.

Fernando says, "If the Russians lose enough money, you won't need to put Renquist on the kill list. The Russians will kill him. Ivashov and his allies won't care about inconveniences, like national borders. They'll hunt Renquist anywhere. They'll contaminate his food with radioactive poison, or gun him down on the streets of London. Renquist's billions won't save him."

Renquist and the bank lobbyists will use every favor, bribe, and threat they can muster. MJ will need a powerful ally if she's going to get Congress to pass None's version of the Jubilee bill.

The stakes have never been higher in Washington. The dollar value on the Jubilee bill dwarfs any existing legislation. Competing Jubilee bills will make or break political careers and which lawmakers win could depend on the smallest information advantage.

MJ dangles the most valuable type of information, secrets. She promises Senator Garfield a secret that no other member of Congress has.

Senator Garfield can't resist the bait. He agrees to meet with her.

MJ marches through the White House at full speed, until she reaches the Oval Office. MJ sits at the Resolute Desk and has Senator Garfield sent in.

She says, "You helped stop the repeal of the minimum wage. You can stand up for what's right again."

Senator Garfield smiles and nods. "I always do the right thing. Bail out the banks or doom the economy. I, for one, am not picking doom the economy."

"Don't listen to lobbyists. Those aren't the actual choices. The peasants sharpen their pitchforks, as we speak. If you don't bail out average

Americans, the pitchforks will come for you. Revolutions are not kind to those in power." MJ can see the writing on the wall. Why can't he?

"I'll take my chances." Senator Garfield taunts MJ with a huge grin. "It's still better than getting shot up, like the last politician that tried to bail out Main Street."

MJ jolts to her feet with murder on her mind. Her rage makes her feel ten times bigger, like she could step on this ant of a man. "My best friend lies dying. You think it's a joke?"

"Dangers lurk on both sides of that fence. I'm not scared of pitchforks." Senator Garfield smiles wider.

"Then you should be scared of me." MJ stares at him with crazed eyes, her breath heavy.

He's a little unsettled, but he plays it off as a joke. "Do you have anything more for me, than threats? What's this secret of yours?"

"This isn't a secret. It's Top Secret." MJ plays a snippet from the warehouse recordings.

Renquist's voice says, "I don't want him unelected, I want him dead."

"Is that?" His smile disappears. Senator Garfield is too scared to ask the full question.

"Yes. It's Alvin Renquist, plotting to kill the President. I've tracked down some of his co-conspirators. Do you know where I last saw them?"

"Uh...no."

"In smoldering piles of ash." MJ stays quiet to let the meaning of her words sink in.

He may not be afraid of pitchforks, but he quickly becomes afraid of MJ.

She talks deliberately, each word marinated in hate and rage. "Anyone who works with Renquist, I will burn to the ground."

"I don't work with him."

"I saw you together at the G20 summit. Don't lie to me."

His words sound desperate. "OK. He's a donor. What do you want from me?"

"Support the Jubilee bill without the bank bailouts, and all past sins are forgiven." MJ tries to make it sound like a generous offer. She doesn't know if he has sins to forgive.

Senator Garfield pauses to weigh his options. "And if I don't."

MJ says, "I gave you the courtesy of a warning. You won't get another."

He pretends everything is alright, as he backs away from MJ, clearly shaken. "Madame President, you have a lot to learn about politics."

"I'll take that as a compliment." MJ grins.

Senator Garfield can't leave fast enough.

MJ returns to the PEOC. All the green dots on the Big Board tell MJ none of the leads panned out.

Next to Fernando, two Air Force pilots salute MJ, as she enters. They're kids, right out of flight school.

"At ease, soldiers. You can salute me when we catch Renquist." MJ gave explicit instructions to skip all formalities, until Renquist's capture. These two must be new.

Fernando says, "Renquist's sub could appear anywhere along the Baltimore shoreline. Given the importance of this mission and the size of the target area, I procured a UAV. It has advanced facial recognition and cell phone surveillance capabilities, SKyIDent, we call it SKID for short."

UAV, Unmanned Aerial Vehicle, a drone. Fernando hates when MJ says drone.

Fernando addresses the young pilots. "Airman Porto, you're pilot. Airman Sanchez, you're sensor operator. Grab your portable ground stations and come with me."

Airmen Morgan and Sanchez put on headsets and grab suitcases from under nearby desks.

MJ follows Fernando into the Executive Briefing Room. She whispers, "Are you sure these pilots are experienced enough? How many missions?"

He holds up one finger.

She grits her teeth.

Fernando looks at the airmen and points at the conference table. "Get set up over there."

He leads MJ to the hospital bed, where Shen keeps None company.

The airmen open their suitcases on the conference table. The suitcase lids provide a sturdy backdrop for the portable UAV ground stations. The airmen flip open the hardened laptops on the left side. On the right side, they flip up another LCD screen and unlock their flight joysticks.

Fernando whispers. "Flying military UAVs over civilian territory is not generally legal. One exception is training missions. So, we'll train Porto and Sanchez."

MJ whispers back, "Can we get more experienced pilots?"

Fernando shakes his head. He whispers, "Less experience means we can trust them. Renquist's people only recruit veteran soldiers."

Six video panels and a TV line the wall that the conference table is shoved up against. A presidential seal adorns the opposite wall.

The airmen plug cables from the portable UAV ground stations into the conference table. Duplicates of the screens from the ground stations take up four of the six video panels.

Airman Sanchez performs a systems check. Cleared for takeoff.

The SKID zips into the air. Airman Porto flies the SKID at cruising altitude, until it reaches Baltimore airspace. Then it descends and patrols the shoreline.

MJ, Fernando, and Shen walk over behind the airmen to watch the SKID do its work.

The four screens show the SKID cockpit view, a map with the SKID's location on a radar overlay, a thermal imaging view, and a steady stream of faces covered with lines and dots.

"What's the plan, once we find Renquist?" MJ asks.

Fernando says, "The FBI Director owes me favors, for some NSA intelligence I gave him. We'll paint the target to guide two FBI helicopters to capture Renquist."

"Can you trust him and his teams?" MJ asks.

Fernando shrugs.

The SKID covers the Baltimore area shoreline and turns around for another pass.

MJ turns on the TV to check for news on the Jubilee Bill. On TV, Senator Garfield makes a passionate speech against bank bailouts.

"We're winning!" Shen thinks about what happened a bit too long. His smile disappears. "What did you say to Garfield? Never mind, I don't want to know."

On the next pass, Airman Sanchez's computer beeps. Facial recognition finds a possible match. Airman Sanchez says, "Facial recognition. Ident triggered."

MJ points at the image. "That's him."

Airman Sanchez says, "We have PID on target."

"Roger, copy that. Coming hard about. I'll recenter on target." Airman Porto turns the SKID around and circles Renquist's reported location.

Fernando says, "We couldn't find him for days. Why now?"

MJ's eyes light up. "He's on the run."

Airman Sanchez checks thermal imaging. "A limo is dismounting personnel. I'm tracking up to ten pax. Driver in limo. Other pax surrounding target."

The SKID descends towards the limo.

Fernando calls for the FBI helicopters on a secure phone. "Helos inbound. ETA ten minutes."

"Woah, they're big!" Airman Sanchez rechecks the readings. "Pax surrounding target, close to seven feet tall."

MJ remember Renquist's bodyguards from the wedding reception. "Eight Andraste bodyguards, his Praetorian Guard. Warn the helos they'll need heavy weapons."

Fernando asks, "Do we have PID on weapons?"

"I'll zoom in for a closer look." Airman Porto heads the nose down. The SKID dives. The altimeter drops thousands of feet in seconds.

Airman Sanchez switches from thermal imaging to standard camera. Details improve as the SKID closes in.

Eight Andrastes carry automatic weapons in a box formation around Renquist.

"We have PID on weapons," Airman Sanchez says.

Renquist pulls out a cell phone, from a small pouch. He looks up.

"Pull up, pull up. He saw you." MJ facepalms. The newbies just gave away their location.

Airman Porto reverses the dive and climbs rapidly. "Negative, Madam President. The SKID is hard to detect."

MJ says, "I saw a cell. We have to tap his phones, to track any co-conspirators,"

"The SKID has an ICOM package," Fernando says. "It should intercept any cell phone traffic."

Video of Renquist and his Andrastes blurs as the SKID gains altitude. A blue dot flashes over Renquist.

A dial tone broadcasts from Airman Sanchez's computer. "He's dialing," Airman Sanchez says.

"Investigate whoever he calls. They're going down with him." Fury rages inside MJ at all Renquist's hidden accomplices.

They all listen intently to the call. It's MJ's voice mail message.

Why is he calling her? MJ checks her personal cell. No bars. The PEOC shielding blocks cell traffic. "Shen, call the White House switchboard. Redirect any calls from Renquist here."

Shen opens a drawer under the table and calls on a secure phone.

MJ's voicemail beeps. Renquist hangs up. With his Andraste escort, he walks 30 meters to the nearby docks and makes another call.

A cheery young woman answers. "White House operator, how may I direct your call?"

"This is Alvin Renquist."

"I'll put you through." Another dial tone replaces the woman's voice.

Fernando gets a status update on the secure phone. "Keep him talking. Helos ETA five minutes."

One of the secure phones rings. MJ picks it up. "Hello."

Renquist laughs. "I knew that was you."

MJ asks, "What do you want?"

"Veto the bill, and you live." He sounds cocky for a man about to be captured.

MJ says, "You can't bargain with me. I have the power to put you on the kill list. Leave American soil, and I'll find you."

"You want to know true power? I got the President of the United States to tie up my loose ends. I made billions, betting against the passengers on that plane."

Fernando pounds the conference table. "I knew it was too easy."

412

MJ's stomach churns. She felt like she was fighting back when she shot down flight NQ836. His words stole victory from her.

Renquist gloats without a hint of fear. "Did Astrid think I didn't know what she was up to? I played her, just like I played you."

Something disturbs the water near Renquist.

Airman Sanchez watches the water closely. "Two-man sub surfacing. I've got helos on my screen, two minutes out."

Fernando says, "Guide in the helos. Lase the target."

"Sparkle on, sparkle on." Airman Sanchez tags Renquist with the laser.

MJ feels lost, unsure of what happened. Every breath Renquist takes is an insult to everything she holds dear.

Renquist enjoys toying with her. "It's time to go. Death, or veto? What's your final answer?"

"Madame President, say something." Fernando eyes the helos on the radar.

The helos won't get there in time. He'll get away. MJ smiles. "I'll show you true power."

MJ shove her way past Airman Porto.

She presses the fire button on his flight joystick.

A missile cam replaces the map screen.

Alarms go off in the PEOC.

The two airmen exchange shocked looks.

"What did you do?" Fernando says, under his breath.

As death zooms closer, Renquist becomes larger in the missile cam. The Andrastes flee the dock.

MJ picks the phone back up. "All that will be left of you is a skid mark."

Renquist points a device at the missile, less than a second before impact.

The missile swerves off, back towards the SKID.

All four screens flash with a bright light, then turn to static.

The missile blew up the SKID. What was that device?

With the SKID gone, only MJ can hear Renquist on the phone.

He whispers, "Who do you think made your drones?" Renquist hangs up.

All the phones in the PEOC ring.

An army colonel from the operations section of the PEOC runs in. "A missile launch was detected, less than 50 miles from the White House."

MJ doesn't have a cover story this time. She looks at Fernando.

The airmen look uncomfortable.

Fernando looks at MJ, then the colonel. "We had an equipment malfunction. The drone was neutralized. No casualties. Send a retrieval team."

MJ and Shen pick up ringing phones and paraphrase the same explanation.

It becomes the official story. The phones go silent.

FBI teams deploy to an empty dock.

There's no sign of the mini-sub. It docked inside Shark Eater and escaped.

If MJ can't kill Renquist herself, she'll get the Russians to do it.

80. **Reboot**

After three days in a coma, Doctor Parker wants to unhook the machines. So much could go wrong. It could be None's last day.

It'll take strength to face what's coming. Shen left for the Presidential Suite to get some proper sleep. He'd never admit it but sleeping in None's bed will make Shen feel closer to his brother.

MJ could only face a day like today with boots on. She wears comfortable clothes she can sleep in: jeans, T-shirt, and a sweater that can double as a makeshift pillow. She knew better than to wear mascara today. On this day of sorrow, MJ left her data glasses on the conference table. All the easier to wipe away tears.

Alone with None, in the Executive Briefing Room, MJ feels her energy draining from her. She's so tired.

The plane. The SKID. Every time MJ fights Renquist, it blows up in her face. It's much harder to do it alone. There's no one to veto giving up.

Alone. The word feels like death kissing her neck.

No. Everything will be alright.

More tears betray her dread.

MJ thought Washington would destroy him. None thought he would leave it a dreamer. They were both right.

She'll cherish every moment she has left with him, even if she collapses in a heap by his bed.

MJ grasps None's hand tightly. "I've got good news. We got the company back. I gave everyone a raise."

She could be a trillionaire in no time if she took Renquist's place, but that's not why None and MJ fought and sacrificed to get the company back. She doesn't need laws to raise the minimum wage. MJ adjusted the Slam Dunk Jobs app settings, giving millions of workers a raise.

Between the Jubilee stimulus and the wage increases, it should buy at least a few years of artificial prosperity.

415

"We did it. We rebooted the economy." MJ forces a smile. The price was too high. She'd spend every dollar bill and print more if it'd bring None back.

From her go bag, MJ pulls out her bottle of Invincible Lips, the lipstick Astrid gave her. She saves it for special occasions. How special will today be? The last day with her best friend?

Blood red, it seems a fitting color. She puts on her war paint and smushes her lips together.

MJ lays her head on None's chest. "I'll kill Renquist with a pen."

The pen is mightier than the sword. All she has to do is sign a piece of paper, to death mark Renquist. MJ fantasizes about all the various ways the Russians will end him after they lose billions.

A secure phone rings.

Reluctantly, MJ leaves None's side to answer the phone.

It's time.

Agent Reynolds has the Jubilee Bill to sign in the Oval Office.

MJ won't be gone long, but she can never get back each lost moment with None.

She wipes away tears and puts on her data glasses.

MJ collects her security detail and marches quickly across the White House. She needs to rush to the Oval Office to sign the Jubilee Bill, before Renquist figures out how to stop her.

When the data glasses come to life, a dozen missed messages from Fernando instantly worries MJ.

Fernando calls again. She answers.

"Why didn't you pick up?" Fernando streams from the front seat of a vehicle.

MJ says, "I was in the PEOC. No reception. Why didn't you use a secure phone?"

"My credentials aren't working. Data glasses are the only thing they didn't know to shut off." Fernando looks in the back and nods.

Ten military types with a weapons arsenal fill what looks to be a van. They wear body armor and helmets. They acknowledge Fernando's go signal and load automatic weapons.

"Is Agent Reynolds with you?" Fernando sounds concerned.

"No. I'm meeting him now."

"Don't. We only searched for ties to Renquist before. Last night we searched financial ties to Ivashov. Zelda and Agent Reynolds are his people. There are half a dozen other compromised agents."

MJ clutches her go bag tightly. She suddenly feels vulnerable around her agents.

She's just down the hall from the Oval Office, where Agent Reynolds tried to lure her. She ducks into the nearby Cabinet Room with her security detail.

Fernando faces forward in the van. The White House gates come up fast. Tires screech, as his van brakes.

Agent Vincent and four security guards fan out with weapons drawn.

Seeing Fernando calms Agent Vincent until he sees the weapons. "Step out of the vehicle."

Fernando steps out of the van with his hands up. "The president's life depends on how you react in the next 60 seconds."

The words sound like a threat to Agent Vincent. He calls in backup. "What's inside?"

"Reinforcements, heavily armed SEALs," Fernando says.

Armed men are the last thing Agent Vincent wants to hear. "Down on the ground!"

Fernando lies flat, face down. "The protective detail has compromised agents. If they reach the Secret Service Andrastes, you'll have war in the White House."

MJ sees only pavement. Running feet stomp around Fernando and the van.

Fernando speaks with urgency. "Agent Vincent, Deploy emergency response teams. Set up a perimeter around the Andrastes."

"You're in no position to give orders," Agent Vincent says.

Fernando tilts up his head. "I have someone on the line who is." He hands out his data glasses.

Agent Vincent puts on the data glasses and attached earpiece. "Hello?"

MJ says, "Give the Secretary of Defense anything he asks for. Accompany him and his men into the White House."

"Yes, Madame President." Agent Vincent hands the data glasses back to Fernando and helps him to his feet. Agent Vincent squeezes in the van with Fernando and his SEALs.

Agent Vincent alerts the Emergency Response Teams, while Fernando's van heads through the gates.

Alarms sound in the White House. MJ tenses up. "What going on?"

"Hide. I'll explain when we get there," Fernando says.

Using situational awareness, MJ makes a mental map of all the exits in the Cabinet Room.

The West wall has two doors, one leading Northwest to the Press Secretary's Office and stairs down. The other goes Southwest towards the Roosevelt Room.

The South door leads to the Presidential Secretary's Office, which leads to the Oval Office, where Agent Reynolds likely waits to spring a trap. Any attack will probably come from there.

417

On the East wall, four French Doors lead out to the Rose Garden. Leafless bushes provide no cover outside, but she could crouch down behind the hedges.

Fernando's van reaches the West Wing driveway.

His men exit the van carrying M4A1 assault rifles with M320 underbarrel grenade launchers attached. One of the SEALs hands assault rifles with grenade launchers to Fernando and Agent Vincent.

Two SEALs carry M32 six-barrel grenade launchers. They open them like oversized revolvers, spin the barrels counterclockwise, and load Hellhound grenades into their M32s.

The armed men sprint in formation towards the West Wing entrance.

MJ hears a barrages of gunfire downstairs, a war zone under her feet. "I need a better hiding place."

Fernando confers with Agent Vincent for suggestions. "There's a Secret Service area below the Cabinet Room. Head there. We're two minutes out."

That's where gunfire's coming from. MJ might not have two minutes. She addresses her security detail. "We need to leave."

MJ's agents check the nearby stairs and the hallway between the Cabinet Room and the Press Secretary's Office.

Under his M4A1, Fernando loads the grenade launcher.

The SEALs load grenades on the run, as they approach the first set of lobby doors.

Two marines hold open the lobby doors. They salute Fernando.

Fernando says, "Clear the area of civilians. Evacuate the building."

The marines nod and follow Fernando's men inside the lobby. The marines break off into the nearby offices to get people out.

Secret Service agents signal MJ to come forward. All clear.

MJ steps through the Northwest door into the hallway. She hears heavy feet stomp towards the stairs.

Two Secret Service agents fire full clips into the stairwell.

Bullets ricochet.

A motor hums for two seconds.

Bullets slice through both agents by the stairs. A hundred tiny holes scatter across the ceiling.

Agents shove MJ back into the Cabinet Room.

Heavy footsteps pound up the stairs.

MJ backs away from the North door. She hears a motor hum again.

A three-second flood of bullets washes away her remaining agents.

Something deadly stalks her, something unseen.

Fear and panic grip MJ.

Last time she came here, she became president. Now MJ needs to run for her life, but her legs don't work.

Focus.

Fernando responds to the sound of the bullets from her video stream. "What happened?"

Focus. MJ ignores Fernando's voice.

Exits. Pick one.

Rose Garden.

MJ crouch runs to the nearest French doors. She opens them and runs outside.

Hide behind hedges? No. Hedges won't stop bullets.

Outside, she turns North. She can double back around the outside wall of the Cabinet Room.

Weapons.

She pauses just around the corner to pull her Decapitator Z9000 weaponized shovel, and Glock 19 from her go bag.

MJ looks back around the corner to see what follows her.

Six tightly grouped gun barrels extend from inside the French doors.

She lingers, unable to look away.

It's a Gatling Gun-style minigun, the kind of guns helicopters carry, the kind that protects the motorcade, the kind that shoots thousands of bullets a minute.

An Andraste comes outside, to look around the Rose Garden. Agent Reynolds' face peers from the transparent helmet. He has that damn smile.

An armored shell hangs over the Andraste. Heavy protective layers cover the ammo feed from the shell to the minigun. It has the look of a turtle from Hell.

Fernando sounds excited. "Woah! A 3,000 round ammo shell!" He cusses under his breath. "I wanted to see the demo, not be in one."

Three other Andrastes with the same equipment fan out to search the Rose Garden.

MJ needs a distraction. She throws her weaponized shovel as far as she can.

The shovel lands behind a hedge with a loud clank.

All four miniguns whirl into action. Flames burst out from the minigun barrels. In a two second burst, 400 rounds shoot up the hedge and take chunks out of the wall behind it.

Time to go. MJ takes a chance they won't expect her to go towards their starting position. MJ sprints down the hallway and peeks around the corner. Gunfire masks the sounds of her running.

It's clear again.

MJ steps carefully over the remains of her security detail. With all the blood, the mushy carpet is slippery. She sneaks past the stairs into the Press Secretary's Office and closes the door behind her. There's one other exit. The closest thing to cover is a curved 90-degree corner desk, the Press Secretary's desk.

On Fernando's stream, the SEALs creep down a hallway towards the Cabinet Room. They split into three squads of four men. Two squads break off left and right, to provide flanking fire for Fernando's team.

Agent Vincent stays with Fernando and the other three SEALs on his squad. They approach the Cabinet Room's Southwest door from a hallway.

The Secret Service had eight Andrastes for the security details. Where are the other four?

The Glock 19 won't penetrate that armor. She clips its holster to her belt.

MJ looks around the Press Secretary's Office for something else to throw, as a distraction.

She flips over a wooden table and kicks off the table legs. They feel like solid weapons in her hands.

SEALs enter the Press Secretary's Office from the other entrance and motion for MJ to get down.

She drops to the floor with the table legs in her hands.

The squad near her takes up firing positions.

MJ wants to join the fight, but she has to survive to sign the Jubilee Bill. She crawls behind the Press Secretary's desk.

Agent Reynolds comes out the Southwest door of the Cabinet Room.

The SEALs concentrate fire on his Andraste from three sides. Hundreds of bullets whittle off tiny pieces of the outer protective layers.

Agent Reynolds seems unfazed. He points the barrels of his minigun right, at a squad of SEALS.

The motor hums. Flames shoot from the minigun barrel, as it sends a hail of gunfire.

Bullets fly over MJ's head. She peeks out from behind the Press Secretary's desk.

Three seconds of gunfire splatters the squad near her. The soldier nearest her falls back into her doorway. Red ooze flows from where his head used to be. His blood seeps into a growing pool nearby.

Just ahead of Fernando, both SEALs with M32 grenade launchers cycle through all six grenades.

Agent Vincent, Fernando, and the SEAL with an assault rifle fire their grenades with a pop.

The projectiles streak down the hallway.

Agent Vincent calls in a heads up. "Cabinet Room under heavy fire."

Blasts leave craters in Agent Reynolds' armor and knock his Andraste back.

The Southwest door disintegrates in a fireball.

Agent Reynolds turns his minigun left, to take out the other flanking squad.

Grenades blow a gaping hole in the Cabinet Room wall.

The barrage of explosions knocks over Agent Reynolds' scorched Andraste. Its minigun finally goes silent.

Fire crackles. Smoke blankets the hallways.

Fernando's switches his data glasses to thermal imaging.

His team reloads their grenades.

Heat and flames glow off the floor and walls near the grenade blasts. MJ sees a ring of fire.

The minigun glows brightly. It points up at the ceiling. Underneath, MJ makes out the heat signature of Agent Reynolds' Andraste, lying on its back.

MJ peeks out at the pool of blood. "They're all dead."

Fernando says, "We're the last squad until reinforcements arrive. Stay hidden."

Agent Vincent screams into his earpiece. "We're dying here! Where are those Emergency Response Teams?"

The glow of three miniguns floats towards the ring of fire in the wall. On the thermal imaging, they remind MJ of ghosts.

Fernando whispers. "Take out the floor. Finish with grenades." He crouches down and braces against the wall.

Three Andrastes step from the Cabinet Room through the flaming hole in the wall.

The M32 grenade launchers fire another volley.

The Andraste miniguns spin up. Flames jet out.

Grenades hit the already weakened floor with glowing flashes.

Minigun bullets arc up into the ceiling, as the war machines fall out of sight, with a loud crash.

A SEAL on Fernando's team slings his assault rifle over his shoulder. He pulls the pins from some grenades and lobs them into the hole.

A swarm of bullets from the floor cuts him down.

Two SEALs with M32 grenade launchers shoot a dozen grenades down the hole.

Bullets shoot through the floors in arcs back and forth across the hallway towards Fernando's squad.

The two SEALs sprint past Fernando and Agent Vincent, to outrun the bullets.

A storm of gunfire zeros in on their steps. The running soldiers get shot through the legs and crotch, their internal organs pulverized. Their bullet-ridden bodies slump to the ground, into the path of yet more projectiles.

Fernando steps backward, one careful step at a time. He whispers, "No sudden moves. Don't give away your position."

Agent Vincent follows his lead. He retreats along the wall, with slow steps.

Fernando straps his assault rifle on his back. He signals to Agent Vincent and points at the M32 grenade launchers. Fernando creeps out towards his fallen comrades.

It's a straight run from the other exit in MJ's room to the hallway where Fernando is. She should make a run for it.

No. A squad of Andrastes stalks Fernando from below. They'll shoot her through the floor.

MJ hears heavy footsteps up the stairs. Her body tenses. They're coming.

The Press Secretary's desk extends to the floor, so MJ can't see anything. She mutes the data glasses and ignores Fernando's feed. MJ listens for every detail.

Heavy thuds move towards her room.

MJ's dead the moment they find her. She grips the table legs in her hands, to feel less defenseless.

A splash.

The pool of blood by the doorway. An Andraste must have stepped in it.

The floor vibrates from the weight of nearby Andraste footsteps.

MJ's heart beats so fast that she's afraid they'll hear it. She holds her breath. Any sound might give her away.

The desk creaks. Something heavy leans on the desktop.

The Andraste looks over the desk. MJ arcs her body with the curve of the desk, to remain out of sight.

Heavy footsteps fade towards the other exit.

Fernando's that way. MJ checks his feed.

Agent Vincent helps Fernando lift up one of the dead SEALs. Fernando slides an ammo belt with grenades off the corpse.

They're both out in the open. MJ has ten seconds to act, or that Andraste will get the drop on him. She can't say a word to warn him, or she'll give away her position.

MJ sneaks from behind the desk.

The Andraste has its back to her.

She glances at the door out. MJ could make her escape.

No.

Save Fernando.

MJ creeps towards the Andraste. She tightens her grip on the table legs. She readies to throw them for a distraction.

A water serpent scabbard dangles off the left of the Andraste's ammo shell.

Her katana. It's Zelda.

Adrenaline shoots through MJ. Instinct takes over.

The Andraste raises a minigun towards Fernando's location.

MJ charges.

Fernando and Agent Vincent load grenades into M32 grenade launchers, unaware of the danger.

The minigun motor hums.

MJ swings both table legs at the minigun, with all her strength and the element of surprise.

The minigun deflects to the right.

Bullets turn the wall near Fernando into Swiss cheese.

The Andraste knocks the table legs out of MJ's hands and turns to face her.

MJ backs away. Stupid instincts. She can't outrun a minigun up close.

Inside the transparent helmet, Zelda stares back with furious eyes. "They're here to kill the president. I'm here for you."

Explosions rock Zelda's Andraste.

Zelda shields the katana from damage like it means something to her.

MJ takes advantage of the distraction, to escape.

She splashes through the pool of blood and flees into the smoky hallway. MJ sprints past two Andrastes checking out the hole in the floor.

Heavy footsteps chase her.

Hums.

Gunfire chews up the walls behind her.

MJ slips on the bloody carpet. With a dive and roll, she clears the minigun field of fire.

Her data glasses fall off, in the roll.

MJ lands on something squishy and decides not to look down.

She stumbles to her feet and emerges from the smoky hallway.

In all black Kevlar armor, Secret Service agents with P90 submachine guns surround her.

The Emergency Response Teams.

MJ points behind her. "Andrastes. Take them out."

She sprints past them, with the Rose Garden on her right.

Hums.

Gunfire.

Explosions.

MJ doesn't look back. She runs. It's all she can do.

She flees for what seems like forever. Her heart pounds like a runaway train.

A presence follows MJ. She senses it but hears nothing.

It's just the fear talking.

MJ stops to catch her breath. She needs a plan, one that doesn't involve charging at a minigun.

The way Zelda spoke sounded personal. MJ would hold a grudge too if someone puked in her mouth and blew her cover as a spy.

Zelda said she was coming after MJ and the president. MJ is president. She must mean None. He's no longer safe in the PEOC.

MJ remembers the Andrastes stored near him. She'll take one and secure None in the other.

With her course of action decided, MJ hurries to the East Wing. She descends into the tunnel system. MJ doesn't stop until she reaches the PEOC.

She types in her code and presses her thumb on the input screen.

Motors slowly open thick steel doors to the PEOC.

A dozen soldiers hide behind desks for cover with automatic weapons drawn.

Behind her.

Heavy footsteps.

Chills run down her spine.

RUN!

She crouch-runs to the right.

A motor hums.

Soldiers open fire.

The minigun splatters heads and splinters tables.

MJ sprints past an open storage locker with empty gun racks, into the Executive Briefing Room. She hurries to None's side.

He lies helpless in the medical bed.

Fernando talked tactics before the speech. He said to absorb damage with the front armor, then turn to retreat. There's nowhere to retreat, and she needs front armor to attack. MJ will absorb damage with her rear armor instead.

Gunfire ceases in the other room.

There's no time. MJ gets inside an Andraste and powers it up.

How can she keep None safe?

She wishes she had Fernando in her ear. Is he still alive?

Zelda enters the doorway. Smoke wisps off the barrels of her minigun.

MJ moves None on his side and hops on the hospital bed, with her back facing Zelda.

The weight of her Andraste drops the bed several inches.

The minigun hums behind MJ.

She cradles None with her Andraste, using her armor to shield him.

The minigun shoots up everything behind them.

None's IV bag ruptures overhead. Liquid sprays down on None and MJ.

The medical equipment monitoring None's heartbeat and pulse flatlines with a loud beep, then goes silent. MJ hopes only the machine died, not him.

The rear armor indicator counts down MJ's life in an easy to read display.

A steady stream of gunfire eats at her protection. MJ's Andraste bleeds in chunks of metal, composites, and exotic materials.

Yellow warning indicators light up.

That means it's time to turn over, but the minigun spews bullets everywhere. A turn would leave None unprotected.

She hopes Zelda has less ammo than she has armor.

Just a few more seconds.

The Andraste screams in flashing red letters. Coolant system failure. Heat levels critical. Rear armor critical.

MJ watches the rear armor indicator fall deep into the red zone.

If this is the end, at least she faces it with None.

The last few armor layers may stop bullets, but she feels each one. Minigun fire pounds her back. Her whole body feels like a giant bruise.

Without heat resistant layers, the energy from impacts gets absorbed in the form of heat. Sweat drips down MJ's face.

The minigun sputters to a stop. Out of ammo.

MJ ignores her body aches and twirls off the bed. She grabs the minigun and slams it in the middle with her Andraste's knee.

Her machine strains with its hydraulics at full power. It takes two seconds to bends the steel gun barrels. Two seconds balancing on one leg. Two seconds vulnerable.

Zelda takes advantage of her awkward position. She knocks MJ onto the ground.

MJ's war machine lands on its back with a thud.

Zelda drops the bent minigun to the floor. She lifts the ammo harness overhead like a weight lifting contest and heaves it at MJ.

MJ scrambles to her feet. She tries to dodge the ammo harness, but it knocks her back down.

The fall brings more red letters. Voice commands offline. Systems overload.

She rolls back to her feet and goes on the attack. MJ tries a karate strike.

Her armor freezes up. She can't move.

With a smirk, Zelda wields the katana. "I was ready to slice and dice you."

She pats her helmet. "Vomit won't save you this time."

Save her? MJ remembers her katana under Zelda's bed. Zelda planned to go *Basic Instinct* on MJ with her own sword. MJ wasn't paranoid, after all.

Zelda moves behind MJ's back.

Her armor is weak there. How can she get Zelda to attack her front?

MJ snorts in contempt. "Stab me in the back. I knew you had no honor. At least, you won't see my dying face."

Zelda returns to the front. She picks up MJ's Andraste and sets it down with its back against the conference table.

There's nothing to fear. Her Andrastes should resist bladed weapons. MJ can wait it out inside until help arrives.

Zelda stabs MJ's Andraste in the gut. It goes in less than a quarter inch.

She pushes with steady power. Zelda monitors the katana thrust closely to avoid tipping MJ over or breaking the blade.

The blade penetrates further, a little at a time. Emotion swells in Zelda's voice. "I'll kill you with the same katana you murdered my partner with."

"I never killed anyone," MJ says. Well, not with a katana, anyway. What is Zelda talking about?

Zelda forces the blade deeper and deeper into MJ's armor. "Don't lie to me. I was there. You didn't hesitate to stab through that bathroom door."

The break-in. Zelda and her partner were the intruders.

MJ remembers the intruder she stabbed in the arm. It was hardly a fatal wound. Zelda exaggerates again as she did about MJ knocking her over.

The blade penetrates halfway through MJ's armor.

The Andraste was supposed to make her invincible. Instead, MJ stands, encased in an armored tomb.

With the heat buildup in her Andraste, it feels like she's being cooked alive. MJ can't move her arms to wipe away the sweat in her eyes.

Between the unbearable heat and pain throbbing through her body, MJ can't concentrate.

Her mind wanders, as she faces doom, a sixteenth of an inch at a time. She feels like Excalibur, stuck in a rock, waiting for King Arthur to free her.

The katana has less than a half inch of armor left to penetrate. Zelda delights in every sign of MJ's suffering.

MJ needs her wits about her. Block out the pain and heat. Focus.

She glances at the undamaged Andraste. That's where she needs to go, but there's not enough power to eject.

What would None do?

He'd turn off non-essential systems.

How?

"Reboot." MJ imagines None saying the word with her.

The Andraste promptly ignores her voice command.

MJ bangs her head on the helmet. "Reboot, damn you."

Something bumps into place. Voice controls online.

A menu comes up on her HUD display, with an error message. "Invalid option: Damn you."

She blinks to clear sweat from her eyes and looks through menu options. "Reboot, Diagnostics Mode."

The messages on her HUD disappear. MJ's war machine goes silent.

Sharp pain stings in MJ's gut. She feels blood trickle from a wound.

The tip of the blade has penetrated the armor. MJ sucks in her stomach.

The helmet display lights back up.

"Eject!"

MJ pushes desperately against the back of her Andraste.

It opens up.

She strains her aching muscles to lift herself out of her Andraste, as the katana blade chases her.

The blade slices deeper until it stabs through to the weakened rear armor.

MJ escapes by inches. She sprints to the undamaged Andraste and powers it up.

Zelda pulls the sword. It's stuck.

There's panic in Zelda's eyes. She uses too much force. The katana shatters near the hilt.

MJ's undamaged Andraste approaches Zelda.

She can't penetrate Zelda's armor, but there's a body inside.

Bodies break.

With nothing left, but adrenaline, MJ has to conserve energy and look for an opening.

This Andraste has its dampening layers. Zelda can punch all she wants.

She lets Zelda hit her repeatedly. MJ inches backward with each blow, as she pretends to be hurt. MJ doesn't have to fake the fatigue or throbbing pain.

After a dozen unanswered blows, Zelda gets overconfident.

There's an opening. MJ kicks at Zelda's left leg.

Zelda lurches forward, as she regains her balance.

With her Andraste's two-ton lifting capacity, and everything she's got, MJ twists Zelda's right arm behind her, into an unnatural position.

Zelda's armor bends further than her bones can.

Crack.

Zelda screams in anguish. Desperate. She squirms out of MJ's grasp. Her right arm falls limp at her side. Zelda turns to run.

MJ presses her advantage and closes in. She sweeps Zelda's legs.

Plummeting towards the ground, Zelda catches herself with her left hand and pushes off, to stumble to her feet. Her right arm flops around.

Weary, MJ watches Zelda flee the room.

Chase down Zelda to finish her off, or check on None? MJ ejects and rushes to None's side.

There's a pulse.

She pulls out his breathing tube. With her ear to None's mouth, MJ hears shallow breathing. He's still alive, but might not be for long, without the machines.

MJ stares into his closed eyes. "Everyone has a warrior inside. Tap into that inner warrior and fight your way back to me."

It's what he should say to MJ in a hospital bed. Being a warrior is her thing, not his. MJ needs to tell him how she feels.

"It's always been us. Everyone else is temporary." The words still aren't enough.

She tried so hard to be fearless. Did MJ bottle up her emotions, along with her fears? Losing him is the fear that trumps all others. She needs to bare her soul to him, while she still has one last chance.

MJ climbs onto the hospital bed and moves None close to her.

Opening up scares her more than fighting for her life.

She clasps his hand. "My whole life I've been afraid to let anyone inside me, but you are inside me. I don't just love you. *I am you.* You're the only one I could truly kiss."

She embraces him closely. It's not close enough.

Tears stream down her cheeks. She worries he will never wake.

If only she could breathe some of her life into him.

She closes her eyes.

MJ nibbles on his lips. Her kisses grow deeper with a hunger, a desire to feel closer, to feel oneness. Her breath joins his.

It's the kiss he waited his whole life for, but he's not conscious to enjoy it.

Her eyes open to give None tender glances. She's never felt closer to him.

MJ confesses every secret she ever kept from him.

None wakes up, disoriented and afraid.

"Say my name, so I know everything is OK."

He struggles to remember her name. "Ma...Ma...Mar..."

MJ sees it in his eyes. He knows her. Say MJ. Say Maria. "Say it. You can. I have faith in you." MJ hopes with every fiber in her being.

"Marjorie?"

Her heart drops. He's not OK. They may face a hard road, but they face it together.

His mind sifts through his strongest memories, trying to remember who she is. "Are you my bro?"

She laughs and smiles back at him. "I'm not your bro. I'm your girlfriend."

MJ stuffs every bit of herself into another kiss. She comes up for air. "Are we clear?"

None beams at her, happy and content. Safe.

Movement.

MJ notices something in her peripheral vision. She looks up.

Darkened by soot and covered in blood, Fernando and Agent Vincent enter with M32 six-barrel grenade launchers.

Fernando barges in like a conquering hero. He carries a blown off section of Andraste armor like a shield in his left hand.

His face sinks when he sees MJ holding None in her arms. He drops the shield and stands frozen.

What can she say to Fernando? She never meant to hurt him. They still have to run the military together.

Agent Vincent keeps his distance, with a concerned look.

MJ can't blame Agent Vincent after she lashed out at him. She waves him over. If she can't do something nice for Fernando, at least she can do something for him.

Concern grows on Agent Vincent's face, as he approaches.

She gives him the gift of redemption. He earned it. "I may have an opening on my security detail. Would you be interested?"

Relieved, Agent Vincent smiles back. "Yes, Madame President." He calls in their location on a secure phone.

Emergency Response Teams enter to clear the area.

Once they signal it's safe, Shen and Doctor Parker come in.

Doctor Parker sprints to None and checks up on him. Her eyes bulge out when she sees the shot up medical equipment.

"I told you she'd be here." Shen gloats he's right. He's even happier to see None awake. He runs to None and hugs him.

MJ points at Shen. "That's your bro."

Shen smiles back at MJ. "Isn't bro your word?" She acknowledges him in a big way by giving it up.

"I have a better word now, girlfriend," MJ says.

Fernando approaches MJ, with hurt and disbelief on his face. "You saved me from Zelda. I meant something to you."

MJ cringes. This is not the conversation she wants to have in front of all these people, especially with None present.

When he can bear her awkward silence no longer, Fernando turns his back to MJ.

The room feels crowded with all the people. MJ scoffs. "Anyone else coming?"

"Some of the joint chiefs are on their way down," an Emergency Response Team agent says.

That can't be good.

Fernando walks to the door. Before he can escape the room, worried generals block his path.

One of them looks at Fernando, then MJ. "Good, you're both here. Madame President, we have a situation."

Before You Go

If you'd be so kind as to share your thoughts on the novel, please review my book. Even a two-line review is a wonderful thing for me to cherish. We live in a world of algorithms and your review will help others find the book.

The adventures of None and MJ will continue in the next book, *Hot Nights and Cold Wars*. To be notified when it's released, join the mailing list on my website **www.pgsundling.com**

In the meantime, read on for the story about the story.

Fact Versus Fiction

Spoiler Alert: the background on this book contains many spoilers. It should not be read before completing the novel.

Originally, I was afraid that many of my ridiculous subplots, like a presidential candidate with his own reality TV show, were too hard to believe. Then Trump ran for president, and the book became more plausible by the day. Even stranger, as the 2016 election cycle progressed, there were a number of strange coincidences. Events similar to plot points in the novel began to happen in real life.

The origins of this book date back decades. This novel originated with a single thought. "What if someone changed their name to None of the Above, and they won the election?" It's a thought thousands have when they first hear about the 'none of the above' concept. It's such a common idea that a California Proposition to add 'none of the above' on the ballot included a clause specifically to prevent that from working.

My path to that idea was less direct. In the days before cell phones could fit in your pocket, everyone had landline phones. The big deal was long distance service. After the government broke up AT&T, the phone company had to give you whatever long-distance provider you wanted. Someone created companies with names like "Whatever" and "I Don't Care." When you got a new phone, they would ask what long distance provider, and if you said "Whatever," you would end up with overpriced phone service. I don't remember the exact year, probably early to mid-80s, there was a warning about these companies on the news.

I had always liked *The Prince and the Pauper* and wondered if there was a similar trick to use mistaken identity to become president. Then I found out about the 'none of the above' concept. Thus, began the journey of this book.

In early versions, the president was a teacher. His major innovation was balancing the budget by legalizing marijuana. As the years passed, this became entirely too plausible, so I needed something more far-fetched.

After a decade of working in technology, in 2004, I shifted the main character to a programmer that improves the education system with software. I gave a speech, that year, on how the educational software would work, as I tried to earn a spot on the reality TV show *American Candidate*. I didn't get on TV, but my efforts helped provide a foundation for the political ideas in the novel.

Inspired by the global economic meltdown in 2008, the book finally had a new crisis for the fictional president to solve.

I accumulated more notes until I worked on a screenplay version in 2012. Before I could complete it, I became a reluctant novelist. I realized that 70% of movies were adapted from other formats, like books, and decided to be in the 70%. Complex and unconventional ideas like mine, tend to work better in novels, or long-form TV. Movies remained my first love, so I decided to write a cinematic novel that felt like a movie.

I focused on this book in 2014 and hoped to complete it by September 2015. If I had succeeded in that timeline, this book would have felt prophetic. About halfway through the book, in 2016, reality veered closer to my fictional timeline.

I wasn't sure it was realistic for viral Internet videos to get an unknown candidate elected as president. Then a candidate named Deez Nuts used viral videos to poll as high as 9% in New Hampshire polls. His support faded after it was revealed that Deez Nuts was a 15-year-old boy who was too young to be eligible for president.

I wondered how realistic it was to have a president at a gun rally, surrounded by armed citizens. Then a petition to make the 2016 Republican convention open-carry gained over 50,000 signatures. The head of Cleveland's police union pleaded with the Governor to suspend Ohio's open-carry laws. Trump could have been surrounded by guns, but the Secret Service stepped in.

A breakdown of the two-party system, like in the book, seems impossible, but there was a moment, in 2016, where it seemed plausible. In Gallop polls, the major party presidential candidates set records for 1st and 2nd most disliked candidates in presidential polling history (unfavorable 61% Trump, 52% Clinton). Many voters didn't vote for a candidate but against a candidate. If none of the above were an option, it would have been a popular one. Between the "Bernie or Bust" Democrats who wouldn't vote for Hillary and the "Never Trump" Republicans against Trump, it seemed like both parties might break apart. In the end, the independents on both sides were brought back in line, and the status quo was maintained.

Early in the book, a group called "Take Back America" tries to unseat all incumbents, Democrats and Republicans alike. Campaigning for candidates from multiple parties seemed unlikely, but a group called "Brand New Congress" is attempting exactly that. They have been fielding Democrats, Republicans, and independents to take down establishment politicians until they can create a brand-new Congress.

In the book, faithless electors change their vote from None, and he has to battle in the California Supreme Court to compensate. While electors have yet to change the course of a presidential election, a record seven faithless electors changed their vote in the 2016 election. Another three tried but had their vote invalidated. The previous record for a living candidate was James Madison all the way back in 1808.

One day, I was in the shower, when I remembered that I had Russians messing with the election in my book. I had written that over a year before, and it was a minor plot point that was easily missed. I had a good laugh over that! Near the end, I point out the full context, for those that might miss it.

The Russians in the book were inspired by news stories about the Russian spy Anna Chapman who was caught in 2010 and deported back to Russia. I had the idea: what if the president was single and did lots of Internet dating and one of them turned out to be a Russian spy? You can't have a very interesting Russian conspiracy with only one Russian, so I expanded it into a major part of the book. Also, since it's a conspiracy, it's not until near the end that its scope is revealed.

Fears of robots and automation bringing mass unemployment have been dismissed in the past, so I wondered if I had gone too far in mentioning it in the book. Then in 2015, the *Rise of the Robots: Technology and the Threat of a Jobless Future* won the Financial Times and McKinsey Business Book of the Year Award.

Early in the election cycle, Pulitzer Prize-winning journalist, Thomas Friedman, stated in the New York Times that his choice for president was None of the Above. I began to worry others would write a similar book before I finished mine.

As I missed deadline after deadline, I tried to motivate myself with an oath. I swore I would not get a haircut until my book was done.

My worries of a competing book increased, when a man in Ontario changed his name to Above Znoneofthe (on ballots: Znoneofthe, Above). He received fewer votes than the leader of the None of the Above party, Greg Vezina.

When I saw an article on Huffington Post that said the Libertarian candidate Gary Johnson was going to petition to change his name to "None of the Above," with his running mate as "Other," I panicked. I took a month leave of absence from my job to expedite the book. It still took well over a year more to finish the novel, but every week the book got better or more complete. Since the book is so late, I have hair down to the middle of my back. Like None, I am an oath keeper.

While researching this section, I found out there was another None of the Above in Indiana. They allow write-in candidates to include nicknames on the official list, so one of the registered candidates for president in Indiana was "Matthew (None of the Above) Roberts."

While many similar events were already in the novel, I did use four items from the 2016 election cycle: "make America great," "fake news," a reference to Trump's murderers and rapists comment, and Trump's wall. Most of those references are in the chapter "Gallows," since that's about when real life began imitating art.

Some similarities between reality and the book, I don't consider coincidences. The best example is the government shutdown fight in the book. There was a government shutdown after I sent the novel to the first set of beta readers. It was just a matter of time before that happened and I expect government shutdown battles to happen again.

The coincidences extended into my personal life as well. I had written the two setup scenes for the subplot I call "the ghost in the machine." None and MJ had both talked to Sarah on the phone, and the next scene in the subplot reveals that Sarah is dead. They've been talking with an AI that mimics her voice. I was excited to write the reveal, but I had to delay it three weeks to work on my taxes.

My story editor spent hours a day talking to a girl online. Their online relationship began to seem serious. I'd become good friends with her too, and they both confided things in me. A few days before I had time to write the next scene, my story editor called and said, "I'm in love with a dead girl." He found her picture on a funeral website!

All those hours he talked to her, a dead girl's picture stared back at him. We both felt this creepy vibe and weren't sure who we were talking to. Should he confront her? What should he say?

At that point, I wrote the reveal that Sarah was dead, without knowing what the truth was about the dead girl my story editor was chatting with. That made the scene feel all the more eerie to me while writing it.

There is a moment in the chapter, "The Money Game," where None hands the Russian president a $100 trillion bill. It's only then that President Orlov internalizes just how artificial money is and lets a handful of cash fly off into the wind. Less than two weeks later, I had that same exact experience.

My story editor and I play a video game called "Gems of War." I was playing the game quickly so that I could rise up the leaderboard. I'm focused on the game when my story editor tells me that there's a bug with one of the virtual currencies and warns me that the developers are banning anyone who takes advantage of the "glory bug." I go to check if I'm affected, and I'm staring at 59 digits of money! So, I'm worried people might think I've cheated, and I stare at all the zeros in shock and awe.

This isn't any game. It may be free to play, but I spent a dozen times more on "Gems of War" than any other video game in my life. I invested over a thousand hours and over $2,600 into earning those virtual currencies. It had real value to me, yet it felt worthless. Endless. Unspendable. The value of money is the biggest mind game there is. You can know it's made up, but without an experience like that, money still has its grip on you. President Orlov's reaction rang true to me.

As I neared the end of the book, I was at the Las Vegas airport. I waited to board my plane home. As I continued writing in the airport terminal, I reached the part of the scene where a plane gets shot down. I got a chill down my spine. With all the coincidences during the writing of this book, continuing the scene on a plane felt like a really bad idea. I didn't write another word until I landed safely back in Los Angeles.

About the Author

I was born in Los Angeles, and other than a few years in Colorado, I've lived here my whole life. As an honorary member of Toastmasters, I gave my first speech on dinosaurs, at the age of seven.

I've loved role-playing games since my childhood. In Dungeons and Dragons, I was drawn to the role of dungeon master because I liked being the storyteller. I created and destroyed many worlds.

I graduated from UCLA with an Aerospace Engineering degree in 1992.

From working in IT at UCLA to coding on billion-dollar e-commerce systems for multiple Fortune 500 entertainment companies, I've worked in technology for almost a quarter century. I was a programmer in a small team that got awarded P.C. Magazine's Editor's Choice for Best Virtual Desktop. I write novels and code software in Los Angeles.

Instead of staring off thoughtfully with a cup of coffee, I thought I'd try something more amusing for my author picture. Wearing my "Meet the Author" shirt, I tried to channel a bit of my cover. After all, every author is a puppet master in their own universe.

Acknowledgements

This book wouldn't be possible without the love and support of my wife Libby and daughter Kathryn, and all the times they sacrificed quality time so I could write. When my daughter grows up, may she enjoy this book as a piece of her father she can carry with her.

My eternal thanks to my lifetime friend and story editor, Curtis W. Ebelherr. This book brought us even closer. I'll always cherish our stream of consciousness editing sessions as you read the book to me. Thanks for not skipping ahead when we had rewrites, even though you were dying to read new chapters.

I'd like to thank Aaron Azevedo, Amy Temprosa, aurora, Brandy, Jessica Brower, Jynifer Smith, Mona Fawn, Morgana, Nick, Rob Ross, and the other dozens of beta readers. This novel would not be as good without your honest feedback.

Of all the beta readers, I saved my utmost gratitude for two that went way beyond the call of duty, each delivering hundreds of corrections and critical feedback.

Thanks to Justin M. Renquist (no relation to the antagonist) for his help. He helped solidify some of the grammar choices in the book and convinced me to use the Oxford comma. The first thing I told him about the book was that he had the same last name as the villain. What a funny coincidence!

Heather Linn gave me insightful, hard-hitting feedback and put up with me when I didn't take bad news well. Her feedback was so extensive that I took time off work so I could spend 11 days just doing rewrites.

For delivering my vision for the cover and author pictures, I'm grateful to rebecacovers. She created a cover worthy of my book and generated the memorable author photos I wanted, despite my subpar source pictures.

My biggest thanks are to you, the reader, for finishing the book. Of all the millions of books and gin joints in the world, you walked into mine. I hope I beat the ham sandwich and brought you at least a few smiles.

Made in the USA
Middletown, DE
26 May 2018